BEYOND THE MEGACITY

New Dimensions of Peripheral Urbanization in Latin America

GLOBAL SUBURBANISMS

Series Editor: Roger Keil, York University

Urbanization is at the core of the global economy today. Yet, crucially, suburbanization now dominates twenty-first-century urban development. This book series is the first to systematically take stock of worldwide developments in suburbanization and suburbanisms today. Drawing on methodological and analytical approaches from political economy, urban political ecology, and social and cultural geography, the series seeks to situate the complex processes of suburbanization as they pose challenges to policymakers, planners, and academics alike.

For a list of the books published in this series see page 417.

EDITED BY NADINE REIS
AND MICHAEL LUKAS

Beyond the Megacity

New Dimensions of Peripheral Urbanization in Latin America

UNIVERSITY OF TORONTO PRESS
Toronto Buffalo London

Toronto Buffalo London
utorontopress.com

ISBN 978-1-4875-0910-1 (cloth)
ISBN 978-1-4875-3972-6 (EPUB)
ISBN 978-1-4875-3971-9 (PDF)

Global Suburbanisms

Library and Archives Canada Cataloguing in Publication

Title: Beyond the megacity : new dimensions of peripheral urbanization in Latin America / edited by Nadine Reis and Michael Lukas.
Names: Reis, Nadine, editor. | Lukas, Michael, editor.
Series: Global suburbanisms.
Description: Series statement: Global suburbanisms | Includes bibliographical references and index.
Identifiers: Canadiana (print) 20210379030 | Canadiana (ebook) 20210379049 | ISBN 9781487509101 (cloth) | ISBN 9781487539726 (EPUB) | ISBN 9781487539719 (PDF)
Subjects: LCSH: Urbanization – Latin America – Case studies. | LCGFT: Case studies.
Classification: LCC HT384.L29 B49 2022 | DDC 307.76098 – dc23

We wish to acknowledge the land on which the University of Toronto Press operates. This land is the traditional territory of the Wendat, the Anishnaabeg, the Haudenosaunee, the Métis, and the Mississaugas of the Credit First Nation.

This work is supported by the Academic Productivity Support Program, PROA VID 2019, University of Chile, and the Fondecyt de Iniciación proyect No. 11150789.

University of Toronto Press acknowledges the financial support of the Government of Canada, the Canada Council for the Arts, and the Ontario Arts Council, an agency of the Government of Ontario, for its publishing activities.

Canada Council for the Arts
Conseil des Arts du Canada

Funded by the Government of Canada
Financé par le gouvernement du Canada

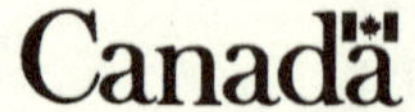

Contents

Figures and Tables

Figures

Tables

BEYOND THE MEGACITY

New Dimensions of Peripheral Urbanization in Latin America

Introduction: Old and New Dimensions of Peripheral Urbanization in Latin America

MICHAEL LUKAS AND NADINE REIS

Over the past two decades or so, representations of Latin American urban peripheries in popular culture, global capitalism, and academic writing have become more complex than ever before. Today, Brazil's favelas[1] are not only feared as dystopian spaces of drug-related violence and police repression, as depicted in the international blockbusters *City of God* (2002) and *Elite Squad* (2007), they are also part of the global tourism industry and represent the latest chic in industrial design. While in its hugely successful series *Narcos* (2015) the online streaming platform Netflix remembers the Medellín of the 1980s and '90s as the headquarters of Pablo Escobar's drug cartel and the deadliest city on Earth, today it is known as one of the world's most creative and innovative centres when it comes to participatory urban planning and urban design (McGuirk, 2014), and is said to be "a miracle of reinvention" (Moss, 2015). In a similar fashion, while Mexico City's district of Ciudad Nezahualcóyotl is depicted by some as a poverty-ridden sector of the world's largest slum (Davis, 2006), for others it "rises from slum to success story" (Wulfhorst, 2016). More recently, the 2020 Mexican movie *I'm No Longer Here* depicts the interweaving of youth counterculture in the poor urban periphery of Monterrey with gang violence and transnational migration flows. Both from the hills of the Monterrey *barrio*, where the main character Ulisis's cumbia crew dances in abandoned construction sites against the backdrop of the high-rise central business district full of entrepreneurial worlding ambitions, and from Brooklyn, where Ulises ends up socially isolated and in absolute poverty, it becomes clear how meshed up the notions of centre and periphery are in practice today. Yet another recent representation of the Brazilian Amazon capital Manaus as the world's COVID hotspot relates back to the classical notion of the periphery as a site of despair. A German news outlet ran a story titled: "'Send us oxygen!' Manaus is living the Corona-apocalypse."[2]

Representations of Latin American cities as "megacities with mega-problems" and apocalyptic spaces of despair stem from the 1960s and '70s, a time when an urban explosion caused by massive rural-to-urban migration left national governments and international observers stunned by the rate of growth and the pace of change.[3] Ever since, the dystopian reading of urban life in Latin America and elsewhere in the Global South has been reproduced in academic discourse and popular culture, perhaps most iconically in Mike Davis's (2006) "Planet of Slums" thesis, or the aforementioned favela blockbusters. Such representations are not "false": in many countries throughout the region the percentage of people living in settlements without functioning urban infrastructure is still very high, as are rates of poverty, which in absolute numbers have been rising almost everywhere. A combination of failed housing policies, drug-related crime, and police repression lead to a worrying perpetuation of social exclusion in many of the Latin American urban peripheries (Perlman, 2011). The recent COVID-19 crisis has impressively demonstrated the enduring vulnerability of the urban population in the Global South, and especially of those social groups that live under precarious conditions in the urban peripheries. Where autoconstruction, overcrowding, and survival economies prevail, lockdown and social distancing are no option. As Bhan et al. (2020) put it, "calls to 'stay at home,' and 'work from home,' as well as the design of mobility-constricting 'lockdowns' as preventive epidemiological practices draw their imagination from the urban arrangements, built forms and economic lifeworlds of the North and of the elite neighborhoods that brought COVID into southern cities." All of this, together with the fact that the pandemic has hit Latin America harder than any other world region so far, show that there is an urgent need to bring forward our understanding of contemporary peripheral urbanization, in Latin America and beyond.

As post- and decolonial urban scholars have pointed out for many years, the problem with the traditional and Eurocentric "megacities mega-problems framework" is that it is simplistic and reductionist, victimizes most of the urban population, and hides socio-spatial, economic, and cultural complexity and diversity from view (Holston, 2008; Caldeira, 2009; Roy, 2011; Varley, 2013). Urban peripheries are not necessarily spaces of despair, criminality, and violence, nor are their shantytowns more informal than many of the public policies that govern them (Caldeira, this volume; Roy, 2011). Of course, they are also and maybe predominantly the sites of everyday life of "ordinary" urban dwellers, of agency and creativity, and there are immense differences within and among them (Roy, 2011; Gilbert & De Jong, 2015). Moreover, and this is only recently starting to be addressed, urban peripheries in Latin America – as elsewhere in the world – are characterized by an astonishing juxtaposition and complex entanglement of poor and rich, "formal" and "informal," of consolidated squatter settlements and high-rise condominiums, dirt roads and super-modern highway systems, land speculation and capital switching, Indigenous ways of life, artificial recreational lagoons, gigantic modernist housing estates, luxury gated communities of impressive scale, business parks surrounded by street vendors, extreme poverty, communal systems of water provision vis-à-vis transnational water companies, extreme wealth, huge informal markets, private bilingual schools, golf courses, the environmental suffering of the poor, and community organizing. Hence, if one of the major tasks of urban theory today – in this "dramatic period of peripheral urbanization that spans the world," as Roger Keil (2018, p. 56) puts it – is to understand how "existing and new forms of peripheral urbanization interlace in a complex pattern of urbanity" (ibid., p. 50), we must look at Latin America in order to move urban theory forward. This is especially due to the maturity of urbanization processes and their long-standing entanglements with global capitalism. Since the colonial invasion, Latin America has been integrated into the world economy as a provider of natural resources and cheap labour. This form of dependent development decisively formed the national urban systems, which have been characterized by high degrees of urban primacy and urban expansion through autoconstruction in the twentieth century. Today the region hosts some of the largest megacities in the world, and 80% of its population live in urban areas. While autoconstruction has seen a recent upsurge, particularly in small- and medium-sized cities, Latin American urban peripheries are at the same time being profoundly transformed by the penetration of finance capital in all aspects of social life.

However, it is not only the scale and scope or the complex layering and processes of hybridization that make peripheral urbanization in Latin America intriguing for the project of shifting the geography of urban theory (Robinson, 2006; Roy, 2009; Brenner, 2014; Keil, 2018); it is also the long-standing and specifically Latin American history of theorizing urbanization "from the margins." The urban explosion that took place between the 1950s and '70s not only reworked the inherited forms of territorial organization in Latin America, but constituted the heyday of theory building on urban peripheries and urbanization, which brought forward a great variety of concepts and theories, perhaps most prominently those of dependent urbanization, informality, and marginality (Jaramillo, 2012; Duhau, 2013). Characteristically, much of that knowledge production – roughly from the 1960s to the 1980s – was deeply intertwined with political activism, planning practice, and public policy, and firmly rooted more in the study of the region's urban peripheries than urban centres (Connolly, 2013). The aim of this book is to reconnect to this Latin American tradition of seeing, considering, and theorizing urbanization *from the margins,* examining how it can be made fruitful for moving urban theory forward.

To do so, we mobilize the notion of *peripheral urbanization*. This allows us to conceptualize this book as an experiment to investigate, from an interdisciplinary perspective, the natures, causes, implications, and politics of current urbanization processes in Latin America, and proceed from there to delineate the elements of a future research agenda on (peripheral) urbanization. The basis of this endeavour is a set of empirical case studies from various countries across the region covering different theoretical and disciplinary approaches from the fields of geography, anthropology, sociology, urban studies, agrarian studies, and urban and regional planning, written by academics, journalists, practitioners, and scholar-activists. What unites these contributions, besides their different ontological and epistemological propositions, is a shift of attention to places, processes, practices, and bodies of knowledge that have often been neglected in the past – the "peripheral" in all its senses. The remainder of this introduction aims to prepare the ground for situating these case studies within the broader history of peripheral urbanization, and for bringing them into conversation with each other in the conclusion of this book. In the following section, we present our approach of mobilizing peripheral urbanization as a theoretical boundary concept. This approach is based on the identification of three major recent advances of urban theory, and an outline of what we understand by, and hope to achieve with, the peripheral urbanization concept. We proceed with a literature review on peripheral urbanization, a concept that emerged in Latin American urban studies in the context of the hitherto unprecedented urban growth rates in the mid-twentieth century.

Recent Debates in Global Urban Theory and Peripheral Urbanization as a Boundary Concept

As Robinson and Roy (2015, p. 181) put it, "urban studies is undergoing a phase of rich experimentation, with a proliferation of paradigms and exploration or invention of various methodologies inspired by the diversity and shifting geographies of global urbanization." Here, scholars increasingly recognize that new understandings of what constitutes the urban, urbanization, and urban theory are needed in the age of global capitalism and a majority urban world. But beyond this common recognition of the necessity of broadening the empirical, epistemological, and methodological scope of urban studies, there is also some considerable degree of debate between different and supposedly incommensurable approaches. In our view, three major theoretical positionalities and scholarly agendas stand out in the last years – namely, planetary urbanization, postcolonial urban studies, and global suburbanisms. It is what we see as a productive tension between these approaches that has inspired the theoretical framing of this book. While we do not have space to dwell on each of these approaches and at the corresponding theoretical debates at length here, it is worthwhile to highlight some key theoretical and methodological aspects of each approach relevant to our understanding of peripheral urbanization.

The first strand of scholarship that inspires our investigation into processes of peripheral urbanization in Latin America is that of the dialectic of the implosion and explosion of urban space under global capitalism, based on Lefebvre's (2003) hypothesis of the complete urbanization of society, and its reconceptualization as planetary urbanization (Brenner, 2014; Brenner & Schmid, 2014; Merrifield, 2013). While the planetary urbanization approach has sparked heated debate and sometimes heavy rejections, especially by postcolonial urban scholars (Peake et al., 2018; Wilson & Jonas, 2018; Angelo & Goh, 2020),[4] it undoubtedly constitutes a milestone in moving urban theory forward. Several aspects are of particular interest to us. First and most importantly, planetary urbanization conceptually relates urbanization processes to the accumulation dynamics of the capitalist world system. In the line of Merrifield's (2018, p. 2) reading of Lefebvre, planetary urbanization is the result of "the closing of the circle of a particular form of capitalism that defines itself less through a model of industrial or agricultural production and more through the production of space. It's a system that now produces planetary geography as a commodity, as a pure financial asset, using and abusing people and places as strategies to accumulate capital. This process quite simply embroils everybody, no matter where." Second, this rescaled global regime of space production unfolds through both concentrated and extended forms of urbanization, through the dialectically related implosion and explosion of space. On the one hand, capital, goods, people, and information keep concentrating in urban centres around the world through agglomeration dynamics and the direction of investment flows; on the other hand, the urban fabric explodes into the former non-urban realm through disjunct fragments and operational landscapes. The notion of extended urbanization allows for decentring and transcending the city and the urban as bounded entities and the only sites of analysis of urban theory. To see through the lens of extended urbanization invites us thus to analyse transformation processes and "spaces far outside cities proper, including agricultural regions, forests, oceans and deserts, in ways that are planetary in scale (Brenner & Schmid, 2014, 2017)" (Angelo & Goh, 2020, p. 2) and focus on "the dissolution of the contradiction of the city and the countryside" (Keil, 2018b, p. 3).

The second major field of contemporary theorizing on the global urban condition comes from authors whose work we sum up under the heading of postcolonial/poststructuralist urban studies. Authors such as Robinson (2006) and Roy (2009, 2011) argue for developing radically new practices and cultures of theorizing, based on a "lively vigilance for difference" against universalizing tendencies in urban theory and specifically claiming the right and indeed the need to begin theorizing "from anywhere" – in conceptual, empirical and geographical terms (Robinson, 2015, p. 187). While they thus share the aim for radically opening up urban theory, they are highly critical of the Marxist method of abstraction as in planetary urbanization, its generally larger scale of analysis, and sometimes also its roots in European intellectual traditions. In her agenda-setting book *Ordinary Cities*, Jennifer Robinson (2006) formulated a powerful critique of the academic and policy related division of labour that sets a handful of "innovative" Western "global cities" against the poor megacities of the "third world." What she proposed instead, and what is a central part of the postcolonial urban studies agenda until today, is to see all cities as ordinary. In methodological terms, this means they must be approached from the perspective of cosmopolitan comparitivism, in other words, placing all cities within the same analytical field, beyond preconceived grand narratives and conceptual binaries. Rather than aspiring for universal theory, postcolonial urban studies draw attention to difference and specificity across particular situations and locations through focusing on everyday life, collective action and agency rather than on any kind of pregiven "structure," constraint or process. What comes to the forefront then are "embodied forms of social difference and the intersection of different forms of oppression" (Angelo & Goh, 2020), going beyond the classical Marxist class or form analysis.

Finally, the third strand of research is the rapidly expanding body of work on global suburbanisms, which is also the editorial context of this book. Most prominently, Keil (2018a) has claimed that Lefebvre's urban revolution is in fact a sub/urban revolution, and that in order to understand the contemporary urban question, research and theorizing has to go where urbanization, urban growth, and territorial transformation mainly take place today – that is, to the urban peripheries, the metropolitan edges, and the global process of peripheral urbanization. In the context of research on global suburbanisms, the lens of comparative urbanism has been applied to gain novel insights into sub/urbanization processes around the world, thereby seeking to decentre the classic view of suburbanization as a distinct North American socio-spatial form and way of life. Rather, as Keil (2018c) explains, "suburbanization is diversifying in morphology, socio-economic and demographic profile, and boundaries between cities and their peripheries are increasingly breaking down." This happens in highly heterogeneous ways around the world. Peripheral urbanization in this sense of suburbanization, post-suburbanization or peri-urbanization – centrifugal movement of people, economic activity and capital – is a major input for global urban theory because it allows for "writing urban theory from the outside in" (Keil, 2018a, p. 46), both on metropolitan and world-systems scales. The aim here is to challenge inherited frameworks of urban theory that throughout the twentieth century have relied on theorizing from the inside out, from city centres to urban peripheries, and from the Global North to the Global South. Furthermore, a key strength of global suburbanisms as heuristic framework for comparative urbanism is that it is open to different types of social theory and epistemological and methodological approaches. In a dialectical way, suburbs (in the cosmopolitan sense of the term) must be understood as "the ultimate capitalist commodity," as "a place of intense use value consumption (albeit realized through immense fetishization of labour practices), and at the same time a prime site for the realization of exchange values, especially in an era of financialization" (Keil, 2017, p. 5); and as "lived spaces of particularity, differentiation and multiple centralities from which the urban fabric has to be understood" (Keil, 2017, p. 13). Taking the global suburbanisms perspective as heuristic thus implies taking structure *and* agency into account, as well as abstraction *and* difference.

In order to contribute to the global suburbanism's endeavour *to write urban theory from the outside in* and to advance articulating the often conflicting approaches of planetary urbanization and postcolonial urban studies, we put the notion of *peripheral urbanization* at the centre of our inquiry. The strength of the *concept* of peripheral urbanization lies in its potential to act as a bridge or, in methodological terms, as a boundary concept. Mollinga (2008, p. 24) defines boundary concepts as "words that operate as concepts in different disciplines or perspectives, refer to the same object, phenomenon, process or quality of these, but carry (sometimes very) different meanings in those different disciplines or perspectives. In other words, they are different abstractions from the same 'thing.'" What Mollinga (2008) describes as the "same thing" in the case of peripheral urbanization refers to a set of socio-spatial processes of peripheralization that occur as distinct from but in relation to processes of centralization, with which they are dialectically related within a world-systemic capitalist totality. All three approaches show a decisive interest in peripheries and processes of peripheralization: global suburbanism lays the focus on metropolitan edges in all its possible forms; planetary urbanization broadens the view to include processes of extended urbanization and thus towards socio-spatial transformation outside of metropolitan regions and even cities in the context of capitalist restructuring; and postcolonial urban studies lead us to engage with the contemporary urban condition from the perspective of the Global South and in the realm of everyday life. What all three approaches have in common is a relational reading of space and the impetus to dislocate the centre (Roy, 2009). Hence, they seek to understand processes that we coin as peripheral urbanization. Peripheral urbanization, then, is a relational and cosmopolitan concept and connects a spatial dimension – as in its addressing of different scales, places, and locales such as urban peripheries, operational hinterlands, and the Global South – to a political, economic, and cultural dimension – which refers to the processes of peripheralization and marginalization, the ubiquitousness of the "New South" (Scholz, 2000), as well as the ubiquitousness of centre and periphery. Importantly, it also is a concept that refers to a process rather than a specific form of settlement (Harvey, 1996; Brenner & Schmid, 2015, p. 165; Keil, 2018a, p. 11).

While this common interest in processes of peripheral urbanization in the three approaches has not been taken up in a systematic way, and as such seems promising, we go one step further and propose to reconnect the notion of peripheral urbanization to the intellectual traditions and new advances of (urban) theory building *in* and *from* Latin America – in the Latin American tradition of theorizing urbanization "from the margins." As we mentioned earlier and will develop below, there is a specific tradition of studying and conceptualizing peripheral urbanization in Latin America, from its interpretation as dependent urbanization in a world-systems perspective and *hábitat popular* in the 1960s and 1970s, to renewed framings of peripheral urbanization as a specific form of space production led by the ordinary residents of the peripheries, as in Caldeira's more recent postcolonial reinterpretation (this volume). Taking into consideration the diverse contributions in this book, we propose, in the conclusion, an agenda for reformulating the concept of peripheral urbanization based on an understanding of peripherality through the concept of the "colonial power matrix" (Grosfoguel, 2011). One of the main insights of decolonial theory is that "what arrived in the Americas in the late fifteenth century was not only an economic system of capital and labor … but a broader and wider entangled power structure that an economic reductionist perspective of the world-system is unable to account for" (Grosfoguel, 2011, p. 8). In line with emerging scholarship on decolonial urban studies, we postulate that the "European modern/colonial capitalist/patriarchal world-system" (Grosfoguel 2007, 2011) is closely related to the production of urban space and the construction of knowledges about the periphery through the binaries of modernity (Farrés/Matarán, 2014, p. 37; Patel, 2016, p. 7).

In order to ground the concept of peripheral urbanization historically, empirically, and theoretically, and to further contextualize the contributions in this book, the following section traces the complex history of the development and theorization of peripheral urbanization in Latin America after 1950.

A Brief History of Performing and Framing Peripheral Urbanization in and from Latin America

This review traces developments in peripheral urbanization, starting from its early phase from the 1950s to the 1970s and the heydays of massive urban growth and the consolidation of the paradigm of "*hábitat popular*" (Connolly, 2013), on to its fading into the background with neoliberalism and the urban studies of globalization from the 1980s to the 2000s, up to the new and more pluralist perspectives emerging over the past two decades with intense debates, for instance, on in/formality, (urban) extractivism, and extended urbanization. As a matter of course, what follows cannot be considered an exhaustive review. It is rather a sketch on what we perceive as the major processes and conceptual developments in and from Latin America on peripheral urbanization in the last decades. In our periodization we relied on existing review pieces of eminent Latin American urban studies scholars such as Schteingart (2000), Connolly (2013), Pradilla Cobos (2013), and Duhau (2016) as well as on the revision of edited books from Latin America and a range of international journals of the Spanish and English-speaking urban studies community.[5] Based on our conceptualization of peripheral urbanization as laid out above, we are specifically interested in research that focuses on the peripherality of urbanization processes in Latin America in a manifold sense, that is, that deals with urbanizing Latin America from world-systemic, socio-spatial, postcolonial, and extended urbanization perspectives.

The Early Phase of Peripheral Urbanization: From the 1950s to the 1970s

In late October 1957, an event took place in Santiago, Chile, that came to epitomize the Latin American peripheral urbanization of the mid-twentieth century. After months of preparation, and supported by Marxist student leaders and communist architects, around five hundred families moved in overnight to occupy a state-owned property in the southern part of the city, throwing stones and resisting attempts by the police to remove them from the site, and setting up tents. By the following afternoon, 2,500 families (around 10,000 people) had arrived in search of a place to live, many of them elderly people and children; and once the police became aware of the presence of these vulnerable individuals, repressive action ceased (Rodríguez Matta, 2015). The settlement of *La Victoria* (meaning "victory" in Spanish) had been born, and with it an important mode of city building based on collective action, self-organization, and autoconstruction (Zibechi, 2012; Caldeira, this volume).

While dynamic urban growth in Latin America dates back to early rural-to-urban migration and foreign immigration towards the end of the nineteenth century, Latin America's true urban explosion began in the 1950s (Almandoz, 2013). Under the second phase of Import Substituting Industrialization (ISI),[6] the increasing technification of agriculture, peasant differentiation, and land concentration caused the displacement of labour from the countryside and triggered large-scale migration to the continents' major cities. In the 1950s and '60s, the "urban question converted into one of the pillars of social struggle" in Latin America (Castells, 1973).[7] All over the continent – from Santiago to Rio de Janeiro, from Bogotá to Mexico City, and from Caracas to Lima – peripheral urbanization through the growth of shantytowns and squatter settlements became a major political issue involving highly variegated forms of social organization, state intervention, and coalition building among emerging urban social movements, political parties, community organizations, and academics.

Beyond the phenomenon of autoconstruction, which was common to all the region's societies and cities, there were important differences in terms of the way in which peripheral urbanization took place. One of these differences lay in the organization of land. Land invasions and squatting on public and private land, for instance, were instrumental in Santiago and Lima, while in São Paulo, Buenos Aires, and Mexico City, migrants' principal means of access to land was through illegal subdivisions controlled by private landlords and developers.[8] These differences affected the degree and forms of collective action present in peripheral urbanization. In Chile, squatting and self-organization led to the formation of a fully fledged squatters' movement which developed into a major political force, playing a decisive role in the election of the country's first socialist president, Salvador Allende (1970–1973). While Chile in the following years became known for its high level of political mobilization in and around urban peripheries, Peru – and especially its capital, Lima – had a major impact on regional policy and theory making thanks to official recognition of the squatter settlements, not as a temporary and undesirable phenomenon but as a mode of city building that was here to stay and should be supported.

In the 1950s and '60s, not all urbanization processes on the urban peripheries or elsewhere involved land invasions, autoconstruction, and political mobilization; rather, it was a time of co-evolution of different modes of city building, combined with diverse urban processes. All over the continent, but with particular intensity in the most rapidly industrializing countries of Mexico, Brazil, and Argentina, urbanization and urban policy became entangled with the modernist and developmentalist ambitions of nationalist governments. For example, in Mexico City in 1957 – the same year that the *La Victoria* settlement in Santiago gave rise to the squatters' movement – a peri-urban shantytown was cleared in order to make room for the construction of one of the world's largest modernist public housing estates, Nonoalco Tlatelolco, comprising 15,000 residential units (McGuirk, 2014, p. 8). This project, as with many others proposed in the region, was inspired by the functionalist ideas of the Congrès Internationaux d'Architecture Moderne (CIAM), which reached its "peak in Costa's and Niemeyer's Brasilia" with the construction of a whole new capital city almost from scratch (Almandoz, 2006, p. 98). Peripheral urbanization thus took place in a hybrid setting of autoconstructed settlements, emerging metropolitan planning, and construction of public housing estates and new highway systems. The building of new transportation and communication infrastructure fuelled the intensification of historical upper- and middle-class spatial strategies that involved "fleeing old-fashioned centers and seeking new styles and landscapes that mirrored their modernizing cosmopolitism" (Almandoz, 2016), thus adding to the complex dynamics of peripheral urbanization in Latin America.

The period from the late 1960s to the early 1980s was a time of prolific theorization on the subject of Latin America's urban peripheries, which aimed to understand the rapid processes of profound urban and territorial restructuring. In her incisive review of the historical literature on peripheral urbanization in Latin America, Connolly (2013) identifies a shift from an initial stage of theory building, based primarily on positivist and phenomenological thinking, to a later stage, where structuralist, Marxist, and political-transformationalist approaches to the urban question began to dominate. The initial stage of research emerged in almost immediate reaction to the eruption of peripheral urbanization in the form of autoconstruction. Academics, activists and professionals alike directed their efforts towards locating, describing, and measuring the new settlements and their inhabitants, and their theoretical-political reactions ranged from shock at the precariousness of living conditions and a view of the settlements as a problem to be eradicated, to admiration of the capacity for collective self-organization and the interpretation of autoconstruction as a permanent phenomenon and solution to the increasingly evident housing crisis. With regard to the latter, Connolly (2013, p. 522) describes how Peru's public policy, between 1957 and 1961, of recognizing informal settlements was "elevated to the level of 'theory' with international reach" in several highly influential publications by British architect John Turner (Turner, 1967, 1968; see also Golda-Pongratz, this volume). An important issue raised in the theoretical debate was the concept of marginality. From the point of view of modernization theory, the sociologist Gino Germani (1973) and the Jesuit priest Roger Vekemans developed a theory of "dualist marginality."[9] In essence, this approach saw the new urban dwellers in Latin American cities as caught in a structural dualism between traditional and modern segments of a modernizing society and thus in a "marginal" position. The view had two important conceptual and political implications for peripheral urbanization: first, that poverty was seen as being rooted in social, psychological, and cultural characteristics of the "marginalized" population itself, which was not yet prepared for modernization; and second, that the societal integration of the urban poor was simply a matter of time, and that marginality and its socio-spatial correlate, the squatter settlements on the urban peripheries, were thus only transitory facets of an inevitable modernization process (Cortés, 2017).

Positivist and phenomenological approaches to peripheral urbanization in general, and in particular those concerning marginality, soon came under heavy criticism from Marxist thinkers whose ideas gained considerable ground in the late 1960s. The concept of peripheral urbanization as such was developed in the context of the world systems and dependency theories, and rapid urbanization in Latin America was analysed in terms of the structural position of "third world countries" in the world economy (Quijano, 1967; Friedmann, 1969; Castells, 1977; Slater, 1978; Kentor, 1981). Anibal Quijano (1967) was one of the first authors to conceptualize the shift from colonial to imperialist-industrial dependency, and how it led to new forms of peripheral urbanization. For Quijano and the other prominent voice in the debate, Manuel Castells (1977), Latin America's historical dependence permeates all institutional, social, and economic aspects of societal development, as well as the internal power structures of Latin American societies, all of which were subordinated to the dominating interests of the metropolitan economies. Around the same time as Lefebvre (2003) developed his thesis on the urban revolution, Quijano (1967) detected the "urbanization of the economic structure" (ibid., p. 25) and the "urbanization of society" (p. 27), whereby "dependent industrialization" (ibid., p. 49) leads to forms of "dependent urbanization" (Castells, 1977, p. 40) whose "logic contains the inevitability of the marginalization of growing sectors of the urban population" (Quijano, 1967, p. 49). The "urbanization of the countryside," through the proliferation of urban products and modes of life, facilitated by new communication technologies and transport infrastructure, pushes the rural population out of their occupational structures and economic relations and towards the urban peripheries of the dependently industrializing urban centre, where rural-to-urban migrants are "again, marginalized and this time, definitively" (Quijano, 1967, p. 40). From this perspective, the strong orientation of urban growth towards one metropolis ("metropolization"), and the fact that urban growth rates exceeded those of the industrial labour market created through ISI ("overurbanization") were conditioned by international economic dependency and the specific accumulation patterns that accompany it (Amin, 1974; Slater, 1978; Walton, 1977; Chase-Dunn, 1975; Kentor, 1981). Castells argued that modernization theory sought to explain "the present rate of urbanization in the 'underdeveloped' countries ... by the initial stage of the process in which they find themselves" (Castells, 1977, p. 40), while in fact, "urbanization in Latin America [was] not the expression of a process of "modernization," but the manifestation, at the level of socio-spatial relations, of the accentuation of the social contradictions inherent in its mode of development – a development determined by a specific dependence within the monopolistic capitalist system" (Castells, 1977, p. 63). While hugely influential in urban studies throughout the region, and forming part of what Connolly (2013) describes as the "Latin American paradigm of the *hábitat popular*," the theory of dependent urbanization was criticized shortly after from a theoretical point of view, mainly because of its extreme structuralism (Nun, 1969; Singer, 1973; Perlman, 1976; Pradilla Cobos, 1982). Causes of "underdevelopment" – and related dependent urbanization processes in the Global South, including Latin America – were above all attributed to the world-systemic level, with internal factors, particularly domestic class relations, often disregarded. Eventually, the economic development of China, India, and the Asian "Tiger" economies in the 1980s made dependency theory in its hitherto form difficult to defend. Thus, the experience of development in the Asian economies, which lifted millions out of poverty within a short period of time, fuelled the breakthrough of the renewed version of modernization theory, as well as the political program that came hand in hand with it in the 1970s and 1980s: neoliberalism.

From the "Hábitat Popular" to Neoliberal Globalization and Socio-spatial Fragmentation: Urban Processes and Debates from the 1980s to the 2000s

By the 1980s, the contexts and processes of peripheral urbanization in Latin America had changed. In demographic and territorial terms, rural-to-urban migration had passed its peak. While in 1930 only 20% to 30% of the population of Latin America lived in cities, by the early 1980s it was already around 70% (Boris, 2001, p. 57), and from that time on, urban growth rates were in decline. However, many cities had doubled or tripled in population over the period, and problems relating to the economic integration of the urban poor and the provision of adequate housing and infrastructure were far from resolved. Since the 1960s, ISI policies had already entered into crisis across Latin America, mainly as a result of historical dependency, leading to the foreign control of the manufacturing sector, causing an increasing outflow of profits abroad and chronic balance-of-payments deficits, increasing inflation and debt (Dos Santos, 1970; Marini, 1973; Gereffi & Evans, 1981). These conditions, along with the often violent and repressive military regimes and authoritarian governments, created the perfect economic, social, and political conditions for the implementation of new accumulation strategies by national and international capital. Following the debt crisis of the 1980s, neoliberalism began to gain ground throughout the region, orchestrated and backed up by the international institutions of the Washington Consensus and their policies of structural adjustment.

With the new macro-economic conditions of economic globalization and the neoliberal policies of liberalization and privatization, modes of city building and space production also changed, and again there were important differences between the various countries in the region. Initially, military regimes – such as that of Augusto Pinochet in Chile, which took power in 1973 – launched programs of violent evictions and the forced removal of shantytowns. For Zibechi (2012, p. 219) this was "a true urban counterrevolution" aimed at destroying the "popular territorial power" embodied in the settlements. However, these repressive measures were gradually replaced by tolerance of illegal occupations and settlements based on autoconstruction. A major influence here was the work of Hernando de Soto (1989, 2000) who claimed that informal businesses and housing in shantytowns were economic assets "that should be revived by the official legal system and turned into liquid capital so people could gain access to formal credit, invest in their homes and businesses, and thus reinvigorate the economy as a whole" (Fernandes, 2002). De Soto's ideas were put into practice not only in Peru, where he advised Alberto Fujimori's (1990–2000) dictatorial government, but in many countries across the region and the world, leading to widespread official recognition of informal settlements, regularization of property titles and land tenure, and slum upgrading programs. So-called sites-and-services programs became a key feature of policymaking by national governments and international institutions such as the World Bank (Davis, 2006, p. 87). The debate surrounding the politics of informal settlement regularization has continued to be a major focus of research on urban peripheries in Latin America to this day, particularly with regard to the impact of regularization on land markets and increasing urban segregation (see for instance Salazar, 2012).

Another policy reform that had a major impact on urban peripheries was the bringing of housing production in line with neoliberal market principles through the installation of demand-side subsidies and credit to allow working class families to purchase housing built by private developers (Gilbert, 2002). In Chile, where the policy was first put into practice, the housing deficit was heavily reduced over the course of the 1990s, virtually eliminating new informal settlements (Salcedo, 2010). Although serious flaws were soon detected in the policy, resulting in deepening patterns of urban segregation, poor quality of housing and the destruction of social capital (Rodriguez & Sugranyes, 2005), the Washington institutions continued to actively promote the model across the region (and other parts of the Global South), transferring them for instance to Brazil, Colombia, Costa Rica, Ecuador, Mexico, Panama, and Peru (Gilbert, 2002, p. 310; Monkkonen, 2012; Janoschka & Salinas, 2017, p. 44; Reis, 2017). The housing policy principles formulated by the World Bank became the blueprint for a whole new mode of peripheral urbanization: "In order to facilitate private sector engagement, governments need to work on property rights development, mortgage finance, targeted subsidies, infrastructure for urban land development, regulatory reform, organization of the building industry, and institutional development" (World Bank, 1993, as cited in Gilbert, 2002, p. 311).

In effect, these policies led to the commodification of housing "and its progressive transformation into an investment asset that may be integrated into globalized financial markets" (Janoschka & Salinas, 2017, p. 44). The same applied to land and infrastructure in the urban peripheries. In several countries in the region, public-private systems for the franchising of highway concessions were introduced, leading to a boom in the construction and modernization of inter- and intraregional transport infrastructure that fuelled outward urban growth (Blanco & San Cristóbal, 2012; Silva, 2011; Lukas & López-Morales, 2017). As Davis and De Durén (2016, p. 3) put it: "Latin American cities began to host degrees of urban sprawl unimaginable in the past, with their extension in space due not merely to population growth and the decline of agriculture but also to the growing neoliberal embrace of formal property rights and newfound public-private sector support for real estate development as a main driver of the urban economy. These trends contributed to more social and spatial inequality, with the poorest populations located in the areas of the city where critical services like transportation, water, and electricity were lacking." Another aspect of socio-spatial inequality relating to the commodification of housing, land, and infrastructure has been the rapid proliferation of gated communities for the growing middle and upper-middle classes, often located in close proximity to the modernized and financially transnationalized road systems (Crot, 2006; De Duren, 2008). "The housing blocks that had accompanied industrialization gave way to the office towers that heralded the service economy. Modernism gave way to postmodernism, and the transparent glass of rationalism became the impenetrable mirrors of a new corporate culture" (McGuirk, 2014, p. 13).

The 1980s saw a crisis of paradigms in Latin American urbanization theory production (Pradilla Cobos, 2013; Duhau, 2016). Marxist political economy, which had inspired the majority of urban research in Latin America during the 1970s and '80s, had lost ground, giving way "to a certain theoretical and methodological eclecticism" (Duhau, 2016, p. 148). However, throughout the 1990s and 2000s, urban research focused strongly on a particular set of questions, described by Duhau (2013) as the "urban studies of globalization." Interest focused now on new economic functions of Latin American metropolises in the context of globalization, and how these led to new socio-spatial orders, morphologies, and inequalities (Roberts, 2005; De Mattos, 2004; De Mattos, 2015; for an overview see Lukas & Durán, 2020). A great deal of inspiration was taken from Sassen's (1991) global city theory. On one hand, Latin American cities were positioned in the hierarchy of the global urban networks according to analysis of the expansion of the Finance, Insurance, and Real Estate (FIRE) sector in cities like Mexico City, Santiago, and São Paulo (Parnreiter, Fischer, & Imhof, 2007). On the other, Latin American cities were seen as undergoing a shift from urban polarization to socio-spatial fragmentation, mediated by processes of globalization and neoliberalization (Caldeira, 2000; Janoschka, 2002a; Borsdorf, 2003). For Mexico City, for instance, Aguilar, Ward, and Smith (2003) considered that "much of the contemporary vibrancy and dynamics of Mexico City's metropolitan development are occurring in 'hot-spots' in the extended periphery, which, to date, have rarely been considered an integral part of the megacity. Yet these areas are also some of the principal loci of contemporary globalization processes." In general, urban research on peripheral urbanization focused on the appearance of "artefacts of globalization" (De Mattos, 1999) such as shopping malls, urban megaprojects, edge cities, and service corridors (Lungo, 2005; Pradilla Cobos, 2013; Kozak, 2013). Prominent empirical examples include Santa Fe in Mexico City, and Avenida Paulista and Puerto Madero in Buenos Aires.

Of particular importance in that context was also work on emerging new patterns and scales of segregation relating to the rapid proliferation of gated communities for the middle and upper-middle classes. Some authors interpreted these as direct expressions of globalization and neoliberalization (Janoschka, 2002a, b; Borsdorf & Hidalgo, 2008); others saw them, in more-fine grained anthropological studies, as resulting from rising inequality, fear, and crime, where "those that won" (Svampa, 2001) were constructing a "city of walls" (Caldeira, 2000). The literature identified spaces such as Barro de Tijuca in Rio de Janeiro (Coy & Pochler, 2002; Coy, 2006; Keil, 2018), Nordelta in Buenos Aires (Janoschka, 2002a) and Piedra Roja in Santiago (Borsdorf & Hidalgo, 2008; Lukas, 2014). In order to understand the political and planning-related dynamics of the links that mediate between the global and the local (Pradilla Cobos, 2013), studies have been conducted of the governance dynamics and actor constellations of regional and local planning in the urban peripheries (Crot, 2006; De Duren, 2008; Lukas, 2014). The fact that gated suburbs often lie immediately adjacent to informal settlements, separated only by physical barriers and walls, has led to a discussion of the limited opportunities of functional and social integration between rich and poor (Bayón & Saraví, 2012). Latin American peripheries have also been addressed in the context of the international discussion on global processes of post-suburbanization (Roitman &

Phelps, 2011; Heinrichs, Lukas, & Nuissl, 2011) whereby some authors studied the influx of North American design ideas (Irazábal, 2006). Herzog (2015, pp. 128–129), for example, observes that in Latin American urban peripheries classical North American design narratives are spreading, "driven by a set of values that include: social exclusivity, an increasing desire for private over public space, fear of crime, a preference for predictable homogenous built environments and a greater emphasis on consumerism and shopping within artificially constructed spaces."

Another important line of research was that on the environmental dynamics related to peri-urbanization and post-suburbanization. Schteingart and Salazar (2005), for instance, studied the urban expansion of Mexico City over protected conservation zone and concomitant environmental degradation. In a similar line, Aguilar (2008) and Aguilar and Santos (2011) analyse how peri-urbanization through "illegal settlements" constitute challenges for land-use policy, again in Mexico, and Ríos and Pírez (2008) address related environmental issues with a view to the installation of gated communities in the urban periphery of Buenos Aires, while Romero and Vásquez (2005) did so for Santiago de Chile.

While new issues such as sustainability, mobility, and identity politics thus were beginning to be added to the urban research agenda (Duhau, 2013), in the 1990s and 2000s the principal focus, however, was on the socio-spatial effects of globalization processes, and an overwhelming majority of studies focused on the Latin American megacities – rather than medium-sized cities or forms of extended urbanization – actually leading to a then much discussed new model of the fragmented Latin American City (Janoschka, 2002**a, b**; Borsdorf, 2003; Borsdorf et al., 2007).

Reframing Peripheral Urbanization: New Perspectives on In/formality, (Urban) Extractivism, and Extended Urbanization in the Twenty-First Century

The leading paradigm of the early 2000s – the urban studies of globalization, with its focus on global city-dynamics and fragmentation through gated communities and megaprojects – continues to be an important line of research today (for an overview, see Lukas & Durán 2020). Segregation, inequality, and security also remain important issues and have extensively been studied (Segura, 2017; Ruiz-Tagle & Romano, 2019; Dammer et al., 2019; Haubrich & Wehrhahn, 2020). However, as the macroeconomic contexts, urbanization dynamics, and leading paradigms in social and urban theory have moved on, there have also been important developments in the processes and studies of peripheral urbanization in Latin America; in other words, classic topics are being revised from different angles (such as informality from a postcolonial perspective) and new issues have been emerging (such as climate change or extended urbanization). In spatial terms, there is much work done on the traditional and still expanding peripheries of the region's megacities, but also – and this is a rather new development – on medium-sized cities and newly emerging settlements and urbanization processes "beyond the city." The notions of the periphery, peripheralization, and peripheral urbanization are broadened, providing new insights on the urban condition in Latin America.

In the remainder of this literature review, we focus on some selected but interrelated issues that we see as particularly important for the understanding of peripheral urbanization as a boundary concept: the increasing penetration and influence of financial capital, actors, and logics in the urban peripheries; new approaches towards the issues of in/formality in theory and urban policy making; urban politics between social movements, urban revolt and the planning for climate change; and extended urbanization and the overcoming of the urban-rural divide in theory and practice. But before that, it is worth making some general remarks on the economic and political landscape over the past two decades.

In very general terms, Latin America in the 1980s and 1990s was characterized by the transition from authoritarianism to democracy, and from the developmental state to neoliberalism. In the 2000s, in many cases the latter was (supposedly) countered by a backlash towards nationalist projects of neo-extractivism. While this transformation led to important effects on peripheral urbanization in the 1980s and 1990s (for instance with a view to urban megaprojects and housing policy as described above), its real scale and scope only began to be understood in the last two decades. A case in point is structural adjustment. With the so-called debt restructuring in the late 1980s and early 1990s, Western lenders enforced a series of structural adjustment measures on Latin American debtor countries. Part of these measures was the opening up of domestic financial markets to foreign investors, banks, and other financial institutions. Massively increased and volatile capital flows, combined with a lack of government regulation of the activities of banks and financial institutions, caused devastating financial and political crises throughout the 1990s, from the tequila crisis in Mexico and the banking crisis in Venezuela in 1994, to the Brazilian financial crisis ("samba effect") in 1999 and the Argentinian breakdown in 2001. In dealing with these crises, the political pathways of Latin American countries in the 2000s can be broadly divided into two categories: on the one hand, those that have continuously conformed to the requirements of Western-dominated multilateral and financial institutions and adapted their economic and financial policies to their needs – for instance Mexico, Brazil, Chile and Peru; and on the other hand, those who, with the help of strong social movements, put left-wing governments to power in the beginning of the 2000s, with the aim of limiting or opposing external dependency, such as Ecuador, Bolivia, and Venezuela, the core countries of what has been called "pink tide" politics. Whereas in the first group of countries, the impact of steadily increased financial integration has had a direct and very visible impact on peripheral urbanization (as we will show below), the second group has not seen urban development shaped by financial investments in the same way.

However, across the region and disregarding the colour of the government, a major political-economic reaction has been the re-organization of space in the context of renewed extractivism and the re-primarization of economies (Gudynas, 2009; Bebbington & Bury, 2013; Burchardt & Dietz, 2014). This includes the massive increase in the extraction of minerals and hydrocarbons and the unprecedented advance of the agrarian frontier for the large-scale industrialized production of monocultures, occurring in the context of global value and supply chains (Bebbington & Bury, 2013; Kay, 2015; Arboleda, 2020a and this volume). In a first phase of neo-extractivism, roughly from 2000 to 2014, the progressive governments of the region were able to increase social expenditure due to the commodity price boom. In countries such as Ecuador and Bolivia, among others, poverty rates began to fall and, through the elaboration of new constitutions, collective and individual rights were consecrated and expanded (Svampa, 2019, p. 32). The region saw the multiplication of national development programs and a multiplication of megaprojects with a view to energy, mineral extraction, oil, and transgenic crops. Moreover, governments embarked on new regional initiatives, most importantly the "Initiative for the Integration of the Regional Infrastructure of South America" (IIRSA in Spanish) and the Plan Puebla Panama (PPP), with the aim of integrating transportation, energy, and telecommunication infrastructures (Arboleda, this volume). Yet, from 2014 on, commodity prices began to fall sharply and have not recovered since then, leading to even more megaprojects due to the increased economic dependence on extractivism. Importantly, in many countries these developments and, in general, shortcomings of full-blown neoliberalism in some countries and neo-developmentalism in others have been accompanied by social mobilizations around social-environmental conflicts as well as urban social movements. Some important instances include the the Zapatista uprising in 1994, the so-called water war against water privatization in Cochabamba in Bolivia in 2000, the píquetero movement in Argentina in 2001, the Forajidos en Ecuador in 2005, the student revolts in Chile in 2006 and 2011, the urban marches in Brazil 2013, and the feminist mobilizations in Argentina in 2016 (Machado & Zibechi, 2016).

Hence, new approaches to understand peripheral urbanization must be contextualized in the economic and political context of financialization, neo-extractivism, and social mobilizations. To begin with, in many Latin American countries, progressing financial integration has led to the financialization of ever larger spheres of society. While neoliberalism and globalization were the major explicatory concepts for urban development in Latin America in the 1990s and early 2000s, more recently there has been a shift in attention towards the concept of financialization. One of the major topics has been the financialization of housing. An important antecedent for housing financialization in Latin America was the creation of tradable mortgage portfolios ("securitization"), which originated in the US in the 1980s and spread rapidly across the world from there, not least with the support of the World Bank (Rolnik, 2013; Soederberg, 2015, pp. 481–482). These policies have substantially transformed the Latin American real estate industry and its business and location strategies. For Mexico, Brazil, and Chile, studies show how the neoliberal policies of flexibilization and liberalization of capital flows have facilitated the emergence of a powerful real estate and development industry (Lukas, 2014; Sanfelici & Halbert, 2015; Janoschka & Salinas, 2017). One of the effects has been a push towards economies of scale, resulting in ever bigger development projects in the peripheries of Latin America´s urban regions. In Brazil, the *Minha Casa, Minha Vida* ("My House, My Life") housing program launched in 2009 resulted in the construction of an estimated 5 million homes (Kowaltowski et al., 2015). In Mexico, in the Metropolitan Area of Mexico City alone, almost 700,000 housing units were authorized in more than 400 residential developments between 2000 and 2015. However, both in Brazil and in Mexico, these housing developments have been harshly criticized, as the residential areas often lacked adequate urban infrastructure, homes were of dismal quality, and their location on the far outskirts of cities caused excessive transportation costs for their inhabitants (Cunha Linka, 2018; Salazar, 2014; Reis, 2017). In Mexico, thousands of homes stand empty today, and many working-class families have been left with debt for homes that are uninhabitable (Reis, 2017; Salazar, Reis, and Varley, this volume).

One of the latest trends, spearheaded by Mexico City, is the financialization of municipal urban infrastructure, including water and wastewater systems, public lighting, and transport, through green municipal bonds (Hilbrandt & Grubbauer, 2020). As argued by Bigger and Webber (2021), the World Bank aims to replicate this model across the Global South, often in order to address problems caused by underinvestment in public goods produced by twentieth century structural adjustment measures and now leading to what these authors call green structural adjustment.

In the past decade, ever more realms of life – such as housing, education, water provision, consumption, and social protection – have been pervaded by the logics of financial accumulation, and public policies and services are reconfigured to extract profits from the low-income population through mass financialization and microfinance (Soederberg, 2013). In this context, Grubbauer (2020) examines how finance capital has recently gained access to the autoconstruction sector through entering strategic partnerships with building-material producers and their networks of building-supply stores. Also with reference to Mexico, Reis (2020) shows how public banking for the poor in the context of the COVID-19 pandemic has served to push forward the World Bank's "financial inclusion" agenda, especially benefitting powerful Mexican financial and corporate elites. According to Gago, the financialization of popular life and popular economy marks one of the latest frontiers of the extractive operations of capital in Latin America (Gago, 2017 also see Mason-Deeze, this volume; for a Marxist angle on the expanded notion of extraction see Arboleda, 2020a, and this volume). Moreover, it is related not only to financialization and neoliberalism "from above" as driven by the state, financial capital, and multilateral institutions as the World Bank, but also to what she calls "neoliberalism from below": A "new type of rationality that … is not purely abstract nor macropolitical but … is embodied in various ways by the subjectivities and tactics of everyday life, as a variety of ways of doing, being, and thinking that organize the social machinery's calculations and affects" (Gago, 2017, p. 2).

Besides the inclusion of urban peripheries and its people into financialized capital accumulation, the meaning and implications of "informal" urbanization remains a very important field of study. As Connolly (2013) points out, already since the 1990s "irregular" settlements have increasingly come to be understood as a permanent part of the contemporary urban order (Connolly, 2013, p. 548; Alfaro et al., 2018). However, much of scholarship of the past 15 years or so has been under the influence of poststructuralist and postcolonial thinking, and there has been increasing attention paid to the active role of "the people" in the production of urban life, politics, and space, both through everyday practices as well as through social movements. Of particular importance, thus, has been the shift towards a more dynamic and actor-oriented understanding of informality as practice and strategy, as well as the highlighting of its diversity across contexts and dimensions and its everyday lived realities and effects (Banks, Lombard, & Mitlin, 2019). Focusing on street vendors in Mexico City, Crossa (2009, 2016) emphasizes the agency and resistance exerted through different informal practices and the continuous re-working of the formal/informal boundary as a strategy of those affected by neoliberal urban policies in Mexico City. Haid and Hilbrandt (2019, p. 3), also working on Mexico City, point out that informality is also used as a strategy by state actors to push their interests and the state's active role in determining informal practices, sites, and activities. Similarly, Mueller (2017) examines the strategic employment of urban informality in the wake of the brutal restructuring of Rio de Janeiro for the Olympic games in 2016, and how this has led to spatial confinement of the urban poor based on territorial stigmatization. Also working on Rio de Janiero, Lanz (2016) dwells on the relations of urban informality, the rise of Pentecostalism and drug gangs in everyday urban life in the favelas and how these are governed. Through their relational mobility lens, Imilán et al. (2020) not only seek to overcome the view on informal housing and urban life as being bound up in stable settlements, but shift the analysis to the twin cities of Alto Hospicio and Iquique in Northern Chile and thus expand the geography of urban theory production beyond the usual suspects of Latin American megacities. In general, perhaps the most important insights of these debates are that "informal" settlements and practices are by no means exceptions or residual spaces to be overcome through development and modernization, but part and parcel of the modes of how urban space is produced and governed in Latin America (Hernández, Kellett, & Allen, 2010), and that they should be understood as a "negotiated mode of urbanization" and part "of a wider strategy of spatial governance" (Gilbert & De Jong, 2015, p. 521). For Teresa Caldeira (this volume), it is thus not "informal" practices as such that are key to urban space production in Latin America. Rather, she emphasizes that the category formal/informal (like others such as legal/illegal) "are always shifting and unstable," and that, therefore, "one needs to set aside the notion of informality (and the dualist reasoning it usually implies) and think in terms of transversal logics to understand these complex urban formations which are inherently unstable and contingent." This insight is key to "de-center urban theory and to offer a bold characterization of modes of the production of space that are different from those that generated the cities of the North Atlantic" (Caldeira, this volume).

Another important field of study with view to in/formality is the fact that Latin America's urban peripheries have (again) become sites for innovative urban interventions, and as a laboratory for urban planning and place making – in fact, not unsimilar to John Turner's time in the 1970s, described above (see also Golda-Pongratz, this volume). For McGuirk (2014, p. 23), Latin America's uniqueness lies in the fact that "no other region of the world has demonstrated the kind of collective effort and imagination that Latin America has in addressing the chronic symptoms of rapid, unplanned urbanization," and the most important recent lesson from Latin America has been "acknowledging the informal as a vital part of the city's ecosystem" (ibid., p. 25). From internationally recognized squatter settlement upgrading programs like favela-bairro (Riley, Fiori, & Ramirez, 2001) to the "Medellín-Miracle" with its cable cars – now also implemented in Rio de Janeiro, La Paz, and other cities around the region – that connect the poor peripheries to the city centre, the new politics and design principles in peripheral urbanization are seen by some as models to be learned from (McGuirk, 2014). However, others see them as urban spectacles and "tales of integration" that overemphasize the role that urban infrastructure as spatial technologies can play in changing the life conditions of the poor (Brand & Davila, 2013; Hernández & Becerra, 2017). What these debates and several related studies have shown is that without doubt Latin America has become a key part of the international urban policy circus based on intense urban interreferencing between multinational organizations, transnational companies, international consultants and experts, and local and national governments (Montero, 2017; Lukas & Brueck, 2018; Jajamovich, 2018; Irazábal & Jirón, 2020; Hilbrandt & Grubbauer, 2020; Silvestre & Jajamovich, 2020; Lukas, this volume; see also the special issue by Jajamovich & Delgadillo, 2020). Another case in point for the increasing transnational visibility and reception of Latin American urbanism (Angotti & Irazabal, 2017) is that the highly influential conferences of the World Urban Forum in 2014 and Habitat III (where the new urban agenda was presented) were held in Medellín and Quito, respectively.

At this point the contemporary complexities of urban politics in Latin America come to the forefront. On the one hand, the urban peripheries, its people, and politicians are targeted by multinational institutions, land developers, real estate firms, banks, and other financial institutions as well as policy brokers. On the other hand, as mentioned above, autoconstruction in the early phase of peripheral urbanization was a form of producing political subjectivities and counter-power from below, which remains the case until today (Zibechi, 2012 and this volume). However, in the context of democratization and changing state-society relationships, there has been a transformation of the way in which urban popular movements have identified: their identity shifted from the *poblador* (founder) and the *vecino* (neighbour) to the *ciudadano* (citizen), as noted by Dosh (2019). For Holston (2009, p. 248) the nexus between democratization and urbanization "[has] made many autoconstructed metropolises strategic arenas for the development of new formulations of citizenship in large measure based on the struggles of residents of the urban peripheries for rights to urban residence, for the right to reside with dignity, security, and mobility" (ibid.). Analysing the mobilization of the urban poor for housing in Santiago de Chile, Pérez (2017) notes that "current urban struggles allow for the rise of a type of urban citizenship through which the urban poor, by conceiving of themselves as city-makers, generate particular understandings of themselves as rights-bearers." Urban social movements today make reference to their belonging to a political community as citizens, and their struggles are not (only) about a mere legal accreditation of rights, but about the exercise of them (Álvarez Enríquez, 2017). However, this process of citizen formation is paradoxical, as Pérez (2017) states for Chile: "Although it is the result of mobilizations aimed at contesting market-based policies, it is permeated by a neo-liberal ethics through which the urban poor legitimate themselves as urban citizens." Moreover, city governments have learned to apply the "right to the city" strategically themselves, while they are in fact practising neoliberal governance (Delgadillo, 2012). Also in those ("pink tide") countries, where social movements have helped progressive parties into power, different studies have shown that once in the government, they did not deliver what they promised in terms of ethno-racial and spatial justice in the sphere of urban policy. For instance, in the case of Ecuador, Durán et al. (2020, and this volume) elaborate on how urban and housing policy under Raphael Correa has reproduced neoliberal urbanism with its inherent violence. For Bolivia, Horn (2018) states that there remains an important gap "between constitutional rhetoric on ethno-racially just cities and actual government practices which often ignore the collective rights of Indigenous peoples" (Horn 2018).

However, it would be incomplete to view urban social movements in Latin America merely through rights- and citizenship-based approaches. Marcelo Lopes de Souza points out that libertarian thought – emphasizing principles such as self-management (*autogestión*), autonomy and a left critique of Marxist principles of verticality, economism, developmentalism, and authoritarianism – have always been present in Latin American emancipatory social movements but have again come to the forefront since the 1990s (Lopes, 2016). This is also true for the perhaps most visible social movement in Latin America's cities in the past decade, the feminist movement, which has emerged with great force all over Latin America, not only as a general societal development but in particular as a response to the increase in feminicides and violence against women. Here, reference is made not only towards the enforcement of rights and the rule of law, but also towards "revitalizing of a widespread, radical, non-state-centric politics (Gago & Gutierrez, 2018). In fact, Verónica Schild argues that what is distinctive about the recent massive women's mobilizations in Latin America is that their gender demands are linked to broader demands for the end of a neoliberal capitalist model of development and its devastating social, economic, and ecological effects (Schild, 2019, p. 24). In this context, authors and activists working in a libertarian tradition stress and foment collective action based on collaboration and the real-world construction of independent, lived alternatives beyond institutional politics and the institutionalization of social movements and their claims (Zibechi, 2012; Gago & Mezzadra, 2015; Streule & Schwarz, 2019). Raúl Zibechi (this volume), calls it the "community approach," "an anti-state mechanism which at the same time fosters freedom through non-hierarchical relationships and the active participation of all members."

Another increasingly important element of urban politics in Latin America that intersects with several of the above-mentioned dynamics is that of planning in the context of changing socio-environmental conditions, triggered by climate change and the exceptionally high disaster risk existing in the region. Authors such as Castro et al. (2015) and Sandoval and Sarmiento (2020) focus on the social (re)production of multiple environmental risks in informal settlements, Rebotier, Pigeon, and Metzger (2019) on the complexities of risk knowledge and management in coastal Ecuador while others analyse post-hazard situations as moments for disaster capitalism and the deepening of neoliberalism, such as Sandoval et al. (2020) and Fuster-Farfán et al. (2020) for Chile. An ever more important area of policy, politics and research is to understand how cities are affected by and reacting to climate change. Hardoy and Romero Lankao (2011), for instance, dwell on challenges and options to mitigation and adaptation responses and highlight the need for what they call a "pro-poor perspective in adaptation actions." Henríquez and Romero (2019) deal with changing urban climate and its effects on the vulnerability of different social groups, and Krellenberg, Welz and Link (2017) with urban climate politics at the local level. Related studies focus on ecosystem services (Perevochtchikova & Rojo Negrete, 2015) and urban green infrastructure planning (Vasquez, 2016) – issues that are becoming central elements of local and regional policy discourses and are also sometimes strategically enacted in order to legitimize the interests of specific power blocs and their urban projects (Anguelovsky, Irazábal, & Conolly, 2019). In line with power-sensitive critical engagements at the intersection of urban politics and socio-environmental relations, there are also studies that approach peripheral urbanization from the perspective of urban or suburban political ecology (Quimbayo & Vásquez, 2016). Coates and Nygren (2020), working on the political ecology of flood and landslide risk in Brazil and Mexico for instance argue "that clientelist governance and state making, including complex forms of political favouritism, create urban hazardscapes, as much as the management of urban disasters acts to reconfigure patron–client relations within 'hazardstates,'" providing a new look on clientelism, an enduring feature of Latin American politics. In another emerging register, Lukas, Fragkou, and Vásquez (2020) argue for the need of a suburban political ecology in order to understand the power dynamics and inequalities inherent in the commodification and financialization of suburban land and water in Santiago's urban periphery, while Arboleda (2016a) brings urban political ecology to the non-city in studying the effects and global metabolic flows of matter, energy, and capital in the neo-extractivist expansion of logistical and financial infrastructures in Northern Chile. This resonates with a tradition of studying questions of urban environmental (in)justice and environmental suffering in Latin America (see Vásquez et al., 2017, for an overview and Swistun, 2018, for an infamous case study based in Buenos Aires).

The above-mentioned tendency of broadening the scale and scope in conceiving and analysing the processes, metabolic flows, and imprints of urbanization is linked to the recognition that urbanization has to be seen as a ubiquitous territorial process that reaches far beyond the limits of the region´s megalopolises and, in fact, beyond the limits of the city itself. The empirical antecedents for this epistemological move in Latin America are twofold: on the one hand, with view to the national and regional urban systems, population growth is now concentrating in small- and medium-sized cities rather than in the megacities. Especially intermediate cities, with a population between roughly 500,000 and 1 million, show the highest growth rates (UN, 2014, p. 15ff.). On the other hand, through the expansion of the extractive frontiers across the region through transnationally organized capital in the agro-industry, mining, fishing, and forestry sectors the urbanization of the rural economies and societies progress at an enormous pace, leading to the emergence of completely new cities in formerly rural areas (Richards & VanWey, 2015; Durán et al., this volume). These observed socio-spatial transformations have brought about a proliferation of new concepts, in search of ways to come to grips with a reality where the nature of the "urban" and the "rural" itself has become questionable, and urban society and space are produced translocally.

From the perspective of urban geography, the increasing intermingling of the "urban" and the "rural" for quite a while has been subject of debates on peri-urbanization (*periurbanización*) – a term that has in fact been far more common in the Latin American academy than "suburbanization" (Gilbert, 2020). Recent literature has especially dealt with the peri-urban areas of the fast-growing medium-sized cities. The peri-urban denotes highly dynamic, neither urban nor rural and difficult to demarcate territories in the urban peripheries. Peri-urban territories are not limited by clear political-administrative boundaries, but characterized by physical, social, and virtual networks along rural-urban flows of people, commodities, goods and services, raw material, and technology (Méndez-Lemus et al., 2017, p. 2). Fast-growing urban settlements are gradually being incorporated into adjacent municipalities with rural activities and identities. In these territories, small farming and local markets co-exist with globalized industrial and consumption areas. Former rural residents are subjected to livelihood transformations through pressures such as increasing land prices and environmental degradation (Méndez-Lemus & Vieyra, 2014; Cruz, 2016), but many households actually combine typically urban and rural production and consumption activities (Lerner et al., 2013). Scholars have also labelled these territories as "rural-urban fringe" (Aguilar, 2008). In this context, there has also been a shift towards understanding urbanization as a multi- or translocal process. For the Bolivian case, scholars point out the multilocal nature of peri-urban spaces, as urban residents keep up continuous relationships with family members and communities in other places. In this sense, the urban and the rural are understood as mutually constituting each other (Cielo & Antequera, 2012).

Related debates on the increasing fluidity of the "urban" and the "rural" in Latin America (Barbieri et al., 2007) have also taken place in rural and agrarian studies. Beginning in the 1990s, the Latin American "new rurality" approach started to question the ways in which rural life and economy have been understood and highlighted the changes that have occurred with neoliberalization, in particular the significance of non-agricultural activities and incomes in the livelihood strategies of the "rural" population (Kay, 2008). The increasing socio-spatial intermingling of the urban and the rural has, for instance, been discussed with regard to the deepening crisis of smallholder agriculture and changing land-use, and conflict over land and water resources (Appendini & Torres, 2008; Ávila Sánchez, 2016, p. 52), food security (Lerner & Appendini, 2011), the merging of the formerly distinct sectors of agriculture and industry through finance capital and rural-urban "classes of labour" (Reis, 2019), and the reformulation of the "agrarian question" in the face of the mutual constitution of spaces of urban mass consumption and the organizational restructuring of agro-industrial hinterlands (Arboleda, 2020b).

A conceptual arsenal that has gained increasing influence is Lefebvre's urban revolution and the related notions of planetary and extended urbanization (Orellana, Link & Noyola, 2016; Link & De Mattos, 2015; Castriota & Tonucci, 2018). The concept of extended urbanization, developed by the Brazilian urbanist and planner Roberto Luís Monte-Mor in the 1980s, is put forward to rethink and overcome the dualities of the urban and the rural, the city and the countryside (Castriota & Tonucci, 2018). Similarly to what was argued during the twentieth-century debate on dependent urbanization (Quijano, 1967; Castells, 1977), Monte-Mór and Castriota (2018, p. 341) see a "double process of industrialization and urbanization of the agrarian world" in Brazil, whereby "the contemporary rural submerges either in industrial processes (like agribusinesses) and their productive logic, or else in urban extensive processes focused on everyday life and the quality of collective reproduction" (Monte-Mór & Castriota, 2018, p. 342). In other regions of Latin America, especially coastal areas and small islands, similar socio-spatial processes are happening in the context of "tourism urbanization" (Brooks, 2018).

Other authors expand "the notion of extended urbanization to the mines and fields in all corners of the earth from where concentrated forms of urbanization (i.e., cities and suburbs) are being provisioned" (Keil, 2018, p. 176). Arboleda (2020a), for instance, sheds light on the operational landscapes of extractivist capitalism that function as metabolic vehicles of planetary urbanization, thus deepening the understanding of "the combined process of metropolitanization and extended urbanization" (Monte-Mór, 2014, p. 112) that increasingly shapes Latin American territorial transformation and its conceptualization. Related studies focus on the extension of infrastructure networks and logistics chains – such as the Initiative for the Integration of the Regional Infrastructure of South America (IIRSA in Spanish) – into regions that hitherto had not been part of urban studies, such as the rainforests of Mexico, Ecuador, Brazil, and Venezuela (Wilson, 2014; Kanai, 2014,

2016; Wilson & Bayon, 2017; Durán et al., this volume; Fonseca, this volume). In the context of what they call "infrastructure scramble," Kanai and Schindler (2018) detect a territorial reconfiguration resulting in uneven and complex urban configurations across vast territories and peri-urban areas.

Summing up this literature review on roughly 70 years of performing and theorizing peripheral urbanization in Latin America, one recognizes an increasing theoretical pluralism in knowledge production along the lines of more and more specialized academic fields and their respective disciplinary paradigms and research foci, especially in the past two decades. The debate on peripheral urbanization in Latin America stems from the mid-twentieth century, when it evolved around theoretical and political-practical discussions about the *hábitat popular* and dependent urbanization. While studies on the urban peripheries of Latin America's megacities have continued until today, peripheral urbanization as theoretical construct faded into background towards the end of the twentieth century, and urbanization processes were very much viewed through the lenses of globalization and neoliberalization. Recently, there has been a renewed interest in urban peripheries and peripheral urbanization, albeit from a range of new perspectives and in the face of new empirical developments, in particular, financialization, and more nuanced understandings of the development of neoliberal capitalism; increased attention towards the agency of residents, social movements, and politics in the production of urban space and life; and a broadened focus on medium-sized cities, and newly emerging settlements and urbanization processes "beyond the city." Theoretically, there has been a revival of Marxist thinking, which is due to an attempt to understand the extended forms of urbanization across the continent that have emerged in the context of renewed extractivism and the urbanization of the countryside. In parallel, studies inflected by postcolonial urban studies are rejuvenating theoretical debates on in/formality, social movements, and political agency from below.

The Structure of the Book

The structure of this book broadly follows the major recent trends identified in this literature review, cutting through the complex field of research on peripheral urbanization in Latin America that has emerged over the past two decades. This divides our contributions into four sections: (1) framing peripheral urbanization in Latin America, (2) metropolitan peripheries under financialization and urban extractivism, (3) community, commoning, and political agency on the urban margins, and (4) extended urbanization between new rurality and operational landscapes. Part 1 presents three theoretical pieces by authors that make important contributions to the concept of peripheral urbanization from different angles. The first contribution is authored by Teresa Caldeira, one of the key thinkers on postcolonial urbanism. Based on her long-standing work as an anthropologist, she uses the notion of peripheral urbanization to analyse autoconstruction as the principal means of urban space production in Latin America and the Global South in general. She argues that peripheral urbanization refers to a general mode of urban space production that (a) operates with a specific temporality and agency, (b) engages transversally with official logics, (c) generates new modes of politics, and (d) creates highly unequal and heterogeneous cities. Second, the writer and activist Raúl Zibechi, well known for his work on urban peripheries as sites of resistance to neoliberal capitalism, reflects on the concept of the community in urban neighbourhoods. He traces the role of community from ancient origins to more recent processes in Mexico and Venezuela and highlights the centrality of collective work in the everyday production of community. The third chapter is by Marxist political scientist Martín Arboleda, who recently made important contributions to urban studies from the perspective of Marxist international political economy. From a planetary urbanization perspective, he highlights the need to rethink natural resource extractivism beyond the mere spatiality of shafts and pits. In particular, he shows how the so-called commodity super-cycle of the early twenty-first century has led to new patterns of uneven geographical development in Latin America, addressing the metabolic interdependence between built and unbuilt environments, city and non-city.

These broader conceptual contributions on peripheral urbanization are followed by twelve case studies from different countries of Latin America, notably Argentina, Bolivia, Brazil, Chile, Ecuador, Mexico, Peru, and Venezuela, which use different theoretical and methodological approaches towards studying peripheral urbanization.

In the second part of the book we revisit the peripheries of the region's megacities and look at the entanglements and complexities of traditional and new dynamics of peripheral urbanization. Clara Salazar, Nadine Reis, and Ann Varley focus on the political and economic processes surrounding the production of large-scale financialized social housing complexes in the megaregion of Mexico City, and investigate the impact of neoliberal changes to housing and agrarian policies on existing modes of urban space production through autoconstruction. They show that the new dynamics of peripheral urbanization, driven by finance capital, have led to highly problematical socio-territorial outcomes, and have actually exacerbated the working class' housing problem. César Cáceres also focuses on peri-urban megaprojects, but with special regard to the residential experience and ways of life of emerging middle-income groups on the outskirts of Santiago, Chile. He reflects upon the "paradoxical happiness" of the peri-urbanites: on one hand, the achievement of higher residential standards in entrepreneurial urban environments leads to the satisfaction of needs and aspirations; on the other, the elevated costs of the entirely privatized services produce increased financial vulnerability. Concluding part 2, Liz Mason-Deese takes an in-depth look at processes of financial extractivism involving poor populations in the periphery of Buenos Aires. Based on a study of the Unemployed Workers' Movement (*Movimiento de Trabajadores Desocupados*) in La Matanza, she describes how the vital popular and solidarity economies developed by the poor and unemployed during the Argentinian crisis of 2001 have been permeated by the operation of financial mechanisms that seek to include these territories in the dominant circuits of capital accumulation.

In the third part of the book the geographical focus of contributions remains on the classical urban peripheries of Latin American megacities, but this time looks at grassroots actions and strategies of resistance. The chapter by Kathrin Golda-Pongratz revisits Lima's *cono norte*, the settlements born in the 1960s through autoconstruction, and where the renowned architect and visioner John Turner studied informal housing production and strategies of self-organization. After describing economic and cultural transformations through the advancement of consumerism and "the power of the malls," which deeply reshaped identities and sub/urban ways of life, the chapter presents scholar-activist interventions aimed at rebuilding community spirit and the reconstruction of urban memory through new territorial practices such as collective walks, ceremonies, and a documentary film. In the next chapter, João Tonucci and Rodrigo Castriota take the case study of Belo Horizonte and explore the upsurge of housing occupations in Brazil, highlighting the new scale of mobilization, the different socio-political dynamics, and new forms of organization. They frame the occupations as "common spaces," neither exclusively private nor public, and address the ambivalence of four interrelated dimensions and elements of commoning: public space, women and social reproduction, agriculture and urban nature, and the question of land. In her chapter on peri-urban Cochabamba, Hannah-Hunt Moeller argues that the nature of Latin American peri-urban spaces is not one of rural-urban continuum, but of distinct spatial affordances enhanced by heritage-based practices. The research proposes that space and its occupants can engender a resistance to singular global forces, as the heterogeneous nature of space is made possible by flows between rural and urban adjacencies. Next, Christina Schiavoni and Ana Felicien explore how communities in Caracas, Venezuela, are blurring the urban-rural divide as they reconfigure territory, relationships, and identities in the attempted construction of food sovereignty. Through the prism of food as a nexus of social relations, they show how the urban-rural divide in Venezuela was constructed over time, and the ways in which it is being deconstructed today. This involves a reconceptualization of both territory and relationships, and a construction of common identities. They argue that it is particularly in the urban peripheries where connections to the countryside, deeply embedded in communities' collective consciousness, are manifesting in everyday practices around food, and being scaled upward and outward through grassroots efforts aimed at blurring the urban-rural divide.

The fourth part of the book looks at new sites and processes of peripheral urbanization in Latin America, beyond the peripheries of the continent's classical megacities. Gustavo Durán, Jonathan Menoscal, and Manuel Bayón take a regional look at the emergence of small and intermediate cities along major routes and in commercial areas, and consider how these are linked to regional and national economic dynamics. Based on the concepts of concentrated and extended urbanization, they analyse how the emergence and expansion of the urban network is related to the fast-growing agro-export industry, and the extent to which it is characterized by the absence of planning and land management, the lack of basic infrastructure, and the production of new forms of informality and vulnerabilities. Michael Lukas analyses the entanglements of territorial and political dynamics in one of the continent's major extractive cities, Antofagasta, in the Chilean Atacama Desert. With the expansion and transnationalization of the copper mining industry, the city has seen processes such as a huge wave of international immigration, a housing crisis, and mineral pollution on hitherto unknown scales, all of which are leading to a range of governance and planning responses. The chapter shows how these new political and planning approaches are not directed towards solving the urban crisis, but rather at the containment of growing social protest movements and the preparation of urban space for the future necessities of the global mining industry. Claudia Fonseca takes us to Yucatán, in southeast Mexico, described in popular discourse as the mystical land of the ancient Mayas, and perceived as geographically remote, rural, and isolated from the rest of the country. Fonseca shows how, since the 1980s, the state has increasingly promoted the maquiladora industry, which has prompted changes in the urbanization process and led to the commodification of nature, racialization of bodies, and the expansion of abstract space. Alisson Flavio Barbieri and Ricardo Ojima lead us to the Brazilian Amazon and the country's northeastern semi-arid region. They argue that urbanization in Brazil has been incomplete in its demographic, economic, and social dimensions, that it reflects the coexistence of a dual society, and that contemporary "urbanization" cannot be understood without an appropriate assessment of rural and agrarian changes that considers how populations have adopted new rural-urban livelihoods and mobility strategies. Finally, Gabriela Torres-Mazuera reflects on the urbanization of the Mexican countryside. Through an anthropological study in Central Mexico, she analyses the deeper meanings and transformations of village life towards what she terms "urbanized rurality." In particular, she argues that the urbanization of the countryside is reflected in a changed role of the state in the countryside, and a transformation of political institutions and policy making on the local and national state levels.

The conclusion of this book revisits the various case studies and perspectives contributed by our authors in order to identify the key trends of peripheral urbanization in Latin America: the inclusion of suburban spaces into the financialized capital cycle; urban peripheries as sites of counterhegemonic politics; and the changing sociospatial patterns of urbanization and the "explosion of space." Based on these trends, we present two propositions on peripheral urbanization, both as a conclusion to this investigation and as possible lines for future research. First, we conclude that urbanization in twenty-first century Latin America must be understood as peripheral in the sense that it is still dependent on the specificity of the region's integration into the capitalist world system. Finally, we argue that further efforts towards an integration of the different strands of scholarly traditions and agendas in critical urban studies is vital today, and present an outlook of how peripheral urbanization as a boundary concept could be used to do so.

ACKNOWLEDGMENT

This chapter is supported by the Academic Productivity Support Program, PROA VID 2019, University of Chile, and the Fondecyt de Iniciación proyect No. 11150789.

NOTES

1 "Favela" is the name given to a specific type of Brazilian shantytown. Many other names are used in other countries, including *barriadas* and *pueblos jóvenes* in Peru, *colonias proletarias* and *colonias populares* in Mexico, *campamentos* and *poblaciones (callampas)* in Chile, and *villas (miserias)* in Argentina.

2 https://www.rnd.de/gesundheit/corona-in-brasilien-manaus-erlebt-die-apokalypse-sauerstoff-schickt-uns-sauerstoff-3CZ4JKBP45D5TMGNBGU4XTG5LI.html. See also: https://www.worldin.news/78953/2021/01/powerful-second-wave-what-triggered-the-apocalyptic-conditions-in-manaus.html.

3 Specific to Latin American urbanization is the fact that explosive peripheral urban growth set in much before the more recent processes in Asia and Africa, and as such has reached a state of greater "maturity" in terms of rural-to-urban-transition.

4 For an overview, see the special issues by Peake et al. (2018) in *Environment and Planning: Society and Space*; and Wilson and Jonas (2018) in *Urban Geography*; see also Angelo and Goh (2020).

5 It is worth mentioning that both of us are of German origin, were trained in European universities and research institutions and have been based at universities and research institutions in Chile (Lukas) and Mexico (Reis) for several years. Due to our geographical and cultural background and our language skills, the review focused mainly on literature in Spanish and English, and much less on the huge amount of Brazilian scholarship in Portuguese and other languages.

6 The first "horizontal" phase of Import Substituting Industrialization (ISI), introduced after the sharp drop of primary commodity prices with the Great Depression in the 1930s, focused on domestic production of consumer nondurables and the local assembly of consumer durables. From the mid-1950s to the 1970s, the second "vertical" phase of ISI attempted to expand import substitution by internalizing all phases in the manufacture of consumer durables (Gereffi & Evans, 1981).

7 By 1950 several countries in the region were mostly urban in demographic terms, with 78% of Uruguay's population living in cities, 65% in Argentina, 58% in Chile, and 53% in Venezuela (Almandoz, 2006, p. 95).

8 While squatting and the occupation of public land was "not very common" in São Paulo, and only 1% of the population lived in squatted settlements in the early 1970s; in Santiago this figure was 20% (Caldeira, this volume).

9 This work on dualist marginality was very much in line with work on the "culture of poverty" by Oscar Lewis (1961), which became particularly influential in the anglophone world.

REFERENCES

Aguilar, A.G. (2008). Peri-urbanization, illegal settlements and environmental impact in Mexico City. *Cities, 25*(3), 133–145. https://doi.org/10.1016/j.cities.2008.02.003

Aguilar, A.G., & Santos, C. (2011). Informal settlements, needs and environmental conservation in Mexico City: An unsolved challenge for land-use policy. *Land Use Policy, 28*(4), 649–662. https://doi.org/10.1016/j.landusepol.2010.11.002

Aguilar, A.G., Ward, P., & Smith, C.B. (2003). Globalization, regional development, and mega-city expansion in Latin America: Analyzing Mexico City's peri-urban hinterland. *Cities, 20*(1), 3–21. https://doi.org/10.1016/S0264-2751(02)00092-6

Alfaro d'Alencon, P., Smith, H., Alvarez de Andres, E., et al. (2018). Interrogating informality: Conceptualisations, practices and policies in the light of the New Urban Agenda. *Habitat International, 75*, 59–66.

Almandoz, A. (2006). Urban planning and historiography in Latin America. *Progress in planning, 65*(2), 81–123. http://dx.doi.org/10.1016/j.progress.2006.02.002

Almandoz, A. (2013). *Modernización urbana en América Latina: De las grandes aldeas a las metrópolis masificadas*. Instituto de Estudios Urbanos y Territoriales, Pontificia Universidad Católica de Chile.

Almandoz, A. (2016). Latin American urbanism, 1850–1950. In *Oxford bibliographies in Latin American studies*. http://www.oxfordbibliographies.com/view/document/obo-9780199766581/obo-9780199766581-0008.xml#obo-9780199766581-0008-div1-0007

Álvarez Enríquez, L. (2017). Ciudadanía, construcción de Ciudadanía y Ciudad. In Eibenschutz, Roberto y Carlos Lavore (Ed.), *La ciudad como cultura: Líneas estratégicas de política pública para la Ciudad de México*. Debate/UAM/CDMX.

Amin, S. (1974). *Accumulation on a world scale: A critique of the theory of underdevelopment*. Monthly Review Press.

Angelo, H., & Goh, K. (2020). Out in space: Difference and abstraction in planetary urbanization. *International Journal of Urban and Regional Research*. Online first: https://onlinelibrary.wiley.com/doi/abs/10.1111/1468-2427.12911

Angotti, T., & Irazabal, C. (2017). Planning Latin American cities: Dependencies and "best practices." *Latin American Perspectives, 44*(2), 4–17. https://doi.org/10.1177%2F0094582X16689556

Anguelovski, I., Irázabal-Zurita, C., & Connolly, J.J.T. (2019). Grabbed urban landscapes: Socio-spatial tensions in green infrastructure planning in Medellín. *International Journal of Urban and Regional Research, 43*(1), 133–156. https://doi.org/10.1111/1468-2427.12725

Appendini, K., & Torrres, G. (2008). ¿Ruralidad sin agricultura? Perspectivas multidisciplinarias de una realidad fragmentada. El Colegio de México.

Arboleda, M. (2016a). In the nature of the non-city: Expanded infrastructural networks and the political ecology of planetary urbanisation. *Antipode, 48*(2), 233–251. https://doi.org/10.1111/anti.12175

Arboleda, M. (2016b). Spaces of extraction, metropolitan explosions: Planetary urbanization and the commodity boom in Latin America. *International Journal of Urban and Regional Research, 40*(1), 96–112. https://doi.org/10.1111/1468-2427.12290

Arboleda, M. (2020a). *Planetary mine: Territories of extraction in late capitalism*. Verso.

Arboleda, M. (2020b). Towards an agrarian question of circulation: Walmart's expansion in Chile and the agrarian political economy of supply chain capitalism. *Journal of Agrarian Change, 20*, 345–363. https://doi.org/10.1111/joac.12356

Arboleda, M. (2020c). From spaces to circuits of extraction: Value in process and the mine/city. *Nexus, 31*(3), 114–133. https://doi.org/10.1080/10455752.2019.1656758

Ávila Sánchez, H. 2016. Periurbanización y gestión territorial: Algunas ideas y enfoques disciplinarios. In A. Vieyra, Y. Méndez-Lemus, & J. Hernández-Guerrero (Eds.), *Procesos urbanos, pobreza y ambiente: Implicaciones en ciudades medias y megaciudades* (pp. 49–70). Universidad Nacional Autonoma de México, Centro de Investigaciones en Geografía Ambiental (CIGA). Morelia.

Banks, N., Lombard, M., & Mitlin, D. (2020). Urban informality as a site of critical analysis. *The Journal of Development Studies, 56*(2), 223–238. https://doi.org/10.1080/00220388.2019.1577384

Barbieri, A.F., Monte-Mór, R.L.M., & Bilsborrow, R.E. (2007, June 11–13). *Towns in the jungle: Exploring linkages between rural-urban mobility, urbanization, and development in the Amazon* [Paper presentation]. The PRIPODE workshop on Urban Population, Development and Environment Dynamics in Developing Countries. Nairobi, Kenya. http://www.cicred.org/Eng/Seminars/Details/Seminars/PDE2007/Papers/Barbieri_MonteMor_Bilsborrow_paperNairobi2007.pdf

Bayón, M.C., & Saraví, G.A. (2012). The cultural dimensions of urban fragmentation: Segregation, sociability, and inequality in Mexico City. *Latin American Perspectives, 40*(2), 35–52. https://doi.org/10.1177%2F0094582X12468865

Bebbington, A., & Bury, J. (2013). Political Ecologies of the Subsoil. In A. Bebbington & J. Bury (Eds.), *Subterranean struggles: New dynamics of mining, oil and gas in Latin America* (pp. 1–16). University of Texas Press.

Bhan, G., Caldeira, T., Gillespie, K., & Simone, A.M. (2020). The pandemic, southern urbanisms and collective life. Essays,

Society & Space, 3. https://www.societyandspace.org/articles/the-pandemic-southern-urbanisms-and-collective-life

Bigger, P., & Webber, S. (2021). Green structural adjustment in the World Bank's resilient city. *Annals of the American Association of Geographers, 111*(1), 36–51. https://doi.org/10.1080/24694452.2020.1749023

Blanco, J., & San Cristóbal, D. (2012). Reestructuración de la red de autopistas y metropolización en Buenos Aires. *Revista Iberoamericana de Urbanismo, 8*, 73–88. http://hdl.handle.net/2099/13035

Boris, D. (2001). Zur Politischen Ökonomie Lateinamerikas: Der Kontinent in der Weltwirtschaft des 20. *Jahrhunderts*. VSA Verlag.

Borsdorf, A. (2003). Cómo modelar el desarrollo y la dinámica de la ciudad latinoamericana. *EURE Revista Latinoamericana de Estudios Urbano Regionales, 29*(86), 37–49. http://dx.doi.org/10.4067/S0250-71612003008600002

Borsdorf, A., & Hidalgo, R. (2008). New dimensions of social exclusion in Latin America: From gated communities to gated cities, the case of Santiago de Chile. *Land Use Policy, 25*(2), 153–160. https://doi.org/10.1016/j.landusepol.2007.04.001

Borsdorf, A., Hidalgo, R., & Sanchez, R. (2007). A new model of urban development in Latin America: The gated communities and fenced cities in the metropolitan areas of Santiago de Chile and Valparaíso. *Cities, 24*(5), 365–378. https://doi.org/10.1016/j.cities.2007.04.002

Brand, P., & Dávila, J. (2013). Metrocables and "social urbanism": Two complementary strategies. In J. Dávila (Ed.), *Urban mobility and poverty: Lessons from Medellín and Soacha, Colombia* (pp. 46–54). DPU-UCL and Universidad Nacional de Colombia.

Brenner, N. (2014). *Implosions/explosions: Towards a study of planetary urbanization*. Jovis Verlag.

Brenner, N., & Schmid, C. (2014). The "urban age" in question. *International Journal of Urban and Regional Research, 38*, 731–55. https://doi.org/10.1111/1468-2427.12115

Brenner, N., & Schmid, C. (2015). Towards a new epistemology of the urban. *City, 19*(2–3), 151–182. https://doi.org/10.1080/13604813.2015.1014712

Brenner, N., & Schmid, C. (2017). Elements for a new epistemology of the urban. In S. Hall & R. Burdett (Eds.), *The SAGE handbook of the 21st century city*. SAGE Publications. http://dx.doi.org/10.4135/9781526402059.n4

Brooks, S. (2018). Growth of touris urbanisation and implications for the transformation of Jamaica's rural hinterlands. In P. Horn, P. Alfaro, & A. Duarte (Eds.), *Emerging urban spaces: A planetary perspective* (pp. 129–148). Springer.

Burchardt, H.-J., & Dietz, K. (2014). (Neo-)extractivism: A new challenge for development theory from Latin America. *Third World Quarterly, 35*(3), 468–486. https://doi.org/10.1080/01436597.2014.893488

Caldeira, T.P.R. (2000). *City of walls: Crime segregation, and citizenship in São Paulo*. University of California Press.
Caldeira, T.P.R. (2009). Marginality, again?! *International Journal of Urban and Regional Research, 33*(3), 848–853. https://doi.org/10.1111/j.1468-2427.2009.00923.x
Castells, M. (1973). *Movimientos sociales urbanos*. Siglo XXI.
Castells, M. (1977). *The urban question: A Marxist approach*. MIT Press.
Castriota, R., & Tonucci, J. (2018). Extended urbanization in and from Brazil. *Environment and Planning D: Society and Space, 36*(3), 512–528. https://doi.org/10.1177%2F0263775818775426
Castro, C.P., Ibarra, I., Lukas, M., Ortiz, J., & Sarmiento, J.P. (2015). Disaster risk construction in the progressive consolidation of informal settlements: Iquique and Puerto Montt (Chile) case studies. *International Journal of Disaster Risk Reduction, 13*, 109–127. https://doi.org/10.1016/j.ijdrr.2015.05.001
Chase-Dunn, C. (1975). The effects of international economic dependence on development and inequality: A cross-national study. *American Sociological Review, 40*, 720–738. https://doi.org/10.2307/2094176
Cielo, C., & Antequera, N. (2012). Ciudad sin frontera: La multilocalidad urbano-rural en Bolivia. *Eutopía, 3*, 11–29.
Cinha, C. (2018). *Shortcomings of Brazil's Minha Casa, Minha Vida programme*. Urbanet. https://www.urbanet.info/brazil-social-housing-shortcomings/
Coates, R., & Nygren, A. (2020). Urban floods, clientelism, and the political ecology of the state in Latin America. *Annals of the American Association of Geographers, 110*(5). https://doi.org/10.1080/24694452.2019.1701977
Connolly, P. (2013). La ciudad y el hábitat popular: Paradigma latinoamericano. In R.R.V. Blanca & E. Pradilla Cobos (Eds.), *Teorías sobre la ciudad en América Latina* (Tomo 1, pp. 505–562). Universidad Autónoma Metropolitana.
Connolly, P. (2017). Latin American informal urbanism: Contexts, concepts and contributions, with specific reference to Mexico. In F. Hernández & A. Becerra Santacruz (Eds.), *Marginal urbanisms: Informal and formal development in cities of Latin America* (pp. 22–46). Cambridge Scholar Publishing.
Cortés, A. (2017). Aníbal Quijano: Marginalidad y urbanización dependiente en América Latina. *POLIS Revista Latinoamericana, 16*(46), 221–238. http://dx.doi.org/10.4067/S0718-65682017000100221
Coy, M. (2006). Gated communities and urban fragmentation in Latin America: The Brazilian experience. *GeoJournal, 66*(1/2), 121–132. https://doi.org/10.1007/s10708-006-9011-6
Coy, M., & Poehler, M. (2002). Gated communities in Latin-American megacities: Case studies from Brazil and Argentina. *Environment and*

Planning B: Planning and Design, 29, 355–370. https://doi.org/10.1068%2Fb2772x

Crossa, V. (2009). Resisting the entrepreneurial city: Street vendors' struggle in Mexico City's historic center. *IJURR, 33*(1), 43–63. https://doi.org/10.1111/j.1468-2427.2008.00823.x

Crossa, V. (2016). Reading for difference on the street: De-homogenising street vending in Mexico City. *Urban Studies, 53*(2), 287–301. https://doi.org/10.1177%2F0042098014563471

Crot, L. (2006). "Scenographic" and "cosmetic" planning: Globalization and territorial restructuring in Buenos Aires. *Journal of Urban Affairs, 28*(3), 227–251. https://doi.org/10.1111/j.1467-9906.2006.00290.x

Cruz, S. (2016). Urbanización y pueblos en la periferia metropolitana de la Ciudad de México. In A. Vieyra, Y. Méndez-Lemus, & J. Hernández-Guerrero (Eds.), *Procesos urbanos, pobreza y ambiente: Implicaciones en ciudades medias y megaciudades* (pp. 125–142). Universidad Nacional Autonoma de México, Centro de Investigaciones en Geografía Ambiental (CIGA), Morelia.

Cunha Linka, C. (2018). *Shortcomings of Brazil's minha casa, minha vida programme*. Urbanet. Online. https://www.urbanet.info/brazil-social-housing-shortcomings/

Dammer, M., Delgadillo, V., & Erazo, J. (2019). La ciudad, espacio de reproducción de las desigualdades. *Andamios, 16*(39), 7–13. https://doi.org/10.29092/uacm.v16i39.672

Davis, D., & De Durén, N. (2016). Cities in Latin America. *The Blackwell Encyclopedia of Sociology*, 1–6.

Davis, M. (2006). *Planet of slums*. London: Verso.

De Duren, N. (2008). *Growth and poverty in the urban fringe: Decentralization, dispersion, and inequality in greater Buenos Aires.* MIT Department of Urban Studies and Planning.

Delgadillo, V. (2012). El derecho a la ciudad en la Ciudad de México ¿Una retórica progresista para una gestión urbana neoliberal? *Andamios, 9*(18), 117–139.

De Mattos, C.A. (1999). Santiago de Chile, globalización y expansión metropolitana: Lo que existía sigue existiendo. *EURE: Revista latinoamericana de estudios urbano regionales, 25*(76), 29–56. http://dx.doi.org/10.4067/S0250-71611999007600002

De Mattos, C.A. (2004). Santiago de Chile: Metamorfosis bajo un nuevo impulso de modernización capitalista. In De Mattos, C. et al. (Eds.), *Santiago en la globalización: ¿Una nueva ciudad?* (pp. 17–46). Ediciones SUR/EURE Libros.

De Mattos, C.A. (2015). *Revolución urbana: Estado, mercado y capital en América Latina*. Colección Estudios Urbanos UC. RIL Editores.

De Soto, H. (1989). *The other path: The invisible revolution in the Third World.* Harpercollins.

De Soto, H. (2000). *The mystery of capital: Why capitalism triumphs in the West and fails everywhere else.* Bantam Press.

Dosh, P. (2019). Urban popular movements in Latin America. In H.E. Vanden & G. Prevost (Eds.), *The Oxford encyclopedia of Latin American politics.* Oxford University Press. https://doi.org/10.1093/acrefore/9780190228637.013.1722

Dos Santos, T. (1970). The structure of dependence. *The American Economic Review, 60*(2), 231–236.

Duhau, E. (2013). La investigación urbana y la metropolis latinamericanas. In R.R.V. Blanca & E. Pradilla Cobos (Eds.), *Teorías sobre la ciudad en América Latina* (Tomo 1). Universidad Autónoma Metropolitana.

Duhau, E. (2016). Social sciences and urban studies: Goodbye to paradigms? *International Journal of Urban and Regional Research,* 147–156. https://doi.org/10.1111/1468-2427.12205

Farrés. Y., & Matarán, A. (2014). Hacia una teoría urbana transmoderna y decolonial: Una introducción. *Polis* [En línea], *37.*

Fernandes, E. (2002). The influence of de Soto's "The mystery of capital." *Land Lines,* 5–8.

Friedmann, J. (1969). The future of urbanization in Latin America: Some observations on the role of the periphery. *Papers in Regional Science, 23*(1), 161–176. https://doi.org/10.1007/BF01941881

Fuster-Farfán, X., Vergara, P., & Imilán, W. (2020). Vivienda sin ciudad, ciudad sin planificación, planificación sin habitantes: APP para la reconstrucción territorial post-desastre. *Revista De Geografía Norte Grande, 77,* 133–156. http://dx.doi.org/10.4067/S0718-34022020000300133

Gago, V. (2017). *Neoliberalism from below: Popular pragmatics and baroque economies.* Duke University Press.

Gago, V., & Gutierrez, R. (2018). Women rising in defense of life: Tracing the revolutionary flows of Latin American women's many uprisings. *NACLA Report on the Americas, 50*(4), 364–368. https://doi.org/10.1080/10714839.2018.1550978

Gago, V., & Mezzadra, S. (2015). Para una crítica de las operaciones extractivas del capital: Patrón de acumulación y luchas sociales en el tiempo de la financiarización. *Nueva Sociedad, 255,* 38–52.

Gereffi, G., & Evans, P. (1981). Transnational corporations, dependent development, and state policy in the semiperiphery: A comparison of Brazil and Mexico. *Latin American Research Review, 16*(3), 31–64.

Germani, G. (1973). *El Concepto de Marginalidad: Significado, raíces históricas y cuestiones teóricas, con particular referencia a la marginalidad urbana.* Nueva Visión.

Gilbert, A. (2002). Power, ideology, and the Washington Consensus: The development and spread of Chilean housing policy. *Housing Studies, 17*(2), 305–324. https://doi.org/10.1080/02673030220123243

Gilbert, L. (2020). Mexico City: Elusive suburbs, ubiquitous peripheries. In J. Nijman (Ed.), *The life of North American suburbs*. University of Toronto Press.

Gilbert, L., & De Jong, F. (2015). Entanglements of periphery and informality in Mexico City. *International Journal of Urban and Regional Research, 39*(3), 518–532. https://doi.org/10.1111/1468-2427.12249

Grosfoguel, R. (2007). The epistemic decolonial turn: Beyond political -economy paradigms. *Cultural Studies, 21*(2–3), 211–223. https://doi.org/10.1080/09502380601162514

Grosfoguel, R. (2011). Decolonizing post-colonial studies and paradigms of political-economy: Transmodernity, decolonial thinking, and global coloniality. *Transmodernity: Journal of Peripheral Cultural Production of the Luso-Hispanic World, 1*(1). https://doi.org/10.5070/T411000004

Grubbauer, M. (2020). Assisted self-help housing in Mexico: Advocacy, (micro)finance and the making of markets. *IJURR, 44*(6), 947–966. https://doi.org/10.1111/1468-2427.12916

Gudynas, E. (2009). Diez tesis urgentes sobre el nuevo extractivismo. In Centro Latinoamericano de Ecología Social (Ed.), *Extractivismo, Política y Sociedad*. Quito.

Haid, C., & Hilbrandt, H. (2019). Urban informality and the state: Geographical translations and conceptual alliances. *International Journal of Urban and Regional Research, 43*(3). https://doi.org/10.1111/1468-2427.12700

Hardoy, J., & Romero Lankao, P. (2011). Latin American cities and climate change: Challenges and options to mitigation and adaptation responses. *Current Opinion in Environmental Sustainability, 3*, 158–163. https://doi.org/10.1016/j.cosust.2011.01.004

Harvey, D. (1996). *Justice,* nature and the geography of difference. Blackwell.

Haubrich, D., & Wehrhahn, R. (2020). The production of (in)security in São Paulo: Changing patterns of daily actions from the perspective of social practices theory. *ERDKUNDE, 74*(2), 85–99. http://doi.org/10.3112/erujunde.2020.02.01

Heinrichs, D., Lukas, M., & Nuissl, H. (2011). Privatisation of the fringes – A Latin American version of post-suburbia? The case of Santiago de Chile. In N. Phelps & F. Wu (Eds.), *International perspectives on suburbanization: A post-suburban world?* (pp. 101–121). Palgrave Macmillan.

Henríquez, C., & Romero, H. (Eds.), 2019. *Urban climates in Latin America*. Springer.

Hernández, F., & Becerra Santacruz, A. (2017). *Marginal urbanisms: Informal and formal development in cities of Latin America*. Cambridge Scholar Publishing.

Hernandez, F., Kellett, P., & Allen, L.K. (2010). *Rethinking the informal city: Critical perspectives from Latin America*. Berghahn Books.

Herzog, L. (2015). *Global suburbs: Urban sprawl from the Rio Grande to Rio de Janeiro*. Routledge.
Hilbrandt, H., & Grubbauer, M. (2020). Standards and SSOs in the contested widening and deepening of financial markets: The arrival of green municipal bonds in Mexico City. *EPA: Economy and Space 2020, 52*(7), 1415–1433. https://doi.org/10.1177%2F0308518X20909391
Holston, J. (2008). *Insurgent citizenship: Disjunctions of democracy and modernity in Brazil*. Princeton University Press.
Holston, J. (2009). Insurgent citizenship in an era of global urban peripheries. *City & Society, 21*(2), 245–267. https://doi.org/10.1111/j.1548-744X.2009.01024.x
Horn, P. (2018). Indigenous peoples, the city and inclusive urban development policies in Latin America: Lessons from Bolivia and Ecuador. *Development Policy Review, 36*(4), 483–501. http://dx.doi.org/10.1111/dpr.12234
Imilán, W., Osterling, E., Mansilla, P., & Jirón, P. (2020). El campamento en relación con la ciudad: Informalidad y movilidades residenciales de habitantes de Alto Hospicio. *Revista INVI, 35*(99), 57–80. https://doi.org/10.4067/S0718-83582020000200057
Irazábal, C. (2006). Localizing urban design traditions: Gated and edge cities in Curitiba. *Journal of Urban Design, 11*(1), 73–96. https://doi.org/10.1080/13574800500297736
Irazábal, C., & Jirón, P. (2020). Latin American smart cities: Between worlding infatuation and crawling provincialising. *Urban Studies*. Online first: https://journals.sagepub.com/doi/abs/10.1177/0042098020945201
Jajamovich, G. (2018). *Puerto Madero en movimiento: Un abordaje a partir de la circulación de la Corporación Antiguo Puerto Madero (1989–2017)*. Teseo.
Jajamovich, G., & Delgadillo, V. (2020). La circulación de conocimientos, saberes y políticas urbanas en América Latina: Introducción. *Iberoamericana, 20*(74), 7–11. https://doi.org/10.18441/ibam.20.2020.74.7-11
Janoschka, M. (2002a). El nuevo modelo de la ciudad latinoamericana: Fragmentación y privatización. Revista de Estudios Urbano Regionales. *EURE, 28*(85). http://dx.doi.org/10.4067/S0250-71612002008500002
Janoschka, M. (2002b). Die Dynamik stadtstrukturellen Wandels in Lateinamerika im Modell der lateinamerikanischen Stadt. *Geographica Helvetica, 57*(4), 300–310. https://doi.org/10.5194/gh-57-300-2002
Janoschka, M., & Salinas, L. (2017). Peripheral urbanization in Mexico City: A comparative analysis of uneven social and material geographies in low-income housing estates. *Habitat International, 70*, 43–49. http://doi.org/10.1016/habitatint.2017.10.003
Jaramillo, S. (2012). Heterogeneidad estructural en el capitalismo: Una mirada desde el marxismo de hoy, Documentos CEDE 010014, Universidad de los Andes – CEDE.

Kanai, J.M. (2014). On the peripheries of planetary urbanization: Extensive consequences to the worlding of globalizing Manaus and its expanding impact. *Environment and Planning D – Society and Space, 32*(6), 1071–1087. *https://doi.org/10.1068*%2Fd13128p

Kanai, J.M. (2016). The pervasiveness of neoliberal territorial design: Cross-border infrastructure planning in South America since the introduction of IIRSA. *Geoforum, 69*, 160–170. https://doi.org/10.1016/j.geoforum.2015.10.002

Kanai, J.M., & Schindler, S. (2018). Peri-urban promises of connectivity: Linking project-led polycentrism to the infrastructure scramble. *Environment and Planning A: Economy and Space, 51*(2), 302–322. https://doi.org/10.1177%2F0308518X18763370

Kay, C. (2008). Reflections on Latin American rural studies in the neoliberal globalization period: A new rurality? *Development and Change, 39*(6), 915–943. https://doi.org/10.1111/j.1467-7660.2008.00518.x

Kay, C. (2015). The agrarian question and the neoliberal rural transformation in Latin America. *EURE Revista Europea de Estudios Latinoamericanos y del Caribe, 100*, 73–83.

Keil, R. (2017). Extended urbanization, "disjunct fragments" and global suburbanisms. *Environment and Planning D: Society and Space, 36*(3), 494–511. https://doi.org/10.1177%2F0263775817749594

Keil, R. (2018a). *Suburban planet: Making the world urban from the outside*. Polity Press.

Keil, R. (2018b). The empty shell of the planetary: Re-rooting the urban in the experience of the urbanites. *Urban Geography, 39*(10), 1589–1602. https://doi.org/10.1080/02723638.2018.1451018

Keil. R. (2018c) After suburbia: Research and action in the suburban century. *Urban Geography, 41*(1), 1–20.

Kentor, J. (1981). Structural determinants of peripheral urbanization: The effects of international dependence. *American Sociological Review, 46*(2), 201–211. https://doi.org/10.2307/2094979

Kowaltowski, D., Granja, A., Moreira, D., Pina, S., Asensio, C., & Castro, M. (2015). The Brazilian housing program "Minha Casa, Minha Vida." *Journal of the Korean Housing Aassociation, 26*(6), 35–42. https://doi.org/10.6107/JKHA.2015.26.6.035.

Kozak, D. (2013). Assessing urban fragmentation: The emergence of new typologies in central Buenos Aires. In M. Jenks, D. Kozak, & P. Takkanon (Eds.), *World cities and urban form: Fragmented, polycentric, sustainable?* (pp. 239–258). Routledge.

Krellenberg, K., Welz, J., & Link, F. (Eds.). (2017). *Cambio Climático, vulnerabilidad urbana y adaptación a nivel municipal: Santiago de Chile y otras ciudades de América Latina*. RIL Editores.

Lanz, S. (2016). The born-again favela: The urban informality of pentecostalism in Rio de Janeiro. *IJURR, 40*(3), 541–558. https://doi.org/10.1111/1468-2427.12362

Lefebvre, H. (2003). *The urban revolution*. University of Minnesota Press.

Lerner, A.M., & Appendini. K. (2011). Dimensions of peri-urban maize production in the Toluca-Atlacomulco Valley, Mexico. *Journal of Latin American Geography, 10*(2), 10–106. https://www.jstor.org/stable/23209586

Lerner, A.M., Eakin, H., & Sweeney, S. (2013). Understanding peri-urban maize production through an examination of household livelihoods in the Toluca Metropolitan Area, Mexico. *Journal of Rural Studies, 30*, 52–63. https://doi.org/10.1016/j.jrurstud.2012.11.001

Lewis, O. (1961). *Antropología de la pobreza: Cinco familias*. Fondo de Cultura Económica.

Link, F., & De Mattos, C. (Eds.). (2015). Lefebvre revisitado: Capitalismo, vida cotidiana y el derecho a la ciudad. *Colección Estudios Urbanos UC*. RIL Editores.

Lopes, M. (2016). Lessons from praxis: Autonomy and spatiality in contemporary Latin American social movements. *Antipode, 48*(5), 1292–1316. https://doi.org/10.1111/anti.12210

Lukas, M. (2014). *Neoliberale Stadtentwicklung in Santiago de Chile: Akteurskonstellationen und Machtverhältnisse in der Planung städtebaulicher Megaprojekte*. Band 125 von Kieler geographische Schriften.

Lukas, M., & Brueck, A. (2018). Urban policy mobilities und globale Produktionsnetzwerke: Städtische Planung in Chile als Legitimationsinstanz extraktiver Industrien. *Suburban – Zeitschrift fuer kritische Stadtforschung, 6*(2/3), 69–90. https://doi.org/10.36900/suburban.v6i2/3.427

Lukas, M., & Durán, G. (2020). The international political economy of cities and urbanization: Insights from Latin America. In E. Vivares (Ed.), *Routledge handbook of global political economy* (pp. 775–795). Routledge.

Lukas, M., Fragkou, M.C., & Vásquez, A. (2020). Hacia una ecología política de de las nuevas periferias urbanas: Suelo, agua y poder en Santiago de Chile. *Revista de Geografia Norte Grande, 76*, 95–119. http://dx.doi.org/10.4067/S0718-34022020000200095

Lukas, M., & López-Morales, E. (2017). Real estate production, geographies of mobility and spatial contestation: A two-case study in Santiago de Chile. *Journal of Transport Geography, 67*, 92–101. https://doi.org/10.1016/j.jtrangeo.2017.09.005

Lungo, M. (2005). Globalización, grandes proyectos y privatización de la gestión urbana. *Urbano, 8*(11), 49–58.

Machado, D., & Zibechi, R. (2016). *Cambiar el mundo desde arriba: Los límites del progresismo*. Ediciones Desde Abajo.
Marini, R.M. (1973/1991). Dialéctica de la dependencia. http://www.marini-escritos.unam.mx/024_dialectica_dependencia.html (4.7.2020).
McGuirk, J. (2014). *Radical cities: Across Latin America in search of a new architecture*. Verso.
Méndez-Lemus, Y., & Vieyra, A. (2014). Tracing processes in poverty dynamics: A tale of peri-urban small-scale farmers in Mexico City. *Urban Studies, 51*(10), 2009–2035.
Méndez-Lemus, Y., Vieyra, A., & Poncela, L. (2017). Peri-urban local governance? Intra-government relationships and social capital in a peripheral municipality of Michoacán, Mexico. *Progress in Development Studies, 17*(1), 1–23.
Merrifield, A. (2013). *The politics of the encounter: Urban theory and protest under planetary urbanization*. The University of Georgia Press.
Merrifield, A. (2018). Planetary urbanisation: Une affaire de perception. *Urban Geography, 39*(10), 1603–1607. https://doi.org/10.1080/02723638.2018.1470809
Mezzadra, S., & Neilson, B. (2017). On the multiple frontiers of extraction: Excavating contemporary capitalism. *Cultural Studies, 31*(2–3), 185–204. https://doi.org/10.1080/09502386.2017.1303425
Mollinga, P.P. (2008). Water, politics and development: Framing a political sociology of water resources management. *Water Alternatives, 1*(1), 24–47.
Monkkonen, P. (2012). Housing finance reform and increasing socioeconomic segregation in Mexico. *International Journal of Urban and Regional Research, 36*(4), 757–772.
Monte-Mór, R. (2014). Extended urbanization and settlement patterns in Brazil: An environmental approach. In N. Brenner (Ed.), *Implosions/Explosions* (pp. 260–267). Jovis.
Monte-Mór, R., & Castriota, R. (2018). Extended urbanization: Implications for urban and regional theory. In A. Paasi, J. Harrison, & M. Jones (Eds.), *Handbook on the geography of regions and territories* (pp. 332–345). Edward Elgar.
Montero, S. (2017). Persuasive practitioners and the art of simplification: Mobilizing the "Bogotá Model" through storytelling. *Novos Estudos CEBRAP, 36*(1), 59–76. https://doi.org/10.25091/s0101-3300201700010003
Moss, C. (2015, September 19). Medellín, Colombia: A miracle of reinvention. *The Guardian*. https://www.theguardian.com/travel/2015/sep/19/medellin-colombia-city-not-dangerous-but-lively
Mueller, F. (2017). Urban informality as a signifier: Performing urban reordering in suburban Rio de Janeiro. *International Sociology, 32*(4), 493–511.
Nun, J. (1969). Superpoblación relativa, ejército de reserva e masa marginal. *Revista Latinoamericana de Sociologia, 5*(2), 178–236.

Orellana, A., Link, F., & Noyola, J. (Eds.). (2016). Urbanización planetaria y la reconstrucción de la ciudad. Colección Estudios Urbanos UC. *RIL Editores.*

Parnreiter, C., Fischer, K., & Imhof, K. (2007). El enlace faltante entre cadenas globales de producción y ciudades globales: El servicio financiero en Ciudad de México y Santiago de Chile. *EURE: Revista latinoamericana de estudios urbano regionales, 33*(100), 135–148. http://dx.doi.org/10.4067/S0250-71612007000300008

Patel, S. (2016). A decolonial lens on cities and urbanisms: Reflections on the system of petty production in India. Asia Research Institute, ARI Working Paper No. 245.

Peake, L., Reddy, R., Tanyildiz, G., Ruddick, S., & Tchoukaleyska, R. et al. (2018). Placing planetary urbanization in other fields of vision. *Environment and Planning D: Society and Space, 36*(3), 374–386. https://doi.org/10.1177%2F0263775818775198

Perevochtchikova, M., & Rojo Negrete, I.A. (2015). The perceptions about payment schemes for ecosystem services: Study case of the San Miguel and Santo Tomás Ajusco community, Mexico. *Ecosystem Services, 14*, 27–36. https://doi.org/10.1016/j.ecoser.2015.04.002

Pérez, M. (2017). "A new Poblador is being born": Housing struggles in a gentrified area of Santiago. *Latin American Perspectives, 44*(3), 28–45. https://doi.org/10.1177%2F0094582X16668318

Perlman, J. (1976). *The myth of marginality: Urban poverty and politics in Rio de Janeiro.* University of California Press.

Perlman, J. (2011). *Favela: Four decades of living on the edge in Rio de Janeiro.* Oxford University Press.

Pradilla Cobos, E. (1982). Autoconstrucción, explotación de la fuerza de trabajo y políticas del Estado en América Latina. In E. Pradilla Cobos (Ed.), *Ensayos sobre el problema de la vivienda en América Latina.* UAM-Xochimilco.

Pradilla Cobos, E. (2013). La economía y las formas urbanas en América Latina. In R.R.V. Blanca & E. Pradilla Cobos (Eds.), *Teorías sobre la ciudad en América Latina* (Tomo 1). Universidad Autónoma Metropolitana.

Quijano, A. (1967). *Dependencia, cambio social y urbanización en Latinoamérica.* ILPES.

Quijano, A. (1971). La formación de un universo marginal en las ciudades de América Latina. In M. Castells & P. Velez (Eds.), *Imperialismo y urbanización en América.* Latina, Gustavo Gili.

Quimbayo, G., & Vásquez, F. (2016). Hacia una ecología política de la urbanización en América Latina. *Ecologia Política, 51.*

Rebotier J., Pigeon, P., & Metzger, P. (2019). Returning social context to seismic risk knowledge and management: Lessons learned from an interdisciplinary research in the city of Esmeraldas, Ecuador. *Cybergeo, 886.* https://doi.org/10.4000/cybergeo.31787

Reis, N. (2017). Finance capital and the water crisis: Insights from Mexico. *Globalizations, 14*(6), 976–990. https://doi.org/10.1080/14747731.2017.1315118

Reis, N. (2019). A farewell to urban/rural bias: Peripheral finance capitalism in Mexico. *The Journal of Peasant Studies, 46*(4), 702–728. https://doi.org /10.1080/03066150.2017.1395856

Reis, N. (2020). The "Bank of Welfare" and Mexico's moral economy. In D. McDonald, T. Marois, & D. Barrowclough (Eds.), *Public Banks and Covid-19: Combatting the pandemic with public finance* (Chapter 8). Municipal Services Project (Kingston), UNCTAD (Geneva) y Eurodad (Brussels), 171–192. Municipal Services Project (Kingston), UNCTAD (Geneva) and Eurodad (Brussels).

Richards, P., & VanWey, L. (2015). Where deforestation leads to urbanization: How resource extraction is leading to urban growth in the Brazilian Amazon. *Annals of the Association* of *American Geographers, 105*(4), 806–823. https://doi.org/10.1080/00045608.2015.1052337

Riley, E., Fiori, J., & Ramirez, R. (2001). Favela Bairro and a new generation of housing programmes for the urban poor. *Geoforum, 32*(4), 521–531. https:// doi.org/10.1016/S0016-7185(01)00016-1

Ríos, D., & Pírez, P. (2008). Urbanizaciones cerradas en áreas inundables del municipio de Tigre: ¿Producción de espacio urbano de alta calidad ambiental? Revista de Estudios Urbano Regionales. *EURE, 34*(101), 99–119. http://dx.doi.org/10.4067/S0250-71612008000100005

Roberts, B. (2005). Globalization and Latin American cities. *International Journal of Urban and Regional Research, 29*(1), 110–123. 10.1111/j.1468-2427.2005.00573.x

Robinson, J. (2006). *Ordinary cities: Between modernity and development.* Routledge.

Robinson, J. (2015). Comparative urbanism: New geographies and cultures of theorising the urban. *International Journal of Urban and Regional Research, 40*(1), 187–199. https://doi.org/10.1111/1468-2427.12273

Robinson, J., & Roy, A. (2015). Global urbanisms and the nature of urban theory. *International Journal of Urban and Regional Research, 40*(1), 181–186. https://doi.org/10.1111/1468-2427.12272

Rodriguez, A., & Sugranyes, A. (2005). *Los con techo: Un desafío para la política de vivienda social.* Ediciones SUR.

Rodríguez Matta, P. (2015). Era lo justo: Producción de periferia en Santiago de Chile en los años cincuenta. *Territorios, 32,* 97–120. https://doi.org /10.12804/territ32.2015.05

Roitman, S., & Phelps, N. (2011). Do gates negate the city? Gated communities' contribution to the urbanisation of suburbia in Pilar, Argentina. *Urban Studies, 48*(16), 3487–3509. https://doi.org/10.1177%2F0042098010397395

Rolnik, R. (2013). Late neoliberalism: The financialization of homeownership and housing rights. *International Journal of Urban and Regional Research, 37*(3), 1058–66. https://doi.org/10.1111/1468-2427.12062
Romero, H., & Vásquez, A. (2005). La comodificación de los territorios urbanizables y la degradación ambiental en Santiago de Chile. *Scripta Nova*, 9.
Roy, A. (2009). The 21st century metropolis: New geographies of theory. *Regional Studies, 43*(6), 819–830. https://doi.org/10.1080/00343400701809665
Roy, A. (2011). Slumdog cities: Rethinking subaltern urbanism. *International Journal of Urban and Regional Research, 35*(2), 223–238. https://doi.org/10.1111/j.1468-2427.2011.01051.x
Ruiz-Tagle, J., & Romano, S. (2019). Mezcla social e integración urbana: Aproximaciones teóricas y discusión del caso chileno. *Revista INVI, 34*(95). http://dx.doi.org/10.4067/S0718-83582019000100045
Salazar, C. (Ed.). (2012) *Irregular: Suelo y mercado en América Latina.* El Colegio de México.
Salazar, C. (2014). Suelo y política de vivienda en el contexto neoliberal mexicano. In S. Guiorguli (Ed.), *Gobierno, Territorio y Población: Las políticas públicas en la mira* (pp. 343–371). El Colegio de México.
Salcedo, R. (2010). The last slum: Moving from illegal settlements to subsidized home ownership in Chile. *Urban Affairs Review, 46*(1), 90–118. https://doi.org/10.1177%2F1078087410368487
Sandoval, V., Gonzalez-Muzzio, C., Villalobos, C., Sarmiento, J.P., & Hoberman, G. (2020). Assessing disaster capitalism in post-disaster processes in Chile: Neoliberal reforms and the role of the corporate class. *Disaster Prevention and Management, 29*(6), 831–847. https://doi.org/10.1108/DPM-01-2020-0005
Sandoval, V., & Sarmiento, J.P. (2020). A neglected issue: Informal settlements, urban development, and disaster risk reduction in Latin America and the Caribbean. *Disaster Prevention and Management, 29*(5), 731–745. https://doi.org/10.1108/DPM-04-2020-0115
Sanfelici, D., & Halbert, L. (2015). Financial markets, developers and the geographies of housing in Brazil: A supply-side account. *Urban Studies, 53*(7), 1465–1485. https://doi.org/10.1177%2F0042098015590981
Sassen, S. (1991). *The global city: New York, London, Tokyo.* Princeton University Press.
Schild, V. (2019). Feminisms, the environment and capitalism: On the necessary ecological dimension of a critical Latin American feminism. *Journal of International Women's Studies, 20*(6), 23–43. https://vc.bridgew.edu/cgi/viewcontent.cgi?article=2145&context=jiws
Scholz, F. (2000). Perspektiven des Südens im Zeitalter der Globalisierung. *Geographische Zeitschrift, 88*(1), 1.20. https://www.jstor.org/stable/27818864
Schteingart, M. (2000). La investigación urbana en América Latina. *Papeles de población* [online]. 2000, *6*(23), 9–25.

Schteingart, M., & Salazar, C. (2005). *Expansión urbana, sociedad y ambiente: El caso de la Ciudad de México*. El Colegio de México.

Segura, R. (2017). Desacoples entre desigualdades sociales, distribución del ingreso y patrones de urbanización en ciudades latinoamericanas: Reflexiones a partir de la Región Metropolitana de Buenos Aires (RMBA). *Revista CS, 21*, 15–39. https://doi.org/10.18046/recs.i21.2278

Silva, E. (2011). Deliberate improvisation: Planning highway franchises in Santiago, Chile. *Planning Theory, 10*(1), 35–52. https://doi.org/10.1177%2F1473095210386067

Silvestre, G., & Jajamovich, G. (2020). The role of mobile policies in coalition building: The Barcelona model as coalition magnet in Buenos Aires and Rio de Janeiro (1989–1996). *Urban Studies*, article first published online, 1–19. https://journals.sagepub.com/doi/pdf/10.1177/0042098020939808

Singer, P. (1973). Urbanización, dependencia y marginalidad en América Latina. In M. Schteingart (Ed.), *Urbanización y dependencia en América Latina*. Ediciones SIAP.

Slater, D. (1978). Towards a political economy of urbanization in peripheral capitalist societies: Problems of theory and method with illustrations from Latin America. *International Journal of Urban and Regional Research, 2*(1–3), 26–52. https://doi.org/10.1111/j.1468-2427.1978.tb00735.x

Soederberg, S. (2013). Universalising financial inclusion and the securitisation of development. *Third World Quarterly, 34*(4), 593–612. https://doi.org/10.1080/01436597.2013.786285

Soederberg, S. (2015). Subprime housing goes south: Constructing securitized mortgages for the poor in Mexico. *Antipode, 47*(2), 481–499. https://doi.org/10.1111/anti.12110

Streule, M., & Schwarz, A. (2019). "Not all spaces are territories": Creating other possible urban worlds in and from Latin America. An interview with Raúl Zibechi. *Geographica Helvetica, 74*(1), 105–111. https://doi.org/10.5194/gh-74-105-2019

Svampa, M. (2001). *Los que ganaron: La vida en los countries y en los barrios privados*. Biblos.

Svampa, M. (2019). *Neo-extractivism in Latin America: Socio-environmental conflicts, the territorial turn, and new political narratives*. Cambridge University Press.

Swistun, D. (2018). Cuerpos abyectos: Paisajes de contaminación y la corporización de la desigualdad ambiental. *Investigaciones Geográficas, 56*, 100–113. http://doi.org/10.5354/0719-5370.2018.51995

Torres-Mazuera, G. (2012). *La ruralidad urbanizada en el centro de México: Reflexiones sobre la reconfiguración local del espacio rural en un contexto neoliberal*. UNAM.

Turner, J. (1967). *Uncontrolled urban settlements: Problems and policies*. United Nations Centre for Housing, Building and Planning, United Nations.

Turner, J. (1968). The squatter settlement: An architecture that works. *Architectural Design, 38,* 350–360.
UN. (2014). *World urbanization prospects: Highlights.* United Nations Department of Economic and Social Affairs. https://esa.un.org/unpd/wup/Publications/Files/WUP2014-Highlights.pdf (August 2018).
Varley, A. (2013). Postcolonializing informality? *Environment and Planning D: Society and Space, 31,* 4–22. https://doi.org/10.1068%2Fd14410
Vasquez, A. (2016). Infraestructura verde, servicios ecosistémicos y sus aportes para enfrentar el cambio climático en ciudades: El caso del corredor ribereño del río Mapocho en Santiago de Chile. *Revista de Geografía Norte Grande, 63,* 63–86. http://dx.doi.org/10.4067/S0718-34022016000100005
Vásquez, A., Lukas, M., Salgado, M., & Mayorga, M. (2017). Urban environmental (in)justice in Latin America: The case of Chile. In R. Holifield, J. Chakraborty, & Gordon Walker (Eds.), *The Routledge handbook of environmental justice* (pp. 556–566). Routledge.
Walton, J. (1977). The international economy and peripheral urbanization. In N. Fainstein & S. Fainstein (Eds.), *Urban policy under capitalism* (pp. 119–123). Sage.
Wilson, D., & Jonas, A. (2018). Planetary urbanization: New perspectives on the debate. *Taylor & Francis, 39*(10), 1576–1580. https://doi.org/10.1080/02723638.2018.1481603
Wilson, J. (2014). The violence of abstract space: Contested regional developments in Southern Mexico. *International Journal of Urban and Regional Research, 38*(2), 516–538. https://doi.org/10.1111/1468-2427.12023
Wilson, J., & Bayon, M. (2017). Fantastical materializations: Interoceanic infrastructures in the Ecuadorian Amazon. *Environment and Planning D: Society and Space, 35*(5), 836–854. https://doi.org/10.1177%2F0263775817695102
World Bank. (1993). *Housing: Enabling markets to work – A World Bank Policy Paper.* The World Bank.
Wulfhorst, E. (2016, October 14). Mexico's Ciudad Neza rises from slum to success story. *Reuters.* https://www.reuters.com/article/us-mexico-slum-neza/mexicos-ciudad-neza-rises-from-slum-to-success-story-idUSKBN12E1F2
Zibechi, R. (2012). *Territories in resistance: A cartography of Latin American social movements.* Oakland. AK Press

PART 1

Framing Peripheral Urbanization in Latin America

1 Peripheral Urbanization: Autoconstruction, Transversal Logics, and Politics in Cities of the Global South

TERESA CALDEIRA

Many cities around the world have been largely constructed by their residents, who build not only their own houses but also frequently their neighbourhoods. They do not necessarily do so in clandestine ways and certainly not in isolation. Throughout the process, they interact with the state and its institutions, but usually in transversal ways. While they have plans and carefully prepare each step, their actions typically escape the framing of official planning. They operate inside capitalist markets of land, credit, and consumption, but usually in special niches bypassed by the dominant logics of formal real estate, finance, and commodity circulation. In the process of house/city building, many make themselves into citizens and political agents, become fluent in rights talk, and claim the cities as their own. In this article, I refer to this mode of making cities as peripheral urbanization.

I use the notion of peripheral urbanization to create a problem-space that allows us to investigate logics of the production of the urban that differ from those of the North Atlantic. I use it as a means of exploring processes of both socio-spatial formation and theory making. My intention is to go beyond the deconstruction of Northern-originated accounts of urbanization articulated by a now long stream of critiques[1] and, in response to Mbembe and Nuttall (2004, p. 352), work with new archives to offer a bold characterization and theorization of a mode of the production of space that is prevalent in cities of the Global South.

Many authors have worked with notions of peripheries and peripheral urbanization to analyse cities in the Global South. I work in conversation with them and cite many of their works throughout this article. At the core of my contribution are two arguments. First, I argue that peripheral urbanization consists of a *set of interrelated processes*. It refers to modes of the production of urban space that (a) operate with a specific form of agency and temporality, (b) engage transversally with official logics, (c) generate new modes of politics through practices that produce new kinds of citizens, claims, circuits, and contestations, and (d) create highly unequal and heterogeneous cities. Second, I argue that peripheral urbanization not only produces heterogeneity within the city as it unfolds over time, but also *varies considerably from one city to another*. Thus, as a model, peripheral urbanization must remain open and provisional to account for variation and for the ways in which the production of the cities it characterizes is constantly being transformed.

It is important to stress that peripheral urbanization does not necessarily entail the growth of cities towards their hinterlands. In other words, *it does not simply refer to a spatial location in the city – its margins – but rather to a way of producing space that can be anywhere.*[2] What makes this process peripheral is not its physical location but rather the crucial role of residents in the production of space and how as a mode of urbanization it unfolds slowly, transversally in relation to official logics, and amidst political contestations.

I definitely do not claim that peripheral urbanization is the only mode of the production of urban space operating in cities of the Global South. Nor do I claim that there is a unified model of "Southern urbanism." In fact, the very different cases I analyse below demonstrate the opposite, that is, that peripheral urbanization unfolds in quite different ways. What I do claim is that peripheral urbanization is remarkably pervasive, occurring in many cities of the south, regardless of their different histories of urbanization and political specificities. I also claim that it is necessary to understand peripheral urbanization as a set of interrelated processes to formulate not only better analyses and theories, but also better urban and planning practices.

I build my arguments by juxtaposing dissimilar cases from a few cities in the Global South. I rely extensively on my own long-term investigations in São Paulo alongside other cases of peripheral urbanization. I have chosen the latter for the following reasons. First, they have been carefully analysed, mostly by scholars based in these cities. Thus, they represent new archives, bringing to the discussion robust knowledge produced from non-North Atlantic perspectives. Usually, these studies circulate within the national context in which they are produced but are not engaged by either Northern scholars or researchers working on other parts of the Global South. To engage these studies, breaking their national barriers and their isolation from the north, is a necessary step to create urban theories that can account for modes of urbanization whose logic is different from that of the industrial cities of the North. It is to take seriously the idea of thinking with an accent (Caldeira, 2000, chapter 1). Second, they are very different from each other, their histories are well known, and their singularities can be clearly contextualized, thus allowing me to consider how different histories may lead to similar outcomes. Third, these cases indicate important transformations in processes of peripheral urbanization, pointing to different futures.

I conceive of my analysis in terms of juxtaposition, a kind of comparison. Several scholars have insisted on the importance of comparison in urban studies, especially as one moves away from North Atlantic models.[3] The main reason to choose juxtaposition is to emphasize the qualitative logic that anchors my approach, a logic standard among anthropologists and ethnographers. To work with the *juxtaposition of dissimilar cases* means to use difference and estrangement as modes of analysis and critique, a perspective that can be traced back to the practices of the European artistic avant-gardes of the early twentieth century, and later to the critique of anthropology in the 1980s (Clifford, 1981; Marcus & Fischer, 1986). This use of juxtaposition operates with a qualitative (analytical) logic that is at odds with a statistical one used by many comparativists (Small, 2009; Yin, 2002). It does not look for the representative, typical, similar, or repetitive. Instead, it proceeds inductively, exploring the differential and internal conditions of cases, looking for a wide range of variation, and aiming at saturation (Small, 2009). The juxtaposition of dissimilar, located, and historicized cases brought together to illuminate one another destabilizes unexamined views and generalizations and opens up new possibilities of understanding.[4]

Thus, in what follows, I juxtapose analyses of Istanbul, Santiago, and São Paulo (with additional considerations about Mexico City and New Delhi), exploring their different configurations to anchor my arguments about peripheral urbanization. I first analyse each of the four interrelated dimensions of peripheral urbanization and then discuss some of the ways in which they have been transformed recently. Throughout the analysis, I build a model based on the exploration of differences, insisting that although we can identify processes of peripheral urbanization in many cities, each iteration is particular and should be analysed as such. The challenge of the analysis is thus to develop a model, peripheral urbanization, that articulates general features while remaining open and provisional to account for the ways in which the modes of operation it characterizes vary and constantly transform.

Agency and Temporality

Peripheral urbanization involves a distinctive form of agency. Residents are agents of urbanization, not simply consumers of spaces developed and regulated by others. They build their houses and cities step-by-step according to the resources they are able to put together at each moment in a process that I call *autoconstruction* (following the Latin American term for it). As Holston (1991) has also shown, each phase involves a great amount of improvization and bricolage; complex strategies and calculations; and constant imagination of what a nice home might look like. Sometimes residents rely on their own labour; frequently, they hire the labour of others. Their spaces are always in the making. Thus, peripheral urbanization also involves a distinctive temporality; homes and neighbourhoods grow little-by-little, in long-term processes of incompletion and continuous improvement led by their own residents. Peripheral urbanization does not involve spaces already made that can be consumed as finished products before they are even inhabited. Rather, it involves spaces that are never quite done, always being altered, expanded, and elaborated upon.

Cities are obviously always being transformed. But peripheries change according to a logic of their own: slowly, progressively, unevenly.[5] For long periods of time, people inhabit spaces that are clearly precarious and unfinished, but with the expectation, frequently realized, that the spaces will improve and one day look like wealthier parts of the city. Peripheral landscapes are marked by constructions and remodelling, resulting in a combination of houses and infrastructure of quite different levels of completion and elaboration. Older areas are better off; newer areas are more precarious in terms of infrastructure, services, and the characteristics of individual houses. As a result of this pattern, peripheral urbanization produces quite heterogeneous landscapes.

The transformations are simultaneously expressed in the conditions of the urban space and the houses. As the neighbourhood grows and the population increases, streets are paved; water, electricity, and sewage arrive; and local commerce expands. With time, façades are improved, houses are enlarged, and, especially, spaces are constantly redecorated. Despite indisputable precariousness and persisting poverty, the processes of transformation of peripheral areas offer a model of social mobility, as they become the material embodiments of notions of progress.[6]

As peripheries improve, however, they may become inaccessible to the poorest residents. Thus, peripheral urbanization is a process that is always being displaced, reproduced somewhere else where land is cheaper because it is more precarious or difficult to access. This presents methodological challenges for the study of peripheral urbanization. To capture its temporality, it has to be studied over time. To capture its continuous reproduction, it has to be studied across space. This is especially clear in Ahonsi's (2002) analysis of Lagos, Duhau's (2014)

Figure 1.1 The autoconstruction of Jardim das Camélias, São Paulo. The pictures were taken from the same terrace over a 35-year period. The photos dates are (1) 1980, (2) 1989, (3) late 1990s, and (4) 2015. Photos by the author.

of Mexico City, and Holston's (2008) and my own (Caldeira, 2000) of São Paulo. Studies looking at one place miss the lateral reproduction of peripheries, getting only the picture of improvement, if focusing on an older area, or precariousness, if focusing on a newer one. Peripheries have to be studied over time and across space. It is only by doing this that one can capture the simultaneous processes of improvement and the reproduction of inequality and precariousness.

Transversal Logics

Peripheries are spaces that frequently unsettle official logics – for example, those of legal property, formal labour, state regulation, and market capitalism. Nevertheless, they do not contest these logics directly as much as they operate with them in transversal ways. That is, by engaging the many problems of legalization, regulation, occupation, planning, and speculation, they redefine those logics and, in so doing, generate urbanizations of heterogeneous types and remarkable political consequences. Cities that have undergone peripheral urbanization are usually marked by significant spatial and social inequality. Nevertheless, because of these transversal engagements, inequalities cannot always be mapped out in simple dualistic oppositions such as regulated versus unregulated, legal residences versus slums, formal versus informal, etc. Rather, these cities exhibit multiple formations of inequality, wherein categories such as "formal" and "regulated" are always shifting and unstable. Thus, one needs to set aside the notion of informality (and the dualist reasoning it usually implies) and think in terms of transversal logics to understand these complex urban formations which are inherently unstable and contingent. Peripheries are improvised. But the fact that there is a significant amount of improvization does not mean that they are either totally unplanned and chaotic or illegal and unregulated. Peripheral urbanization does not mean an absence of the state or planning, but rather a process in which citizens and governments interact in complex ways. While residents are the main agents of the production of space, the state is present in numerous ways: it regulates, legislates, writes plans, provides infrastructure, polices, and upgrades spaces. Quite frequently, though, the state acts *after the fact* to modify spaces that are already built and inhabited.

Many peripheral and autoconstructed spaces involve a substantial amount of planning. Nezahualcóyotl or Ciudad Neza in the Mexico City Metropolitan Region, for instance, was built on a perfect grid, as were many other peripheral areas all over Latin America. In the 1960s, it was laid out by a developer as 150,000 uniform but irregularly parcelled out housing lots that were then built up by individual residents (Duhau, 2014, p. 151; 1998, pp. 145–146). In fact, countless neighbourhoods in peripheries have been laid out and sold by developers acting according to clear plans and legal standards. In general, however, their creation, commercialization, and development involve several layers of irregularity – sometimes quite difficult to untangle – and many turns of negotiation between all the agents involved in the process, including the state. In the case of the Mexico City metropolitan region, a common source of irregularity is the creation of urban developments on *ejido* land – that is, land that, distributed to landless peasants to hold in common after the Mexican Revolution, is illegal to sell, lease, or rent.[7]

There is broad agreement in the literature about housing in the Global South that both irregularity and illegality are the most common means through which the poor settle in and urbanize cities. Frequently, illegality and irregularity are the only options available for the poor to become urban dwellers, given that formal housing is not affordable and public housing is not sufficient. The conditions of irregularity regarding land tenure and construction vary widely.[8] They range from swindling of private property to lack of official permits; from corruption in the allocation of communal land to failure to follow municipal codes; from disputes over parcelling of tribal land to the appropriation of ecologically protected areas for private constructions. In addition to irregularities related to land, there are others associated with construction. Autoconstructors and developers try to follow city ordinances and regulations but usually do this unevenly. In spite of the fact that there is a good amount of irregularity regarding land tenure, land use, and construction in peripheral areas, they are neither necessarily illegal nor invaded. Although there are many squatter settlements in peripheral areas built on invaded land,[9] a significant proportion of autoconstructors actually pay for and have claims to their lots. In São Paulo, for example, 80% of peripheral residents have legitimate claims to property ownership (Holston, 2008).

Legality is a complex issue, however. James Holston (2008) shows that, in Brazil, illegality is a common condition of the urbanization patterns of the poor because it is a dominant means of rule in general, not because the poor have a tendency to marginality and precariousness. Moreover, when there is illegality, it is usually not because the residents have taken land that is not their own, but rather because they have been swindled by developers or because legislative changes have made illegal what used to be acceptable. Tuna Kuyucu (2014) shows that, in Istanbul, the use of legal ambiguity and administrative arbitrariness by public and private actors has been fundamental for the constitution of a private property regime that anchors a capitalist market of land and the displacement of *gecekondu* settlements.

Whatever the case, what these and other studies demonstrate is that the legal status of neighbourhoods built by peripheral urbanization is frequently subject to transformation. Residents bet on the possibility of legalization and regularization and most frequently either succeed in seeing it happen or live with the consequences of ongoing irregularity. Thus, there is a temporality related to legalization as well. In the early 1970s, Nezahualcóyotl was the object of the first large-scale land regularization project by the Mexican federal and state governments (Duhau, 2014, p. 151). Now, most Latin American countries have large programs for the regularization of urban land and the question of land policy is central to the research agendas of several institutions (Smolka & Mullahy, 2010). However, studies have also shown that the state itself is responsible for the creation and recreation of irregularity and illegality, as it passes laws and master plans that alter the status of lands and buildings, making the irregular into the regular and vice versa (for Delhi, see Bhan, 2016). A single zoning law can render a whole area irregular or legalize it overnight. Obviously, these shifts engender intense political struggles, as they involve immediate repercussions in terms of the profitability of real estate and the dislocation of residents.

Processes of regularization help to illuminate both the constant exchange between autoconstructors and the state and the logic of planning that it embeds. São Paulo's case illustrates this point. During the period of largest expansion of the city in 1950–70, private entrepreneurs laid out developments in distant areas, leaving vast stretches of empty land between new developments (Caldeira, 2000; Camargo et al., 1976). They created dirt roads and bus lines to connect the city to the new settlements, usually irregular, that they sold to workers trying to avoid exorbitant rent. The expectation of both residents and developers, almost always realized, was that the state would follow up and install the necessary infrastructure. When this happened, developers began to sell the land they had initially bypassed, benefitting from the valorization brought about by the improved infrastructural conditions and previous urbanization. The same happened as a result of governmental programs of land regularization/legalization. These practices initiated cycles of land development/regularization/valorization that entangled the state, investors, and citizens. In the process, state planners and agencies acted routinely after the fact in a way that benefitted private developers, improved neighbourhoods, and consolidated the rights of residents. We can recognize the same logic in programs of land regularization and slum upgrading.

Finally, the transversal logics of peripheral urbanization appear in the modes of consumption and credit. Autoconstruction involves substantial consumption related to the acquisition of both building materials and appliances, furniture, and decorative items. But autoconstructors typically lack access to credit from institutions, such as banks and the state, to finance the acquisition of land or construction of homes. There is credit, but it comes from developers, merchants, and popular department stores, usually with very high interest rates. Everything happens in a kind of alternate market that specializes in the needs of the urban poor and bypasses some dominant official logics. Maintained in this market, urban land and housing remain relatively cheap and accessible.

Experiments in Politics and Democracy

The perennially shifting conditions in the peripheries, and their accompanying configurations of both improvement and the reproduction of inequality, have entangled the agents that produce urban space in complex political relationships. Peripheral urbanization generates new modes of politics through practices that produce new kinds of citizens, claims, and contestations. These politics are rooted in the production of urban space itself – primarily residential urban space – and its qualities, deficiencies, and practices (Caldeira, 2015). The unstable conditions of tenure, the skewed presence of the state, the precariousness of infrastructures, the exploitation by developers and merchants, the constant abuse of residents by the institutions of order, and several processes of stigmatization and discrimination against residents make peripheries spaces of invention of new democratic practices. Sometimes, the political relationships that develop in these spaces are quite clientelistic. In many other cases, however, they become what James Holston (2008) calls "spaces of insurgent citizenship." In many parts of the world, social movements and grassroots organizations from peripheries have created new discourses of rights and put forward demands that are at the basis of the rise of new citizenships, the formulation of new constitutions, the experimentation with new forms of local administration, and the invention of new approaches to social policy, planning, law, and citizen participation (Holston, 2009). They have also metamorphosed over time and been an important presence in the protests that have filled the streets of several cities in recent years, such as Istanbul (El-Kazaz, 2013; Tugal, 2013), São Paulo (Caldeira, 2013, 2015), and Cairo (Ismail, 2014).

These movements have certainly generated changes in the everyday lives and qualities of urban space, as they have forced their improvement. But they have also produced deep political transformations. They have taken the materiality of spaces of inhabitation to anchor movements that create new political subjectivities and expose the inequalities that sustain the reproduction of peripheral urbanization and the limits of dominant political arrangements. In the process, they have also opened new spaces of experimentation that have substantially transformed the character of the public sphere in their societies.

Heterogeneity

Thus, peripheral urbanization is a process through which residents engage in modes of production of space that constitute themselves as simultaneously new kinds of urban residents, consumers, subjects, and citizens. Areas produced by peripheral urbanization are dynamic, creative, and transformative, as AbdouMaliq Simone (2004, 2010, 2014) has insistently demonstrated. These engagements, however transformative they may be, obviously are not able to erase the gap that separates peripheries and their residents from other spaces and social groups – a gap that has been constantly recreated and frequently increased, both in terms of the disparity of quality of urban spaces and the income and resources of social groups. Peripheries are, undoubtedly, about inequality. They are poor, precarious, discriminated against, and frequently violent. Nevertheless, they are not homogeneously poor and precarious; and usually are much better than they were in the past. Their inherent dynamism and pattern of transformation complicate accounts of socio-spatial inequality.

Over time, peripheral urbanization generates heterogeneous urban spaces. This is especially clear in Latin America, where this mode of producing the city has been prevalent for over 50 years. Several studies demonstrate that, as they grow and improve, the social and spatial heterogeneity of peripheral neighbourhoods gradually and persistently increases. But because each area has a dynamic of its own – and because service and infrastructure reach them at different times and depending on various conditions, including the level and persistence of political organization of their residents – the result is that what used to be relatively homogeneous areas 50 years ago are now quite heterogeneous. São Paulo is a good example, as demonstrated by the contrast between an analysis produced in 1977 by Seplan, the secretary of economics and planning of the state of São Paulo, and two recent analysis using data from the 2010 census by Nery at NEV/USP (2014) and Marques (2014). While the Seplan study revealed a homogeneous and vast periphery, the more recent studies reveal a city that is a kind of patchwork. Since at least the middle of the twentieth century, peripheral urbanization has offered poor residents of metropolises of the Global South opportunities to inhabit these cities by maintaining alternative markets and spaces in which housing and urban life are precarious but affordable. Although this mode of urbanization continues to prevail in several cities, in others it is starting to show its limits. Simultaneously, some governmental interventions in the housing market intended to expand low-income housing have both increased the availability of housing for the poorest and affected the dynamics of autoconstruction. In the second part of this article, I discuss some cases that point to transformations in processes of peripheral urbanization. I am guided by the following question: What conditions permit the continuing improvement of peripheral urban spaces while simultaneously preserving the ability of the poor to inhabit these spaces and the metropolises they build? It seems that two conditions are fundamental: the strong organization of residents and the engagement of states that are compelled to commit to principles of social justice.

Creation of New Land Markets

Let me start with Istanbul. While it is a textbook case of peripheral urbanization, it also clearly exemplifies the current trend towards the erasure of the types of space and housing built over the last half century through autoconstruction.[10] As the population of the city increased exponentially, starting in the 1950s, new migrants were incorporated through a double mechanism – employment and housing – which entailed their integration in local social and political networks. Migrants arriving in the city settled in what became known as *gecekondu*. These squatter settlements were built in what were once distant but are now centrally located areas of the city. The settlements were illegal, since they were typically invasions of what used to be public land. As the new urbanites settled in these areas and constructed their houses step-by-step, they engaged in networks that simultaneously enhanced their chances of getting a job, provided a local basis of support, and ultimately guaranteed the legalization of their lands. This happened especially from the 1970s onwards, when populist regimes organized these neighbourhoods, exchanging votes for the provision of infrastructure and land titles. As a consequence, explains Keyder (2005, pp. 126–127), "the lifecycle of a squatter neighborhood was such that after a few elections it could become an area of multiple-storey apartment buildings ... Urban politics was the natural arena in which immigrants engaged; they elected and supported politicians who could credibly promise local returns. Migrants became citizens through their allegiance to the space of residence."

This story of success, however, also had contradictory effects. As those who had occupied public lands obtained titles of either private ownership or user's rights (*tapu tahsis* deeds; see Kuyucu, 2014: 10), a good amount of land that had been kept outside of the capitalist market was incorporated into it. The first consequences of this shift were felt by newer generations of migrants to the city. Unable to find easy land on which to reproduce the same process of peripheral urbanization, they became renters in the two- and three-storey buildings owned by the earlier generation of migrants. This brought about a significant antagonism between the two groups, changing the social dynamics in *gecekondu* neighbourhoods (Balaban, 2011; Keyder, 2005).

But this comprised only the first wave of commodification of land in Istanbul. The second and more aggressive, which has come more recently, is still unfolding. As Istanbul has joined the ranks of cities with plans to become "world-class," and as the pressure from real estate developers for access to land has increased, the state has abandoned its previous tolerance for the occupation of public land and its practice of privatizing this land in electoral negotiations. Instead, it has actively contributed to the destruction of these areas and their transfer to private developers. Several processes are involved. On the one hand, land is now sold for a profit to bigger developers in a process of enclosure of urban space (Balaban, 2011), property transfer (Kuyucu & Ünsal, 2010), and population displacement. Moreover, as upscale high-rise developments radically change the landscape of the city, developers arrive at the old *gecekondu* areas to buy out the residents of desirably located neighbourhoods.

But there are other and more aggressive modes of conversion of *gecekondu* areas. A crucial one, backed by new legislation, consists of large state-led projects to designate these areas as "urban transformation zones" to be cleared for the sake of conversion to other high-end and high-profit uses. This designation allows state agencies to relocate *gecekondu* families to housing in more remote areas. Analysing the unfolding of these transformation zones, Kuyucu (2014) brilliantly shows how flexible and ambiguous modes of land occupation are transformed into legal property to be used for real estate development through processes that operate by exploiting legal ambiguity and state arbitrariness embodied in the new legislation. Although conversion in this case clearly legalizes the illegal, as is common in processes of peripheral urbanization, now the legal conversion no longer benefits autoconstructors, who are forced away from the areas they have previously occupied.[11]

Thus, peripheral urbanization in Istanbul has assured the social mobility and integration of one generation of migrants. If for a moment this incorporation was guaranteed by a configuration that kept urban lands outside of a strong capitalist market and the state on the side of new poor urbanites, the transformations of this arrangement – the state's unwillingness to pursue popular housing alternatives, promotion of projects to displace available *gecekondu*, and stronger push for the incorporation of land into an upscale real estate market – are not only transforming and destroying old peripheries, but also constituting a new regime of private property (Kuyucu, 2014) and making urban land inaccessible to new migrants, who either become renters or squat in more remote areas, reproducing the periphery elsewhere. The protests of 2013 expressed in important ways the culmination of citizens' dissatisfaction with these "urban transformations" and the project of city they embed (El-Kazaz, 2013; Tugal, 2013).

Although I do not have space to analyse other cases of the constitution of land markets, I want to mention that New Delhi shows tendencies strikingly similar to Istanbul (Bhan, 2016; Ghertner, 2015). In Delhi, as in other metropolises of the Global South, processes such as evictions of previously accepted settlements; enclosure or privatization of public or commonable land; the aggregation of urban land in peri-urban fringes, sometimes through violent means, and its incorporation into formal property markets; and the use of state force to clear land for private development undermine the possibilities of the continuation of peripheral urbanization. These processes provoke market-driven displacements and foreclose alternative markets and spaces for housing the poor.

Expansion of Social Housing

The case of Santiago, Chile, another metropolis shaped by peripheral urbanization, also reveals deep changes in this mode of the production of space. In particular, it has been transformed through repression and the adoption of neoliberal policies that have made Chile into a "best practice" case for the provision of social housing. Thus, while strong state interventions are at the basis of these changes in practices of peripheral urbanization, they are quite different from those in Istanbul. In Santiago, peripheral urbanization was not stopped by either interventions in markets of private property or ambitions of world-class status, but rather by a type of state policy, initiated by the military regime, that transformed housing for the poor into an affordable commodity (although of very low quality) to be acquired legally in areas supplied with services and infrastructure.

While Santiago grew immensely in the second half of the twentieth century, like Istanbul and São Paulo, its rate of population growth has slowed to less than 1% per year.[12] It is no longer a precarious city, as distribution of urban infrastructure, from water to sewage to electricity, has become basically universal. In the 1960s, the occupation of public land in distant areas was the predominant mode of access to housing for the poor, and the government did not take long to improve these areas. The government of Eduardo Frei (1964–70) initiated large programs of land regularization and housing construction, which were intensified by socialist president Salvador Allende (1970–73). Organized invasions of public land multiplied and both governments "increased their support of the organized poor for squatting operations" (Salcedo, 2010, p. 94) at the same time as they tried to accelerate housing production. Thus, during this period, the state was sympathetic to peripheral urbanization and acted in favour of legalizing the illegal, improving infrastructure, and supporting the organization of the new urban residents, who became important political forces sustaining the regime.

Expectedly, the scenario changed radically with the establishment of the military dictatorship in 1973. Social movements were severely repressed and land invasions came to a halt. The military regime decided to stop illegal occupation of land and treat housing as a private matter to be addressed by the market. Unable to ignore the housing deficit, however, it adopted a model in the mid-1970s based on capital subsidies and the production of low-income housing by the private market. The government started giving capital subsidies to families, ones who could prove that they were poor but able to save, and directing them to housing units built en masse by private real-estate developers. It also started to produce houses to be built "progressively" by owners. This meant building rudimentary bathrooms and kitchens and asking the owners to do the rest, in a kind of officially organized autoconstruction. The goal of this program was to eradicate illegal occupations by providing legal and serviced housing. Years later, this program would become the model supported by the World Bank and other international agencies (Gilbert, 2002, p. 311).

Ironically, though, the best results of this neoliberal program were obtained after democratization in 1990, when the model was perfected to reach the poorest sectors of society, which had originally been ignored by it. In this new version, houses were completely subsidized and families would receive the title deed for free, with no bank loans to repay (Salcedo, 2010, p. 95). The state also intensified the production of housing for the poorest sectors. The program became a huge success and is indicated as one of the main factors that helped the proportion of poor people in the Santiago Metropolitan Region drop from 33% to 10% between 1990 and 2006 (Rodríguez & Rodríguez, 2009, p. 15). In 1990, the housing deficit in Santiago was significant; around 20% of the population lived in shantytowns. Through the new housing policies, however, more than a million Chileans have moved from shantytowns to new housing units where they have become property owners (Salcedo, 2010, p. 90). Even critics of the policy agree that the majority of the poor (at least 80%) are now legal property owners and only around 1.5% of the population continues to live in shantytowns (Salcedo, 2010: 91). As one study puts it, home ownership today is "the minimum decent condition" (Salcedo, 2010, p. 111).

However, the narration of this success story in the recent literature comes with an enumeration of problems. One of the most obvious is intensified spatial segregation, as the new housing complexes for the poor have been built disproportionately in distant localities and away from jobs. Additionally, apartments are very small (less than 40 m^2), poorly constructed, and isolated, causing residents to complain bitterly.

The literature analysing the new housing borrows the imagery from a well-known dystopian repertoire to describe spaces of anomie, deterioration, mental illness, lack of sociability, drug addiction, alcoholism, and violence (see Ducci, 1997). One interesting finding of these studies, however, is that residents who moved to houses and were able to transform them are relatively happy. Many managed to transform even apartment units by expanding into common areas or adding improvised balconies. Real discontent exists among those who cannot transform their units and feel trapped in residences for which there is no clear real estate market. What is missed is the possibility of autoconstruction.

The Chilean case shows a clear intervention of the state to solve the housing deficit for the poor and change the character of peripheral urbanization. This intervention is hybrid. Although it started as a neoliberal policy to create property owners who bought residences produced by private developers for the market, its decisive success came from modifications provided by a state with a strong commitment to eradicating poverty and reaching the poorest who had otherwise remained at its margins. Chile is not the only country to have taken decisive steps to enlarge the social capacity of the state and reinvent its social welfare system. Many others – from Costa Rica and Colombia to South Africa, Mexico, and Brazil – have borrowed directly from Chile's housing model (Gilbert, 2004a). All these cases introduce new dimensions to peripheral urbanization, affecting both its transversal logics and the types of spaces produced. In the context of these programs, the production of residential spaces includes not only infrastructure and services, but also legal and regularized housing. Moreover, with official subsidies and loans, housing for the poor exits the alternative circuits of financing and production to join the world of banks, real estate developers, and official mechanisms of financing. In countries that adopt these procedures, the housing market for the poor changes substantially, as the programs introduce a level of formality and capitalization previously unknown. Consequently, they transform the house into another type of commodity: no longer a space to be autoconstructed and improved over time, but rather one, often limited and low quality, to be consumed as a finished product. What is not transformed, however, is spatial segregation, as the new housing complexes are invariably built in far away areas, consolidating class separation and spatial inequality.

Maintenance of a Popular Land Market

São Paulo, like Santiago, exploded in the 1950s during a period of rapid industrialization, but now grows very little, less than 1% per year.[13]

The large majority of the migrant workers who came to the city autoconstructed their houses in the peripheries, usually in places very far from the centre. Squatting and occupation of public land were not very common; in the early 1970s, only 1% of the population lived in *favelas*, squatted settlements, compared to 20% in Santiago. Most of the people who moved to the peripheries bought a piece of land and, as a result, approximately 80% of the residents of the peripheries today have claims of ownership to their lots and houses. Thus, although a market of land existed in São Paulo since at least the 1950s, it was a kind of segregated market. The portion of this market accessible to the poor involved not only long distances from the centre and a lack of infrastructure, but also several layers of illegality and irregularity – from the direct swindling of land, to the lack of registered deeds and permits, to the absence of infrastructure required by law. The risk of buying a property with precarious title was part of what made it affordable to the poor.

As in many other cities, the conditions of irregularity and the precariousness of urban infrastructure were at the basis of efforts by residents in the peripheries to organize social movements. These movements simultaneously pressed the state to improve infrastructure and services in the peripheries and enlarged their claims for democracy. In the context of democratic transition, the state directly engaged these movements and invested heavily in infrastructural improvements. By the early 2000s, water, electricity, and asphalt were almost universally distributed throughout the city, as in Santiago. However, what makes São Paulo's case especially interesting is the way in which organized movements and different governments at the local, state, and federal levels have engaged in a process of experimentation with urban policies that has focused simultaneously on the regularization of land in the peripheries; the prevention of the displacement of their inhabitants; the reproduction of improved spaces for low-income housing; and, more recently, the provision of social housing through the program *Minha Casa, Minha Vida*, inspired by the Chilean model. These policies have incorporated the improved spaces of the peripheries into the regularized city and expanded home ownership for the poor.

As is well known, one of the main achievements of Brazilian social movements has been to influence the 1988 Brazilian Constitution and a radically innovative urban legislation, the 2001 *Estatuto da Cidade* (see Caldeira & Holston, 2005). In 2002, the city of São Paulo approved a new master plan according to this federal legislation. The plan has many remarkable aspects, but one of its most interesting dimensions is a clear attempt to address the issue of illegality and irregularity in the peripheries of the city. Many of the authors of São Paulo's master plan had a long history of advocacy for the peripheries and lobbying for progressive urban legislation and new modes of municipal administration that included popular participation. Having a deep knowledge of the conditions in the peripheries, and sometimes being organizers of social movements and NGOs themselves, many planners were clear that they wanted to legislate for what they termed "the real city." This meant legislating for the city with its complex patterns of urbanization combining all gradations of spaces, bringing the whole heterogeneous and unequal city into a framework of legality and regularity, improving the peripheries without either making them unaffordable to the poor or displacing their residents.

In the final plan, which resulted from significant controversy, the city was treated as being formed by discordant and unequal process of urbanization that required different types of planning instruments. To deal with the peripheries, the 2002 plan used a figure called Special Zones of Social Interests (ZEIS). Inside of these vast zones, most of them peripheral, legislators granted exemptions in land-use standards in the interest of promoting low income housing, accepting as normal substandard practices that would be unacceptable in the better off parts of the city (Caldeira & Holston, 2014). Planners were acutely aware that if they did not create specific norms for peripheral neighbourhoods, it would be hard to control land speculation and the expulsion of poor residents. Thus, the peripheries had to be legalized and improved. But in a way they had to remain peripheral, as a specific market of land not viable for large real estate developments. One way of doing this was to limit the size and prohibit the combination of lots inside of ZEIS. In other words, in their campaign to continue to improve and legalize the peripheries without displacing their residents, urban reformers officially legalized inequality in the production of space.

This plan has now been replaced by another from 2014 that reinforces some of the same principles. Despite the fact that the new plan is too new for us to assess its effects, and that the old one has been the subject of several criticisms and many of its aspects have never been enforced (Caldeira & Holston, 2014), it is evident that São Paulo has not seen the same level of market-driven displacement or undermining of the rights of the poor as in Istanbul or New Delhi, for example. Since the 1970s, the eviction of favelas in São Paulo has been rare, and recently slum upgrading has become routine. Although inequality is still the unmistakable mark of Brazilian cities, the quality of the urban environment has significantly improved, and residents in the peripheries have been able to remain in the neighbourhoods they built, in large measure due to their constant organizing.

Although autoconstruction has been the dominant mode of urbanization and housing for the poor in Brazilian cities, in 2008 the federal government introduced a massive housing project, in line with those in Chile and Mexico, called *Minha Casa, Minha Vida*. It provides subsidies for poor families to acquire housing produced by private developers. In its first five years, it has supported 557 developments in the São Paulo metropolitan region, resulting in the construction of around 110,000 units, with 34% of them for the very poor (Marques, 2014). As in Santiago, the program offers legal housing in serviced areas. Nevertheless, most of the developments for the lowest income group are in the peripheries – not only of the city, but of the broader metropolitan region. Thus, although illegality and irregularity are not being reproduced, the separation is, as the developments are usually located in remote locations, reinforcing a pattern of spatial and social segregation (LabCidade, 2014).

In sum, São Paulo is a case in which peripheral urbanization has always involved a market of land, but one that has been segregated and in which affordability has been associated with distance, precariousness, illegality, and irregularity. It is also a case in which citizens and the state have become deeply involved in simultaneous experiments of regularization of land and democratization that have shaped not only the city, but also Brazil.

Temporality and Politics in the Production of Cities and Theories

Peripheral urbanization consists of a set of interrelated processes that entangle citizens and states in the production of cities of great heterogeneity and dynamism. At the core of the way in which these cities constantly change are the everyday efforts of residents to autoconstruct their dwellings and improve their neighbourhoods. But central to the whole process are also what I called transversal logics and the modes of political engagements that they generate. I have argued that peripheral urbanization engages transversally logics of legal property, formal labour, state regulation, and market capitalism. Actors in this way of creating cities – residents, government officials at various levels, developers, speculators, activists – engage with each other not necessarily outside of mainstream logics, but rather by taking them transversally as matter of negotiation and transformation. The unspoken assumption of all involved is that precariousness, irregularity, and illegality may constitute the present condition under which they urbanize vast areas of the city, but are not permanent conditions. They are rather a matter of struggle, negotiation, and especially of transformation – in short, of politics.

The discussions above show that the institutions of the state are crucial in creating conditions for urbanization, regularization, legalization, and the incorporation of the poor in the city. But it is also clear that the state does not act in the favour of autoconstructors out of its own heart. It does so when organized citizens are able to force it in their favour, maintaining their disturbing presence in public spaces and demanding changes to institutions, legislation, and ways in which the state operates and formulates policies. When this organizing becomes impossible or is weakened, policies may turn in the opposite direction, and dispossession and displacement of autoconstructors may result. Attempts to understand the dynamics of processes of peripheral urbanization must then necessarily dissect the transversal logics at play in each situation and the tense political engagements they generate.

At the core of peripheral urbanization is also a certain temporality: there is constant transformation. Of course, cities are always being transformed. But when the people involved in producing them agree that the present conditions – of the house, the neighbourhood, or their legal status – are provisional and are constantly engaged in changing them, the attention to processes of transformation is critical. In fact, the cases I analysed indicate not only that the conditions in peripheral areas always change, but also that in several cities this mode of producing the city may have changed recently in ways that test the limits of the notion of peripheral urbanization. This happens when the state sides with the formation of capitalist markets of land at the cost of displacing autoconstructors instead of incorporating them, or when massive projects of social housing generalize the possibility of the acquisition of legal and finished housing by the poor, thus eclipsing autoconstruction.

Peripheral urbanization is a widespread process throughout the Global South, but it shapes cities unevenly. Although similar processes of urbanization can be identified in Istanbul, Santiago, and São Paulo, their histories are quite different, as are their present configurations and their tendencies of transformation. They are dissimilar cases. Their juxtaposition and the exploration of the tensions and variations that exist among them illuminate each other and open up new possibilities of understanding.

The three points I have just highlighted – the role played by transversal logics and their unpredictable outcomes; the constant processes of transformation; and the dissimilarity among cases – put specific demands on analyses of peripheral urbanization. Research of these processes should capture the instability of formations of legality and regulation and a significant amount of improvization, experimentation, and contestation shaping the relationships among all involved, from residents to agents of the state. Research should also identify emergent conditions, configurations in the making, and constant transformation. Finally, research should explore significant dissimilarity among cases. The analysis I provided in this article pursues these paths using both qualitative logic and historical investigation. It approaches the set of processes that constitute peripheral urbanization simultaneously in their internal historical transformations *and* across their dissimilar formations. If the model of peripheral urbanization presented here is powerful, it is because it tries to bring to the forefront the double instability of this mode of producing cities in the Global South: it is structured by ambiguity and contestation; and it is always being transformed.

ACKNOWLEDGMENTS

This paper has been in the making for a while. Its first incarnation was presented at the conference "Peripheries: Decentering Urban Theory" that James Holston and I organized at UC Berkeley in 2009, sponsored by Global Metropolitan Studies and in collaboration with the Center for Contemporary Culture of Barcelona. Another version was presented at Stanford University at the Workshop "Decentering Urban Theory: Peripheries, Urbanization and Popular Participation" in 2013. Finally, a later version was presented as a LSE Cities Lecture in October 2014. I thank the participants in these events for their comments, especially James Holston, James Ferguson, Thomas Blom Hansen, Rick Burdett, and Austin Zeiderman. I also thank Ayfer Bartu Candan, Asher Ghertner, Carter Koppelman, Miguel Pérez, Alex Werth, my students who have engaged with me over the years in discussions about peripheries, and my colleagues and collaborators Gautam Bhan, Kelly Gillespie, and AbdouMaliq Simone. Finally, I would like to thank three anonymous reviewers for their insightful comments.

This article has been published previously in *Environment and Planning D: Society and Space, 35*(1): 3–20, 2017.

NOTES

1 Some of the main references in a now well-known argument that mainstream urban theories are inadequate to analyze the types of urbanization in the Global South include Robinson (2002), Roy (2009), McFarlane (2008), Mbembe and Nuttall (2004), and Watson (2009).

2 Nor does peripheral refer to macro relations of uneven development, as in world system theory.

3 For example: McFarlane (2010), Robinson (2011, 2016), and Ward (2010).
4 Simone's analysis in *For the City Yet to Come* (2004) is a brilliant example of this practice of thinking through dissimilar cases.
5 Sometimes I use the term "peripheries" to refer to spaces produced by peripheral urbanization.
6 See Bonduki and Rolnik (1979), Caldeira (1994), Holston (1991), and Holston and Caldeira (2008).
7 On Mexico and its various layers of land irregularity, see Azuela (1987), Castillo (2001), Duhau (1998, 2014), Gilbert and de Jong (2015), and Jones and Ward (1998).
8 See, for instance, Bhan (2016), Holston (2008), Payne and Durand-Lasserve (2012), and Varley (2002).
9 Although the terminology is usually not very precise, residences built on invaded land to which residents have no claim to ownership are usually referred to with special words, such as *favelas, invasões, campamentos,* etc.
10 My discussion of Istanbul is based especially on the works of Balaban (2011), Candan and Kolluoglu (2008), El-Kazaz (2013), Keyder (2005), Kuyucu (2014), Kuyucu and Ünsal (2010), and Tugal (2013).
11 One might be tempted to use the label "gentrification" to describe the processes of land requalification going on in Istanbul. The Turkish scholars who have been analyzing these processes tend not to use this term. Instead, they talk about enclosure of urban space (Balaban, 2011), property transfer (Kuyucu & Ünsal, 2010), institutionalization of private property regimes (Kuyucu, 2014), and social exclusion (Keyder, 2005). Thus, although they do not address the discussion directly, it seems that they would be in agreement with Ghertner's (2014) argument that gentrification theory should not be extended into the Global South. In fact, the assumptions of this theory are based on the Euro-American cases of postindustrial societies with well-established private property regimes. When transferred to different contexts in which the main dynamic may be exactly the constitution of such regimes, the notion of gentrification may "reduce the analytical clarity" of the studies (Ghertner, 2014, p. 1556).
12 My discussion of the Chilean case is based especially on Ducci (1997), Gilbert (2002, 2004a, 2004b), Murphy (2014), Rodríguez and Sugranyes (2005), Rodríguez and Rodríguez (2009), Salcedo (2010), and Tapia Zarricueta (2010).
13 My analysis of São Paulo is based mainly on my own research over the last 30 years, as well as on my work in collaboration with James Holston, and on his own research. It is a work constructed in constant dialogue with our Brazilian colleagues, especially Nabil Bonduki, Eduardo Marques, Marcelo Nery, Regina Prosperi Meyer, and Raquel Rolniki. In the last 40 years, Brazilian social scientists and urbanists have generated a phenomenal production

elucidating the processes of peripheral urbanization throughout Brazil. I deeply regret only been able to cite a minimal fraction of this production here.

REFERENCES

Ahonsi, B. (2002). Popular shaping of metropolitan forms and processes in Nigeria: Glimpses and interpretations from an informed Lagosian. In O. Enwezor, et al. (Eds.), *Under siege: Four African cities: Freetown, Johannesburg, Kinshasa,* Lagos (pp. 129–152). Ostfildern-Ruit: Hatje Cantz Publishers.

Azuela, A. (1987). Low income settlements and the law in Mexico City. *International Journal of Urban and Regional Research, 11,* 523–542. https://doi.org/10.1111/j.1468-2427.1987.tb00065.x

Balaban, U. (2011). The enclosure of urban space and consolidation of the capitalist land regime in Turkish cities. *Urban Studies, 48,* 2162–2179. https://doi.org/10.1177%2F0042098010380958

Bhan, G. (2016). *In the public's interest: Evictions, citizenship and inequality in contemporary Delhi.* Delhi: Orient BlackSwan.

Bonduki, N., & Rolnik, R. (1979). *Periferias: Ocupação do Espaço e Reprodução da Força de Trabalho.* São Paulo: FAU-USP.

Caldeira, T.P.R. (1994). *A política dos Outros – O Cotidiano dos Moradores da Periferia e o que Pensam do Poder e dos Poderosos.* São Paulo: Brasiliense.

Caldeira, T.P.R. (2000). *City of walls: Crime, segregation, and citizenship in São Paulo.* Berkeley: University of California Press.

Caldeira, T.P.R. (2013). São Paulo: The city and its protests. *Kafila.* Available at: https://www.opendemocracy.net/en/opensecurity/sao-paulo-city-and-its-protest/ (Accessed 30 July 2015).

Caldeira, T.P.R. (2015). Social movements, cultural production, and protests: São Paulo's shifting political landscape. *Current Anthropology, 56,* pp. 126–136.

Caldeira, T.P.R., & Holston, J. (2005). State and urban space in Brazil: From modernist planning to democratic interventions. In A. Ong & S.J. Collier (Eds.), *Global assemblages: Technology, politics, and ethics as anthropological problems* (pp. 393–416). London: Blackwell.

Caldeira T.P.R., & Holston, J. (2014). Participatory urban planning in Brazil. *Urban Studies, 52,* 2001–2017. https://doi.org/10.1177%2F0042098014524461

Camargo, C.P.F.D., Cardoso, F.H., Mazzuchelli, F., Moisés, J.A., Kowarick, L., Almeida, M.D., Singer, P.I., & Brant, V.C. (1976). *São Paulo 1975: Crescimento e pobreza.* São Paulo: Loyola.

Candan, A.B., & Kolluoglu, B. (2008). Emerging spaces of neoliberalism: A gated town and a public housing project in Istanbul. *New Perspectives on Turkey, 39,* 5–46.

Castillo, J. (2001). Urbanisms of the informal: Transformations in the urban fringes of Mexico City. *Praxis: Journal of Writing and Building, 2*, 102–111.

Clifford, J. (1981). On ethnographic surrealism. *Comparative Studies in Society and History, 23*(4), 539–564.

Ducci, M.E. (1997). Chile: El lado oscuro de una política de vivienda exitosa. *Revista EURE, 23*, 99–115.

Duhau, E. (1998). *Hábitat popular y política urbana*. Mexico City: Universidad Autonoma Metropolitana.

Duhau, E. (2014). The informal city: An enduring slum or a progressive habitat. In B. Fischer, B. McCann, & J. Auyero (Eds.), *Cities from scratch: Poverty and informality in urban Latin America* (pp. 150–169). Durham, NC: Duke University Press.

El-Kazaz, S. (2013). It's about the park: A struggle for Turkey's cities. *Jadaliyya*. http://www.jadaliyya.com/pages/index/12259/it-is-about-the-park_a-struggle-for-turkey%E2%80%99s-citie (Accessed: 12 November 2018).

Ghertner, A. (2014). India's urban revolution: Geographies of displacement beyond gentrification. *Environment and Planning A, 46*, 1554–1571. https://doi.org/10.1068%2Fa46288

Ghertner, A. (2015). *Rule by aesthetics: World-class city making in Delhi*. New York, NY: Oxford University Press.

Gilbert, A. (2002). Power, ideology and the Washington consensus: The development and spread of Chilean housing policy. *Housing Studies, 17*, 305–324. https://doi.org/10.1080/02673030220123243

Gilbert, A. (2004a). Helping the poor through housing subsidies: Lessons from Chile, Colombia and South Africa. *Habitat International, 28*, 13–40. https://doi.org/10.1016/S0197-3975(02)00070-X

Gilbert, A. (2004b). Learning from others: The spread of capital housing subsidies. *International Planning Studies, 9*, 197–216. https://doi.org/10.1080/1356347042000311776

Gilbert, L., & de Jong, F. (2015). Entanglements of periphery and informality in Mexico City. *International Journal of Urban and Regional Research, 39*, 518–532. https://doi.org/10.1111/1468-2427.12249

Holston, J. (1991). Autoconstruction in working-class Brazil. *Cultural Anthropology, 6*, 447–465. https://doi.org/10.1525/can.1991.6.4.02a00020

Holston, J. (2008). *Insurgent citizenship: Disjunctions of democracy and modernity in Brazil*. Princeton, NJ: Princeton University Press.

Holston, J. (2009). Insurgent citizenship in an era of global urban peripheries. *City & Society, 21*, 245–267. https://doi.org/10.1111/j.1548-744X.2009.01024.x

Holston, J., & Caldeira, T. (2008). Urban peripheries and the invention of citizenship. *Harvard Design Magazine, 28*, 18–23.

Ismail, S. (2014). The politics of the urban everyday in Cairo: Infrastructures of oppositional action. In S. Parnell & S. Oldfield (Eds.), *The Routledge handbook on cities of the Global South* (pp. 269–280). London: Routledge.

Jones, G., & Ward, P. (1998). Privatizing the commons: Reforming the ejido and urban development in Mexico. *International Journal of Urban and Regional Research, 22*, 76–83. https://doi.org/10.1111/1468-2427.00124

Keyder, C. (2005). Globalization and social exclusion in Istanbul. *International Journal of Urban and Regional Research, 29*(1), 124–134.

Kuyucu, T. (2014). Law, property and ambiguity: The uses and abuses of legal ambiguity in remaking Istanbul's informal settlements. *International Journal of Urban and Regional Research, 38*, 609–627. https://doi.org/10.1111/1468-2427.12026

Kuyucu, T., & Ünsal, Ö. (2010). "Urban transformation" as state-led property transfer: An analysis of two cases of urban renewal in Istanbul. *Urban Studies, 47*, 1479–1499. https://doi.org/10.1177%2F0042098009353629

LabCidade (2014). Ferramentas para a avaliaçaõ da inserçaõ urbana dos empreendimentos do MCMV. Available at: http://www.labcidade.fau.usp.br/arquivos/relatorio.pdf (Accessed: 9 March 2015).

Marcus, G., & Fischer, M. (1986). *Anthropology as cultural critique: An experimental moment in the human sciences*. Chicago: Chicago University Press.

Marques, E. (2014). *A Metrópole de São Paulo no Século XXI – Espaços, Heterogeneidades e Desigualdades*. São Paulo: CEM-UNESP.

Mbembe, A., & Nuttall, S. (2004). Writing the world from an African metropolis. *Public Culture, 16*, 347–372.

McFarlane, C. (2008). Urban shadows: Materiality, the "Southern city" and urban theory. *Geography Compass, 2*, 340–358. https://doi.org/10.1111/j.1749-8198.2007.00073.x

McFarlane, C. (2010). The comparative city: Knowledge, learning, urbanism. *International Journal of Urban and Regional Research, 34*, 725–742. https://doi.org/10.1111/j.1468-2427.2010.00917.x

Murphy, E. (2014). In and out of the margins: Urban land seizures and homeownership in Santiago, Chile. In B. Fischer, B. McCann, & J. Auyero (Eds.), *Cities from scratch: Poverty and informality in urban Latin America* (pp. 68–101), Durham, NJ: Duke University Press.

Núcleo de Estudos da Violência da Universidade de São Paulo (NEV/USP) (2014). *A delimitação de áreas chave para estudos longitudinais em São Paulo (SP), Brasil*. São Paulo: NEV.

Payne, G., & Durand-Lasserve, A. (2012). *Holding on: Security of tenure: Types, policies, practices, and challenges* [Paper presentation]. The Special Rapporteur on Adequate Housing, UN.

Robinson, J. (2002). Global and world cities: A view from off the map. *International Journal of Urban and Regional Research, 26*, 531–554. https://doi.org/10.1111/1468-2427.00397

Robinson, J. (2011). Cities in a world of cities: The comparative gesture. *International Journal of Urban and Regional Research, 35*, 1–23. https://doi.org/10.1111/j.1468-2427.2010.00982.x

Robinson, J. (2016). Thinking cities through elsewhere: Comparative tactics for a more global urban studies. *Progress in Human Geography, 40*, 3–29. https://doi.org/10.1177%2F0309132515598025

Rodríguez, A., & Rodríguez, P. (2009). *Santiago, una ciudad neoliberal*. Quito: OLACCHI.

Rodríguez, A., & Sugranyes, A. (2005). *Los con techo: Un desafío para la política de vivienda Social*. Santiago: Ediciones SUR.

Roy, A. (2009). The 21st-century metropolis: New geographies of theory. *Regional Studies, 43*, 819–830. https://doi.org/10.1080/00343400701809665

Salcedo, R. (2010). The last slum: Moving from illegal settlements to subsidized home ownership in Chile. *Urban Affairs Review, 46*, 90–118. https://doi.org/10.1177%2F1078087410368487

Secretaria de economia e planejamento do estado de São Paulo (Seplan) (1977). *Subdivisão do Município de São Paulo em Areas Homogêneas*. São Paulo: Seplan.

Simone, A. (2004). *For the city yet to come: Changing African life in four cities*. Durham, NC: Duke.

Simone, A. (2010). *City life from Jakarta to Dakar: Movements at the crossroads*. London: Routledge.

Simone, A. (2014). *Jakarta: Drawing the city near*. Minneapolis: University of Minnesota Press.

Small, M.L. (2009). "How many cases do I need?" On science and the logic of case selection in field-based research. *Ethnography, 10*, 5–38. https://doi.org/10.1177%2F1466138108099586

Smolka, M.O., & Mullahy, L. (2010). *Perspectivas urbanas: Temas críticos en políticas de Suelo en América Latina*. Cambridge: Lincoln Institute of Land Policy.

Tapia Zarricueta, R. (2010). Vivienda social en Santiago de Chile: análisis de su comportamiento locacional, período 1980–2002. *Revista Invi, 73*, 105–131. http://dx.doi.org/10.4067/S0718-83582011000300004

Tugal, C. (2013). Occupy Gezi: The limits of Turkey's neoliberal success. *Jadaliyya*. Available at: http://www.jadaliyya.com/pages/index/12009/occupy-gezi_the-limits-of-turkey%E2%80%99s-neoliberal-succ (Accessed: 25 August 2013).

Varley, A. (2002). Private or public: Debating the meaning of tenure legalization. *International Journal of Urban and Regional Research, 26*, 449–461. https://doi.org/10.1111/1468-2427.00392

Ward, K. (2010). Towards a relational comparative approach to the study of cities. *Progress in Human Geography, 34*, 471–487. https://doi.org/10.1177%2F0309132509350239

Watson, V. (2009). Seeing from the South: Refocusing urban planning on the globe's central urban issues. *Urban Studies, 46*, 2259–2275. https://doi.org/10.1177%2F0042098009342598

Yin, R. (2002). *Case study research: Design and methods.* Thousand Oaks, CA: Sage.

2 Urban Community and Resistance

RAÚL ZIBECHI

Since time immemorial, humanity has used community as a means of dealing with challenges and disasters, and it is through the profound interpersonal relationships involved in community ties that mankind has found ways to survive. In recent decades, the idea of community has been undergoing a process of reinvention, not just in order to protect it, but also with a view to emancipation. This change is considered by activists to be not only vital in times of collective crisis and tragedy, but also a way of breaking the chains of capital.

The history of urban communities is a long one, stretching as far back as the early cities of Europe. "Auzolan" was the name given by the founders of the Basque city of Vitoria to the concept of community labour around the year 1400. It was work performed "to carry out certain tasks for the good of the community or for the creation or maintenance of communal infrastructure or equipment" (Ayllu, 2014, p. 25). Collective labour was a means to satisfy the material needs of the group, and at the same time strengthened the collective feeling of community, with auzolan work generally ending in a group celebration attended by all involved.

Studies that have pieced together the history of Vitorian neighbourhoods have concluded that this organizational approach to resolving collective problems represented "the transfer to the city of a form of organisation which had been in use since ancient times in villages and hamlets ... and to which, as a result, they were "naturally" accustomed (Ayllu, 2014, p. 36).

There is no documentary evidence to suggest when these practices first emerged, although it is known that they were present in numerous cities and their peripheral neighbourhoods during the Middle Ages. There were no by-laws or regulations in place to manage such practices, but their essential nature was similar in all cases. In the absence of any state provision, auzolan was driven by the needs of each community, and collective ability dictated the way in which these would be met. According to the residents themselves, it was a form of "spontaneous" (according to current language) or "natural" organization and action (Ayllu, 2014).

Methods of self-organization (which is certainly not a term that would have been used at the time) and collective culture were brought from rural areas to the cities by migrating peasants, propagating the solidarity that characterized the sprawling families of the countryside. Here, greater importance was ascribed to the collective than to the individual. However, far from being official ways of working, they were at most customs – common sense which did not need to be made into law. Moreover, the transformation of common sense into written law would only mark the beginning of its disappearance, of its neutralization, as part of the autonomy of these communities was derived from their invisibility to government.

Centralization of the Spanish state brought about the destruction of this parallel form of power, as the elite began to develop a new political culture. However, political cultures do not simply vanish. The Concejos Abiertos, or open councils, survive to this day (319 in Álava and 348 in Navarre), the remnants of these communities' ancient decision-making systems. Despite the advent of modernity and consumerism, they endure in *reuniones de portal*, or community meetings, during which residents discuss the problems of the buildings in which they live. However, it is during moments of difficulty, crisis, or tragedy that the true potential of collective labour becomes apparent.

In August 1983, when the city of Bilbao suffered flooding that laid waste to entire neighbourhoods and caused the deaths of 39 people, the auzolan rose again in all its strength in an event which came to be known as "the flood of solidarity." In the absence of an effective state response, thousands of people offered their support to the families that had lost their homes, assisting in the clean-up and rebuilding of the areas worst affected (Asociación de Familias de Rekaldeberri, 1983).

Urban communities and their collective practices can be seen even in today's vast Latin American cities. The Acapatzingo community and other urban communities that have worked together to secure housing for their members demonstrate that the traits of community may be seen even in such vast population centres as Mexico City. The piece of land on which the Acapatzingo community now exists – La Polvorilla, in the borough of Iztapalapa in eastern Mexico City – was progressively occupied from 1994 onwards following the removal of "intermediaries" who had been taking advantage of those moving in. The land was eventually purchased in 1998, and the construction of homes began in 2000 (Lao & Flavia, 2009; Zibechi, 2009).

Acapatzingo is an excellent example. By mid-2013 it included two nurseries producing food, a community radio station run by teenagers, and spaces for children, the elderly, and young people, including a skating and cycling park, two basketball courts, and an open-air theatre. Construction of a health centre and pre-, primary, and secondary schools is in its initial phase. Entry to the area is regulated by a vigilance committee, in which each of the families participates on a rotating basis, and the police are not permitted access except under special circumstances, and with the authorization of the community.

Construction of the neighbourhood was a lengthy process in which cultural and subjective aspects played a more important role than housing and the physical space itself, and the result was the transformation of an abandoned wasteland into a beautiful neighbourhood. But how did a group of individuals succeed in forming this thriving community? Residents maintain that a change in subjectivity came about once they found themselves able to rise above "our own fears, combat the trauma instilled in us from childhood, and make a clean break from selfishness and apathy" (FPFVI-UNOPII, 2008). This challenge is both individual and collective, while being neither entirely one nor the other. Polarity between the two is removed, fading without confusion or the disappearance of either. But how is this possible? The answer is through occupation of land and the conversion of settlements into open-air schools, where "meetings, demonstrations, guard duty and the working day foster collectivism and concern for each other" (FPFVI-UNOPII, 2006). In short, I exist by sharing with him and with her; I am not in isolation, but as part of the collective. It is through everyone else that my individuality may grow, strengthen and be.

This stark change in subjectivity seen among members of the movement is firmly rooted in Mexican colonies, in which a community culture exists "naturally." The Front (*Frente Popular Francisco Villa Independiente*) has managed to rescue this phenomenon and prevent its loss at the hands of market forces, political parties, and the patriarchy. "The principal aim is for the rescue and conservation of community culture to extend to whole families, in order that children and young people may grow up in an environment shaped by these values, subversive as they are in a society such as ours" (FPFVI-UNOPII, 2006).

The foundation of community here is always the same: the "brigades," each formed of 25 families. They represent not only the basis of territorial organization, but the very essence of the community itself. At brigade meetings, each family has a vote. Issues of importance, such as the rules of the settlement, are discussed, and each brigade then puts forward its own arguments until all are in agreement. The brigade intervenes in the event of conflict – which may even include intra-familial disturbances – and depending on the seriousness of the issue may request the involvement of the vigilance commission or even the general council. Once a month, each brigade takes responsibility for the security of the settlement, although the concept of vigilance, rather than being traditionally control-based, focuses on collective self-protection, and as a result is strongly educational in nature.

The vigilance commission is also responsible for maintaining the settlement border, and for dictating who may enter and who is forbidden to do so. This is perhaps the most important aspect of autonomy, a condition which implies establishing a physical and political boundary that differentiates internal and external space, ensuring that the autonomous body does not dissolve into its surroundings. This is a reflection of living systems, which create a perimeter to delineate the territory in which interactions occur, enabling the constituent parts to function as a single unit (Maturana & Varela, 1995).

Another example is the Centro Integral Cooperativo de Salud (CICS) health centre run by Cecosesola (Lara State Centre for Social Services) in the city of Barquisimeto in Venezuela. It is a network of 50 agricultural cooperatives, small-scale agro-industries, and providers of services including health, transportation, funerals, savings and loans, food distribution, and homeware supplies. The health centre receives around 180,000 people each year, and is run according to community principles.

One of its most interesting features is the collective approach to management of such a large centre: all of the employees associated with the network of cooperatives are involved in its running, and this is even increasingly the case with the doctors. Management is important: although health services are provided on a non-commercial basis, the administrative approach has an important role to play in determining the dominant culture, and in theory would have the potential to allow the market and the state to regain control of projects born in opposition to such a reality.

My intention is to highlight the creative and emancipatory aspects of human relations present in the community approach. In the cases of the Acapatzingo community and the Cecosesola network of productive communities, the "community approach" represents an anti-state mechanism which at the same time fosters freedom through non-hierarchical relationships and the active participation of all members. It is important to keep in mind that community may function in any sector of society, including in areas involving advanced technologies, such as modern health centres. Contrary to popular belief, community is not limited merely to rudimentary activities in rural areas.

At this point, it seems appropriate to reflect on what is meant by the concept of community. Latin America is home to a seemingly endless number of community institutions, but it is common even for native and peasant communities to be defined in terms of the central importance of property or so-called common goods. Common ownership of land, assembly-based management, the election of authorities by all members, and a shared culture and world view are all clearly key aspects of community and are routinely placed at the heart of the concept. However, while common goods, both material and immaterial, are in effect the foundation upon which a community institution is built and supported, alone they are insufficient to ensure its sustained operation.

I propose that community be considered in terms of the social relationships that exist within it, viewing it not as a static unit, but instead focusing on how it flows, on the day-to-day life of its members, and on the capacity of their relationships to construct and reproduce it. I believe that community is kept alive through collective work, which is a creative activity that renews and strengthens the community. In this light, *minga* and *tequio* can be interpreted not as negative human activity working contrary to capital or the state, but the means by which members of a group produce community. They are the expression of heterogeneous social relations, without which the concept of community could not be applied.

Community does not simply exist. It is constructed and renewed on a daily basis by the collective effort of men and women, of children and the elderly who, by working together, produce community. To reduce it to an institution – made once and for ever; established – obscures the fact that it is collective effort that gives life, sense, shape, and background to that community. Let us therefore say we produce community instead of simply we are a community. One of the central pillars of any living community is assembly, and this should be considered not only as a common good, but as the essence of "good living."

The difference is substantial. By placing collective property at the heart of community, as the cornerstone of production of common material goods, we achieve nothing more than a recreation of the economic rationale that dominates Western critical thinking. Conversely, by heeding the Zapatistas and placing collective work at the centre, we remove all essentialist temptation and shine the spotlight on two key aspects: human relationships, and the living and tangible nature of collective work, which is capable of generating greater value than the abstract concept associated with the value of change and in turn with property.

REFERENCES

Asociación de Familias de Rekaldeberri (1983). *Más allá del barro y las promesas*. Bilbao: Revolución.

Ayllu, E. (2014). *Las vecindades Vitorianas*. Vitoria: Ned Ediciones.

Cecosesola. (2009). *Hacia un cerebro colectivo*. Barquisimeto: Cecosesola.

FPFVI-UNOPII. (2006). *Ponencia para acto en la casa de Dr. Margil en Monterrey*. November.

FPFVI-UNOPII. (2008). *Una construcción con esfuerzo colectivo*. México D.F.: mimeo.

Lao, W., & Flavia, A. (2009, January 6). *El frente popular Francisco Villa Independiente no es solo un proyecto de organización, es un proyecto de vida* [Interview with Enrique Reynoso]. *Rebelión*. http://www.rebelion.org/noticia.php?id=78519 (Accessed: 15 December 2013).

Maturana, H., & Varela, F. (1995). *De máquinas y seres vivos*. Santiago: Editorial Universitaria.

Zibechi, R. (2009, December). *Interview with Enrique Reynoso*. México D.F.

3 Planetary Urbanization and the Commodity Super-Cycle in Latin America[1]

MARTÍN ARBOLEDA

Introduction

The book *Implosions/Explosions: Towards a Study of Planetary Urbanization*, edited by Neil Brenner (2014), has as its cover an image of the desolate, colossal, and gloomy tar sands of Alberta, Canada. These barren landscapes are the result of a historical turning point in the intensity and scale of resource extraction, and have come to constitute the universal epitome for the massive socio-ecological plunder that underlies fossil fuel-powered, modern urban life. Such aesthetic – which became popularized by the work of photographers like David Maisel, Garth Lenz, and Edward Burtynsky – conveys without ambiguities the philosophical and political urgency of transgressing the dominant epistemology of urban studies in which cities are considered the only morphological embodiment of urbanization. With his visionary 1970 work *The Urban Revolution*, Henri Lefebvre began to lead the way in such direction because, for him, the concentration of population that economic growth and industrialization demands corroded the borders of a traditionally self-contained urban form, making of urbanization a boundless phenomenon (Lefebvre 2003/1970). This led him to describe capitalist globalization as a generalized "explosion of spaces" that is fundamentally underpinned by contradictory, yet interwoven processes of homogenization and fragmentation of territories (Lefebvre, 2009/1979, 2009/1980, 1991/1976, 2003/1970; Brenner, 2000).

Primary commodity production is arguably one of the main driving forces of the explosion of spaces that Lefebvre so presciently described. Indeed, the fluctuations of international commodity markets in recent years attest to the frantic pace at which resource extraction is changing the face of the planet (Toro, 2012; Klare, 2008; IMF, 2011, 2014; ECLAC, 2013; World Bank, 2009, 2012; in Latin America: Gudynas,

2010; Bebbington, 2012). Since Latin American countries possess some of the world's largest mineral and oil reserves, as well as investment-friendly regulatory frameworks, the region has become the main destination for capital allocations in world mineral prospecting, accounting for almost one third of total mining investment in 2010, with USD 180 billion (ECLAC, 2013). The budget for exploration in the region increased more than fivefold during the 2003–2010 period, going from USD 566 million to USD 3 billion (ibid), making several Latin American countries recipients of substantial flows of foreign direct investment (FDI) (Cancino, 2012). This triggered a massive wave of infrastructure, energy, and mining undertakings across the whole regional territory in order to facilitate resource extraction at a very large scale.

On the basis of recent discussions on the notion of planetary urbanization (Brenner & Schmid, 2013, 2015; Brenner, 2014 for an overview), this chapter proposes to rethink mineral extraction beyond the mere spatiality of shafts and pits. This, the chapter argues, requires understanding the mine as *produced by* and *entangled in* processes of urban metabolic transformation unfolding at various spatial scales. The chapter argues that the expansion of primary commodity frontiers in Latin America has come to manifest itself in terms of a contradictory tension between spatial homogenization – in the form of multiscalar state spatial strategies – and territorial fragmentation – in the form of fixed capital allocations and state-led spatial segregation. When considered jointly, these contradictory movements allow us to grasp the patterns of uneven geographical development that are concomitant to the so-called commodity supercycle – one of the most persistent and wide-ranging resource extraction booms in modern history.[2] This dialectic of sociospatial transformation, it is worth noting, has assumed a more advanced configuration as a result of the new geo-economic order wrought by processes of industrial upgrading taking place across East Asian economies, and especially by the rise of China (Charnock & Starosta, 2016; Khanna, 2016; Arboleda, forthcoming).

Increasing spatial separation between manufacturing and extraction following a new international division of labour has pushed mining, oil, and agro-industrial companies to embrace connectivity, speed, and homeostasis as central organizational principles (Arboleda, forthcoming). This has led some commentators to argue that instead of being world factories or industrial powerhouses, the so-called Asian Tigers are more adequately conceptualized as "logistics empires" (Cowen, 2014; Khanna, 2016). And, indeed, recent critiques of existing studies of "extractivism" have pointed out that it is important to supersede the idea of the mine or extraction site as a self-contained object, and grasp its imbrication within wider processes and scales of territorial transformation that include manufacturing, finance, urbanization, and logistics, among others (Gago & Mezzadra, 2015; Mezzadra & Neilson, 2017). The approach that this chapter proposes, then, intends to shed light into the present context of global territorial change, where the deployment of spatial technologies and logistical infrastructures for the swift mobilization of raw materials across space has led to a more advanced configuration of metabolic interdependence between built and unbuilt environments, city and non-city. The chapter begins with a brief discussion of Lefebvre's notion of implosion/explosion, and of its reinterpretation by critical urban theorists. The remaining sections reflect on the splintered, geographically uneven territorial configurations that have emerged in Latin America as a result of the commodity supercycle.

The Implosions and Explosions of Capitalist Urbanization

Throughout his major works on space, written since the late 1960s (Stanek, 2011), Lefebvre describes a simultaneously amplified and exploded urban reality, as the irruption of industry and its pursuit of labour markets and raw materials implied an immeasurable extension of the city beyond city limits (Lefebvre, 2003/1970). For Lefebvre, it is the massive concentration in urban areas concomitant to the process of capital accumulation (implosion) which ultimately leads to the expansion of multiple infrastructural networks across space (explosion). In this process, space as a whole inserts itself into the modernized mode of production, being utilized to produce surplus value. For that reason, the urban fabric, with its multiple networks of communication and exchange becomes a part of capital (Lefebvre, 2009/1979). In this context, the spatial arrangement of a city, a region, a nation, or even a continent increases productive forces in a similar way as do the machinery and equipment in a factory (ibid, p. 188). What is most problematic about this, Lefebvre (2009/1978) adds, is that space adopts a hierarchy that corresponds to that of social classes, assuming the form of a collection of ghettos. These ghettos are not simply juxtaposed, as they are "hierarchized in a way that represents spatially the economic and social hierarchy" (ibid, p. 244).

Recently, an emerging strand of critical urban thought has reclaimed this processual view of the urban to move beyond the entrenched epistemology of mainstream urban studies in which the urban is thought to be embodied only in cities, usually determined as such by population thresholds and densities (Brenner, 2014; Brenner & Schmid, 2013

Monte-Mór, 2014; in Latin America: Kanai, 2014; Wilson & Bayón, 2015). Transcending such epistemologies and ideological visions in studies of resource extraction is something of uttermost political and analytical urgency, because despite the fact that extraction sites may not have the densities and population thresholds of large urban areas, they nonetheless become the recipients of infrastructures, capital, and migratory flows that transform them completely, superseding any clear distinction between city and country. Authors in this tradition have proposed to consider the twenty-first-century global urban condition in terms of a dialectic between *concentrated and extended* forms of urbanization (Brenner 2014). The former corresponds to densely settled zones (i.e., cities, metropolitan regions, megacity regions, and so forth), whereas the latter would include infrastructures for energy, tourism, telecommunications, and transportation, as well as extraction sites and landfills, among many other places that both result from and facilitate the dynamics of urban agglomeration (ibid). The process of extended urbanization that collapses the erstwhile urban/non-urban divide, as the following sections explain, is deeply rooted in the contradictory tension between spatial homogenization and territorial fragmentation.

The Tendency to Homogenization

The process through which the uneven spatialities of the urban fabric are produced is profoundly contradictory, as it falls within a schema that Lefebvre referred to as "homogeneity-fragmentation-hierarchization" (2009/1980; 2003/1970). First of all, there is a tendency to homogenization in which capitalism produces a space that is a "reflection of the world of business on the national and international level" (Lefebvre, 2009/1979, p. 187). According to Harvey (2006/1982), spatial integration – understood as the linking of commodity production in different places through exchange – is a precondition for the accumulation and circulation of capital. This homogenizing process is thus achieved by reducing physical barriers to the movement of commodities and capital (ibid, p. 375). As Smith (2008/1984; Harvey, 2006/1982) noted, such spatially integrated systems require a tightly woven constellation of institutional apparatuses. In other words, and as Lefebvre (2009/1980) suggested, the homogenization of space is not only produced by relations of production, because insofar as it is a political product, it also implies a network of strategies and administrative and governance controls.

To begin with, international financial institutions (IFIs) have been officially endorsing the shift to resource-intensive economies in Latin

America ever since the structural adjustment reforms of the 1980s and 1990s. The World Bank Group published in 1996 a watershed document titled *A Mining Strategy for Latin America and the Caribbean*, in which "friendly" policy recommendations were made with the purpose of implementing reforms to attract transnational mining investment.[3] In the years that have followed the report, the World Bank and other multilateral organizations have sponsored programs to reform mining codes, ease profit repatriation, reduce royalty rates, and support geological surveying to create more incentives for companies to invest (Bebbington et al., 2008, 2008a; Rudas & Espitia, 2013; Infante, 2011; Padilla, 2010). As a result, most mining codes in the region were reformed to reflect this new approach (Fuentes, 2012; Toro, 2012; Pardo 2013). Reforms of mining and land regulations have also coincided with a parallel tendency to negotiate and sign bilateral Free Trade Agreements (FTAs) with direct impact on resource extraction. Usually tailored to boost investment, these legal instruments include tax exemptions, litigation agreements, compensations, and other dispositions to attract and protect transnational extractive companies (see CooperAcción, 2012; Toro, 2012).

This tendency to homogenization has not been circumscribed to legal formulations and trade agreements only, as it has also encompassed specific spatial manifestations via large-scale territorial planning and infrastructure programs. Since commodity exchange needs reducing to a minimum the physical barriers to the movement of goods and money through space (Harvey, 2006/1982; Smith, 2008/1984), achieving region-wide spatial integration across national borders has become a key priority. In contexts like this, Harvey (2006/1982) has noted, the effect is the creation of a hierarchy of means – market, institutional, and state – for the production, modification, and transformation of spatial configurations to the built environment (p. 397). The Initiative for the Integration of the Regional Infrastructure of South America (IIRSA in Spanish), which was launched at a South American presidential summit in 2000, is among the most ambitious and promethean endeavours to transform the subcontinent into a spatially integrated system (Smith, 2008/1984) for an export-led regional economy (Zibechi, 2006; Martínez & Houghton, 2008; Razeto et al., 2009). According to Zibechi (2006), the IIRSA is a multi-sectorial process that is aimed at developing and integrating transport, energy, and telecom infrastructures throughout the region, ordering geographical space in the form of terrestrial, fluvial, and air transport networks; oil, gas, and water pipelines; waterways, sea and inland ports, and power lines among others. Funded by three multilateral banks[4] (Razeto et al., 2009), it encompasses over 500 infrastructure projects distributed along 10 "Integration and Development" axes, and has an estimated cost of USD 75 billion.

The Plan Puebla Panama (PPP), launched in 2001 at a projected cost of USD 20 billion, and with the aim of integrating transportation, energy, and telecommunication infrastructures from Mexico to Panama, was intended to be the Mesoamerican counterpart of the IIRSA. According to Wilson (2011), both projects are underpinned by the New Economic Geography (NEG), an offshoot of neoclassical economics that incorporates questions of geography and location so as to make sense of processes of uneven development. The NEG outlook – in which space is conceptually reduced to transport costs – has been implemented by multilateral agencies in order to incorporate remote regions of the world to transnational circuits of capital via large-scale infrastructure projects like IIRSA and PPP (ibid). The PPP, however, was officially terminated in 2008, after seven years of struggle from civil society organizations, especially peasant communities and Zapatista groups in Southern Mexico (Wilson, 2014). After its official abandonment, the PPP was replaced by the Mesoamerica Project for Integration and Development, a more modest infrastructural integration program.

In sum, and although the social geographies that have emerged from these institutional arrangements could be regarded as a reflection of the world of business at a regional scale and perhaps hint at the disruptive nature of the commodity boom, focusing on the national and international scales alone would yield an incomplete picture of the unevenness of the operational landscapes of extended urbanization that ceaselessly proliferate throughout the region. It is precisely for that reason that Smith (2008/1984) considered that dependency theory, centre-periphery theory, world systems analysis, and the various approaches to uneven development fail to capture the full extent of the geographies of capitalism and, one might add, of the geographies of capitalist urbanization as well. Consequently, the following subsection constitutes a descent into the complex world of resource extraction on the ground, because it is only in conjunction with the local scale that one can visualize the way in which primary commodity production weaves together the thickening, fractured, and geographically uneven fabric of urbanization.

The Tendency to Fragmentation

For Neil Smith, the necessity for accumulation leads to a continuous investment in fixed capital in the form of facilities such as railways, factories, warehouses, power stations, and so forth (Smith, 2008/1984).

Since these facilities need to be geographically immobilized as a precondition for accumulation, there is a spatial (machineries and infrastructures) and social (labouring processes) concentration of capital, which is the main determinant of the tendency to spatial fragmentation (ibid). Resource extraction (especially minerals) is among the most spatially immobile economic activities, because besides being eminently place-specific, it requires vast amounts of investment on fixed capital in the form of machinery and infrastructures. As Bebbington et al. (2008) have argued, the extractive industries are by definition a "point source activity" (Bridge, 2009), in the sense that the geographic unevenness of geological formations results in the fact that some areas become territories of extraction while others do not. To the extent that extraction generates significant socio-environmental effects, geological unevenness invariably translates into territorial and social difference (Bebbington et al., 2008a).

Mining entails, as a first step, the removal of large quantities of soil by using explosives in order to access the minerals that are buried underground, profoundly affecting communities, ecosystems and water sources (Padilla, 2012). Also, large quantities of poisonous chemicals – most notably cyanide – are required for the process of lixiviation, which is central to mining and consists of separating minerals from rocks (Padilla, 2012; Peña, 2013). Since processes of mineral extraction are developed at colossal scales, the material footprint of large-scale mining is gigantic, as each mine requires between 460 and 1,060 litres of water per gram of mineral and produces between 50 and 140 million tons of solid waste per year, on average 40 times more waste than any Latin American megacity during the same time period (Cabrera & Fierro, 2013; Ramírez & Ibagón, 2012). In recent years, leakages and malfunction of tailings dams have led to severe environmental destruction and social crisis, with the failure of the Bento Rodrigues dam in the Brazilian state of Minas Gerais in 2015 being perhaps the most illustrative example. It is estimated that 60 million cubic metres of toxic sludge from mining waste were poured into the Doce river, causing mass flooding, 17 deaths, and the loss of livelihoods for thousands of villagers (Mendes, 2017).

It should be noted that contemporary agribusiness works in similar ways as open-cast mining, both in terms of its material footprint and of its impact on local communities (Gudynas, 2010; Otero & Lapegna, 2016). Agro-industrial monocultures not only rely on the use of highly toxic herbicides and fertilizers, the most controversial of which is glyphosate. They are also land-intensive, and this has become an important driver of processes of deforestation and mass displacement of local communities. It is estimated that 900,000 peasants in Paraguay have been displaced by landlords and state forces during the last decade in order to enable the expansion of capital-intensive monocultures (Carneri, 2017). In Argentina, estimates suggest that around 200,000 rural families have been displaced by transgenic soybean production (Feeney-McCandless, 2017).

Because of the explosives, soil removal, and use of chemicals among other processes, once a given territory has been used for open-cast mining (or monoculture agribusiness), other economic activities like subsistence farming, livestock, or even real estate become unfeasible (Ramírez & Ibagón, 2012). It has been argued that concessions grossly exaggerate the effects of extraction on the landscape because only a small proportion of areas granted for extraction is converted into actual mines or wells (Bebbington, 2012). Concession holders therefore exert absolute power over the territories handed over by the state, producing classic enclave economies that, according to Bridge (2009), are at the same time deeply integrated into the global economy and also fragmented from national space. Since concessions only give their holders the right to the subsoil, rights to the surface have to be acquired through market transactions, negotiation, or compulsory purchase (Bebbington, 2012), thus triggering ongoing processes of primitive accumulation, where corporations constitute semi-sovereign spaces within territories of extraction.

Furthermore, and because of its manifold negative impacts, extraction projects elicit protest and revolt from communities and activists alike (Bebbington et al., 2008), making armed private security, militarization, and violence the background of these emerging operational landscapes. Activists are typically subject to surveillance and intimidation, and in countries like Colombia and Brazil they usually become victims of selective killings, sometimes at the hands of "death squads" or paramilitary groups[5] (Duque, 2012; Karmy & Salinas, 2008; Bermúdez, 2012 for the relationship between extractivism and low-intensity warfare). In a context of planetary urbanization, Merrifield (2014b) contends that war no longer comprises grandiose campaigns by troops, but as the Latin American case demonstrates, "is rather a micro-everydayness of peacetime intervention, a dogged affair in which the police and the paramilitary play interchangeable roles" (p. 41). A recent report by Global Witness revealed that 2019 had been the most dangerous year on record for defenders of the environment, as 212 Indigenous land and environmental activists were murdered (Global Witness 2020).

Also, since resource extraction requires large amounts of fixed capital allocations, the microeconomic effects of the irruption of vast capital flows in mining districts form a breeding ground for inequality and dispossession. These flows of circulating money tend to create microeconomic distortions resulting in artificial price increases across most goods and services – especially rents, which have been said to increase over 300% in some mining areas[6] (see for example Colombia Solidarity Campaign, 2013, p. 92). In Antofagasta, one of the largest mining cities in Chile, for example, the cityscape has been aggressively splintered between the zones that most directly partake in the supply chain of extraction, and those that aspire to do so but have been left behind. The wealthy quarters of the city are interspersed by shopping malls and opulent residential units boasting swimming pools, tennis courts, and expensive cars. Opposing the small islands of gated communities and conspicuous consumption tailored for the steady, salaried workers of the mining industry, are Antofagasta's dark peripheries of informal, precarious, and subcontracted work. Faced with soaring rents, a considerable part of these workforces has been left with no choice but to inhabit the city's swelling shantytowns and precarious settlements, informally referred to as campamentos (Thodes Miranda, 2016; Techo Para Chile, 2016, p. 60).

In sum, the processes of ecological plunder and social dispossession that result from the expansion of primary commodity frontiers in Latin America are well documented (Padilla, 2012; Soliz et al., 2012; Idárraga, 2012; Reyes & León, 2012; Duque, 2012; Vargas, 2013; Cabrera & Fierro, 2013; Bebbington, 2012; Hinojosa & Hennermann, 2011; Hinojosa, 2011), thus figuring as key drivers of the process that Caldeira (2016) has aptly termed "peripheral urbanization," and which informs this book. These processes of sociospatial transformation, it should then be noted, have also laid the foundations for new forms of territorial struggle that are pushing the frontier of political possibility in important ways. In a well-known passage of the Communist Manifesto, Marx and Engels highlight how, with the advance of industry, the bourgeoisie inadvertently replaces the isolation of workers with their union through association and in so doing, lays the foundations for the revolutionary movement of the proletariat. The spatial technologies and logistical apparatuses harnessed to expand the scale of mineral production, then, have enabled new modalities of encounter and mobility. The workers directly employed by the mining industry and their contractors, as well as Indigenous and peasant communities, have been transformed by the frenzied movement of these infrastructural systems, becoming an arrogant force that not only is oppositional but invents new modes of inhabiting the world (Teubal, 2009; Zibechi, 2012; Arboleda, 2017).

Conclusions

This chapter has cast light into how an emerging geography of advanced industrialization has resulted in new, more advanced modalities of material interdependence between city and non-city, natural resources, and urbanization. As Asian Tigers become consolidated as the world's main buyers of raw materials, complex logistical systems have begun to weave together mines in Latin America with an expanding constellation of cities, factories, ports, and other auxiliary infrastructures of industrialization across East Asia and other parts of the world. This context of global sociospatial change, I have argued, requires conceptual frameworks and theoretical experimentations that are sensitive to the ways in which natural resource extraction and capitalist urbanization interact with each other in mutually transformative ways. This chapter has revisited Henri Lefebvre's notion of planetary urbanization as well as some of its recent appropriations by critical urban theorists to make sense of the variegated urban and territorial morphologies that are rapidly changing the face of Latin America as a result of a voracious international demand for raw materials. On that basis, I have illustrated how the geo-economic dynamics of the commodity supercycle have come to manifest themselves in the form of a dialectical movement of homogenization-fragmentation of territories at multiple spatial scales. This contradictory movement, the chapter showed, accounts for the production of deeply fractured and sclerotic territorial formations in the midst of unprecedented material integration in the space economy of global capitalism.

NOTES

1 This chapter is a revised and abridged version of a longer article titled "Spaces of Extraction, Metropolitan Explosions: Planetary Urbanization and the Commodity Boom in Latin America," originally published in 2015 in *International Journal of Urban and Regional Research, 40*(1), 96–112.

2 The specialized literature tends to refer to a "commodity boom" in order to define the overarching increase in the price of raw materials that began in the early 1990s and lost momentum in 2014. However, this major world-historical event is most adequately understood as a commodity "supercycle," because despite periods of decline (notably in the aftermath of the 2008 financial crisis, and in 2014), demand has not returned to pre-1990s levels. In fact, 2017 has signaled a new iteration of price increases, also connected to China's booming manufacturing sector as the central driving determination.

3 Newspaper article published in *El Tiempo* on 24 December 2013, authored by mining expert Guillermo Maya.
4 The Inter-American Development Bank, the Corporación Andina de Fomento (known in English as the Development Bank of Latin America), and the Fondo Financiero para el Desarrollo de la Cuenca de la Plata (FONPLATA).
5 Interview with a member of Colombia Solidarity Campaign, 3 May 2013, and interview with a member of Amnesty International, 20 June 2013.
6 Interview with a member of Consejo de Defensa Huasco, 28 November 2013, with an official from the Planning Secretariat of Vallenar, Chile, 3 December 2013, and 2013 report "La Colosa, Una Muerte Anunciada," by Colombia Solidarity Campaign.

REFERENCES

Arboleda, M. (2017). La naturaleza como modo de existencia del capital: Organización territorial y disolución del campesinado en el superciclo de materias primas de América Latina. *Anthropologica, 38*, 145–176. http://dx.doi.org/http://doi.org/10.18800/anthropologica.201701.006

Arboleda, M. (forthcoming). *Planetary mine: Territories of extraction in the fourth machine age*. London and New York: Verso.

Bebbington, A. (2012). Underground political ecologies. *Geoforum, 43*(6), 1152–1162. https://doi.org/10.1016/j.geoforum.2012.05.011

Bebbington, A., Bebbington, D., Bury, J., Lingan, J., Muñoz, J., & Scurrah, M. (2008a). Mining and social movements: Struggles over livelihood and rural territorial development in the Andes. *World Development, 36*(12), 2888–2905. https://doi.org/10.1016/j.worlddev.2007.11.016

Bebbington, A., Hinojosa, L., Bebbington, D., Burneo, M.L., & Warnaars, X. (2008). Contention and ambiguity: Mining and the possibilities of development. *Development and Change, 39*(6), 887–914. https://doi.org/10.1111/j.1467-7660.2008.00517.x

Bermúdez, R.E. (2012). Impactos de los grandes proyectos mineros en Colombia sobre la vida de las mujeres. In C. Toro, J. Fierro, S. Coronado, & T. Roa (Eds.), *Minería, territorio y conflicto en Colombia*. Bogotá: Universidad Nacional de Colombia.

Brenner, N. (2000). The urban question as a scale question: Reflections on Henri Lefebvre, urban theory and the politics of scale. *International Journal of Urban and Regional Research, 24*(2), 361–378. https://doi.org/10.1111/1468-2427.00234

Brenner, N. (2014). Introduction: Urban theory without an outside. In N. Brenner (Ed.), *Implosions/explosions: Towards a study of planetary urbanization*. Berlin: Jovis.

Brenner, N., & Schmid, C. (2013). The urban age in question. *International Journal of Urban and Regional Research, 38*(3), 731–755. https://doi.org/10.1111/1468-2427.12115

Brenner, N., & Schmid, C. (2015). Towards a new epistemology of the urban? *CITY, 19*(2–3), 151–182. https://doi.org/10.1080/13604813.2015.1014712

Bridge, G. (2009). *The hole world: Spaces and scales of extraction.* New Geographies, 2, 43–48.

Cabrera, M., & Fierro, J. (2013). Implicaciones ambientales y sociales del modelo extractivista en Colombia. In L.J. Garay (Ed.), *Minería en Colombia: Fundamentos para superar el modelo extractivista.* Bogotá: Contraloría General de la República de Colombia.

Caldeira, T. (2016). Peripheral urbanization: Autoconstruction, transversal logics, and politics in cities of the Global South. *Environment and planning D: Society and space,* published online ahead of print 18 July 2016.

Cancino, A (2012). La dudosa fortuna minera de Suramérica: Los países andinos Colombia, Chile y Perú. In C. Toro, J. Fierro, S. Coronado, & T. Roa (Eds.), *Minería, territorio y conflicto en Colombia.* Bogotá: Universidad Nacional de Colombia.

Carneri, S. (2017). La codicia por la tierra en Paraguay. *El País.* https://elpais.com/elpais/2017/02/07/planeta_futuro/1486488199_675583.html (Accessed: 16 October 2017).

Charnock, G., & Starosta, G. (Eds.). (2016). *The new international division of labour: Global transformation and uneven development.* London: Palgrave Macmillan.

Colombia Solidarity Campaign. (2013). *La Colosa, a death foretold: Alternative report about the Anglo Gold Ashanti gold mining project in Cajamarca, Tolima, Colombia.* London.

CooperAcción. (2012). *Los TLC, las inversiones y la expansión minera en América Latina.*

Cowen, D. (2014). *The deadly life of logistics: Mapping violence in global trade.* Minneapolis: University of Minnesota Press.

Duque, M. (2012). Minería: Yacimientos explosivos: Las bonanzas y el conflicto, la historia se repite. In C. Toro, J. Fierro, S. Coronado, & T. Roa (Eds.), *Minería, territorio y conflicto en Colombia.* Bogotá: Universidad Nacional de Colombia.

ECLAC (Economic Commission for Latin America and the Caribbean). (2013). *Natural resources within the Union of South American Nations: Status and trends for a regional development agenda.* United Nations.

Feeney-McCandless, I. (2017). Por una vida digna: Science as technique of power and mode of resistance in Argentina. *Alternautas, (Re)Searching Development.* Available at: http://www.alternautas.net/blog/2017/22/9

/por-una-vida-digna-science-as-technique-of-power-and-mode-of -resistance (Accessed 16 October 2017).

Fuentes, A. (2012). Legislación minera en Colombia y derechos sobre las tierras y territories. In C. Toro, J. Fierro, S. Coronado, & T. Roa (Eds.), *Minería, territorio y conflicto en Colombia*. Bogotá: Universidad Nacional de Colombia.

Gago, V., & Mezzadra, S. (2015). Para una crítica de las operaciones extractivas del capital: Patrón de acumulación y luchas sociales en el tiempo de la financiarización. *Nueva Sociedad, 255*, 38–52.

Global Witness. (2020). *Defending tomorrow: The climate crisis and threats against land and environmental defenders*. Global Witness, London.

Gudynas, E. (2010). Agropecuaria y nuevo extractivismo bajo los gobiernos progresistas de América del Sur. *Territorios, 5*, 37–54.

Harvey, D. (2006). *The limits to capital*. London and New York: Verso. (Original work published 1982)

Hinojosa, L. (2011). Riqueza mineral y pobreza en los Andes. *European Journal of Development Research, 23*, 488–504.

Hinojosa, L., & Hennermann, K. (2011). *La dimensión ambiental de las dinámicas territoriales rurales en contextos de expansión de las industrias extractivas. Working Paper No. 64*. Programa Dinámicas Territoriales Rurales – Centro Latinoamericano para el Desarrollo Rural.

Idárraga, A. (2012). El devenir de la minería trasnacional en Colombia: incertidumbres en torno a la relación ambiente, trabajo y salud. In C. Toro, J. Fierro, S. Coronado, & T. Roa (Eds.), *Minería, territorio y conflicto en Colombia*. Bogotá: Universidad Nacional de Colombia.

Infante, C. (2011). *Pasivos ambientales mineros: Barriendo bajo la alfombra*. Observatorio Latinoamericano de Conflictos Ambientales.

International Monetary Fund. (2011). *Regional economic outlook: Western hemisphere*.

International Monetary Fund. (2014, September). *Quarterly review of commodity markets*.

Kanai, J.M. (2014). On the peripheries of planetary urbanization: Globalizing Manaus and its expanding impact. *Environment and Planning D: Society and Space, 32*, 1071–1087. https://doi.org/10.1068%2Fd13128p

Karmy, J., & Salinas, B. (2008). *Pascua Lama: Conflicto armado a nuestras espaldas*. Santiago de Chile: Editorial Quimantú.

Khanna, P. (2016). *Connectography: Mapping the global network revolution*. London: Weidenfeld & Nicholson.

Klare, M. (2008). *Planeta sediento, recursos menguantes: La nueva geopolítica de la energía*. Santiago de Chile: Ediciones Urano.

Lefebvre, H. (1991). *The production of space*. Cambridge: Wiley-Blackwell. (Original work published 1976)

Lefebvre, H. (2003). *The urban revolution*. Minneapolis: University of Minnesota Press. (Original work published 1970)

Lefebvre, H. (2009a). Space and the state. In N. Brenner & S. Elden (Eds.), *Henri Lefebvre: State, space, world*. Minneapolis: University of Minnesota Press. (Original work published 1978)

Lefebvre, H. (2009b). Space: Social product and use value. In N. Brenner & S. Elden (Eds.), *Henri Lefebvre: State, space, world*. Minneapolis: University of Minnesota Press. (Original work published 1979)

Lefebvre, H. (2009c). Space and mode of production. In N. Brenner & S. Elden (Eds.), *Henri Lefebvre: State, space, world*. Minneapolis: University of Minnesota Press. (Original work published 1980)

Martínez, G., & Houghton, J. (2008). La IIRSA: O el mega-ordenamiento de los territorios indígenas. In J. Houghton (Ed.), *La tierra contra la muerte: Conflictos territoriales de los pueblos indígenas en Colombia*. Bogotá: Ediciones Anthropos.

Mendes, C. (2017). *Victims of Brazil's biggest environmental disaster still fighting for compensation*. Reuters. https://www.reuters.com/article/us-brazil-disaster-compensation/victims-of-brazils-biggest-environmental-disaster-still-fighting-for-compensation-idUSKBN1D7203 (Accessed 17 November 2017).

Merrifield, A. (2014b). *The new urban question*. London: Pluto Press.

Mezzadra, S., & Neilson, B. (2017). On the multiple frontiers of extraction: Excavating contemporary capitalism. *Cultural Studies*. Advance online publication 17 March 2017. https://doi.org/10.1080/09502386.2017.1303425

Monte-Mór, R. (2014). Extended urbanization and settlement patterns in Brazil: An environmental approach. In N. Brenner (Ed.), *Implosions/explosions: Towards a study of planetary urbanization*. Berlin: Jovis.

Otero, G., & Lapegna, P. (2016). Transgenic crops in Latin America: Expropriation, negative value and the state. *Journal of Agrarian Change, 16*(4), 665–674. https://doi.org/10.1111/joac.12159

Padilla, C. (2010). *Minería: ¿Desarrollo o Amenaza para las comunidades en América Latina?* Observatorio de Conflictos Mineros de América Latina.

Padilla, C. (2012). Minería y conflictos sociales en América Latina. In C. Toro, J. Fierro, S. Coronado, & T. Roa (Eds.), *Minería, territorio y conflicto en Colombia*. Bogotá: Universidad Nacional de Colombia.

Pardo, L.A. (2013). Propuestas para recuperar la gobernanza del sector minero colombiano. In L. J. Garay (Ed.), *Minería en Colombia: Fundamentos para superar el modelo extractivista*. Bogotá: Contraloría General de la República de Colombia.

Peña, G. (2013). *Manifiesto del Agua*. Bucaramanga: Comité Para la Defensa del Agua y el Páramo de Santurbán.

Ramírez, D., & Ibagón, N. (2012). Exportando minerales para importar alimentos. In C. Toro, J. Fierro, S. Coronado, & T. Roa (Eds.), *Minería, territorio y conflicto en Colombia*. Bogotá: Universidad Nacional de Colombia.

Razeto, C., Soto, D., & Marconi, A. (2009). *Pascua Lama, IIRSA: Acumulación por desposesión: El imperio contraataca*.

Reyes, P., & León, I. (2012). Las nuevas guerras justas y la política (neo) extractivista global: Nuevos escenarios de biopoder. In C. Toro, J. Fierro, S. Coronado, & T. Roa (Eds.), *Minería, territorio y conflicto en Colombia*. Bogotá: Universidad Nacional de Colombia.

Rudas, G., & Espitia, J.E. (2013). Participación del Estado y la sociedad en la renta minera. In L.J. Garay (Ed.), *Minería en Colombia: Fundamentos para superar el modelo extractivista*. Bogotá: Contraloría General de la República de Colombia.

Smith, N. (2008). *Uneven development: Nature, capital and the production of space*. Athens: The University of Georgia Press. (Original work published 1984)

Soliz, F., Maldonado, A., & Valladares, C. (2012). Las actividades extractivas minan los derechos de los niños y las niñas en las fronteras. In C. Toro, J. Fierro, S. Coronado, & T. Roa (Eds.). *Minería, territorio y sonflicto en Colombia*. Bogotá: Universidad Nacional de Colombia.

Stanek, L. (2011). *Henri Lefebvre on space: Architecture, urban research and the production of theory*. Minneapolis: University of Minnesota Press.

Techo Para Chile. (2016). *Catastro de Campamentos 2016: El número de familias en campamentos no deja de aumentar*. Techo. http://www.techo.org/paises/chile/wp-content/uploads/2016/09/Catastro-Nacional-de-Campamentos-2016.pdf (Accessed 24 January 2017).

Teubal, M. (2009). Peasant struggles for land and agrarian reform in Latin America. In H. Akram-Lodhi & C. Kay (Eds.), *Peasants and globalization: Political economy, rural transformation and the agrarian question*. London and New York: Routledge.

Thodes Miranda, E. (2016). Segregación socioespacial en ciudades mineras: El caso de Antofagasta, Chile. *Notas de Población, 102*, 203–227.

Toro, C. (2012). Geopolítica energética: Minería, territorio y resistencias sociales. In C. Toro, J Fierro, S. Coronado, & T. Roa (Eds.), *Minería, territorio y conflicto en Colombia*. Bogotá: Universidad Nacional de Colombia.

Vargas, F. (2013). Minería, conflicto armado y despojo de tierras: Impactos, desafíos y posibles soluciones jurídicas. In L.J. Garay (Ed.). *Minería en Colombia: Fundamentos para superar el modelo extractivista*. Bogotá: Contraloría General de la República de Colombia.

Wilson, J. (2011). Colonising space: The new economic geography in theory and practice. *New Political Economy, 16*(3), 373–397. https://doi.org/10.1080/13563467.2010.504299

Wilson, J. (2014). The violence of abstract space: Contested regional developments in Southern Mexico. *International Journal of Urban and Regional Research, 38*(2), 516–538. https://doi.org/10.1111/1468-2427.12023

Wilson, J., & Bayón, H. (2015). Concrete jungle: Planetary urbanization in the Ecuadorian Amazon. *Human Geography, 8*(3), 1–23. https://doi.org/10.1177%2F194277861500800301

World Bank. (2009). *Global economic prospects.*

World Bank. (2012). *Global economic prospects: Managing growth in a volatile world.*

Zibechi, R. (2006). *IIRSA: La integración a la medida de los mercados.* Programa de las Américas del IRC.

Zibechi, R. (2012). *Territories in resistance: A cartography of Latin American social movements.* Oakland, Edinburgh, Baltimore: AK Press.

PART 2

Metropolitan Peripheries under Financialization and Urban Extractivism

4 Large-Scale Housing in Peripheral Urbanization: Persistence and Change in the Production of Urban Space in the Mexico City Megaregion

CLARA SALAZAR, NADINE REIS, AND ANN VARLEY

Introduction

The main objective of this chapter is to explore how the neoliberal reforms of housing and land polices in Mexico have affected the production of urban space in the Greater Mexico City Area. As in many other Latin American countries, the production of urban space in Mexico has largely been driven by the process of *autoconstrucción*, by which future residents build their own "self-help" dwellings.[1] Around 70% of urban space in Mexico had been created "informally" up to the 1990s (Duhau, 1998). In the Greater Mexico City Area, autoconstruction has to a large extent taken place on agrarian property (*ejidos*) in the peripheries of urban agglomerations.[2] Urbanization on ejidos has happened outside of state planning processes, without state supervision of construction processes or building standards, and on land without basic urban infrastructure. Up to 1992, ejido land was tied to the rights holder and could not legally be sold on the formal land market. However, land transmission has been a reality in ejidos as long as they have existed.

Ejido land has been subdivided and sold through informal markets among the population, albeit without legal documentation guaranteeing property rights (Varley, 1985; Duhau, 1998; Clichevsky, 2000; Salazar, 2012). Although some have argued that the lack of property rights makes land cheap, this is not necessarily the case for the poor, as it may still be relatively expensive in relation to their precarious conditions of urban space (Abramo, 2009). Nevertheless, it has been the only viable option for a large part of the population in order to satisfy the vital need of a roof over their heads. This is also because the buyer pays by installments without interest, via "naked contracts" drawn up on the basis of relations of trust and solidarity (Lomnitz, 1973; Di Virgilio, 2015). Over time and acting "after the fact" (Pírez, 2016; Caldeira, this volume), the state has provided these so-called informal settlements with infrastructure, and they have become the object of land titling programs (Varley, 1987; 2017; Calderón, 2004; Clichevsky, 2006a; 2006b; Salazar, 2012).

Significant transformations in the production of urban space have been taking place in Mexico since the 1990s, and they are most in evidence in the urban periphery. These transformations have quantitative-spatial and qualitative dimensions. On the one hand, urban growth has expanded in medium-sized cities, giving rise to the emergence of a polynucleated metropolitan megaregion in Central Mexico. The growth of the Mexico City Metropolitan Zone (MCMZ) has been overtaken by the growth of cities in the surrounding states, including Puebla (2.9 million inhabitants), Querétaro (1.3 million), Cuernavaca (983,000), Pachuca (557,000), and Tlaxcala-Apizaco (540,000). The MCMZ has already merged with the Toluca Metropolitan Zone (TMZ) (2.2 million). With a total population of 29.9 million, the megaregion as a whole incorporates around one quarter of Mexico's population (Romo Viramontes & Velázquez Isidro, 2017, p. 15).

On the other hand, there have been two policy reforms, implemented in the context of the Washington Consensus structural reform prescriptions of the 1990s, with significant implications for urban space production: the housing policy reform and the agrarian "counter-reform" (Calva, 1993; Valtonen, 2000; Navarro-Olmedo et al., 2016). The housing policy reform was based on the idea that the market would be more efficient in providing social housing than the government, and that governments should create more space for a private and profitable housing market. In addition, based on the thinking of Peruvian economist Hernando de Soto, whose ideas became a key ingredient of the neoliberal reform agenda of the 1980s and '90s, people were believed to be poor because they lacked clearly defined property rights. If people became the legal owners of their properties, they could take out mortgages for the construction or improvements of homes. They could also sell the house legally and take advantage of this. According to de Soto's rationale, the establishment and enforcement of legal property rights was key to poverty reduction. In Mexico, these ideas materialized with the deregulation and financialization of social housing policy as we explain later in this chapter.

With regards to the agrarian counter-reform, a Constitutional reform in 1992 laid the foundation for the commodification of ejido land (Cornelius & Myhre, 1998; Appendini, 2001; Assies, 2008), allowing land rights holders to sell land to private investors and partner up with joint ventures. Since 1992, *ejidatarios* (the official members of ejido communities) have been legally allowed to sell ejido land. Hence, the agrarian reform coincided with increasing pressure for access to land by developers for the construction of large-scale housing projects (*conjuntos urbanos*) in the context of the privatization and financialization of social housing policy.

This chapter aims to investigate the impacts of these reforms on the production of urban space in the Greater Mexico City Area. As observed by Connolly (2013), the general tendency in the literature on the Latin American city has been a disappearance of the *hábitat popular* as an object of theoretical reflection[3] as compared to the period between 1950 and 1990. Instead, the focus has shifted to more general concerns regarding the city and the urban order as a whole. In her recent work, Teresa Caldeira has assigned a key role to the process of autoconstruction in cities of the Global South, by elevating it to the status of *the* mode of urban space production in Latin America and the Global South in general. Caldeira uses the concept of "peripheral urbanization" to denote a mode of urban space production that "(a) operate[s] with a specific temporality and agency, (b) engage[s] transversally with official logics, (c) generate[s] new modes of politics, and (d) create[s] highly unequal and heterogeneous cities" (Caldeira, this volume). In particular, Caldeira argues that the specificity of peripheral urbanization is the vital role of residents in the production of urban space. In this chapter, we aim to reflect on the question to which extent neoliberal policy changes challenge this notion of peripheral urbanization – a question that Caldeira herself poses (Caldeira, this volume). What has been the impact of neoliberal changes in housing and agrarian policies on modes of urban space production through autoconstruction? And how do these processes affect the conceptualization of "peripheral urbanization" as a process that is primarily carried out by residents and their complex entanglements with the state (Caldeira, this volume)? Based on empirical research, including statistical analysis of primary and secondary data and qualitative interviews, in two study sites in the Mexico City megaregion – the Mexico City Metropolitan Zone (MCMZ) and the Toluca Metropolitan Zone (TMZ) – our contribution draws three main conclusions. First, we show that private capital has become a major actor in urban space production in the peripheries of Mexico City and Toluca. After around 20 years of private involvement in social housing policy, millions of new homes have been constructed. However, the living quality in the new neighbourhoods is often worse than in self-built areas. Moreover, the strong demand for land by developers in the MCMZ has caused dramatic increases in land prices, affecting the ability of the poor to afford living space in the area. Hence, contrary to neoliberal assumptions, this has not solved the housing problem but exacerbated it for many of the poor.

Second, and in consequence of the first point, areas of autoconstruction have shifted within the megaregion. The poorer segments of the population must search for housing in cheaper areas of the megaregion, such as the periphery of Toluca. Their demand for land for autoconstruction is met by a higher supply of land due to the destruction of smallholder agriculture through neoliberal agrarian policy and the high demand for water by developers. In summary, considering the failure of the private sector to supply adequate housing, the poor are still key agents in the production of urban space on ejido land.

Third, a crucial finding from our study is that the strategies of private developers resemble those of autoconstruction. They acquire comparatively cheap land in the peripheries without urban land use status and public infrastructure, while the state creatively adapts or disregards existing regulations in order to provide the housing projects with legal status. As they fail to deliver necessary infrastructure, it is then up to the citizens and the state to turn the new residential areas into liveable urban spaces. Finally, we conclude that the neoliberal age has brought about new modes of urban space production in Mexico, but first and foremost, it has contributed to the consolidation of peripheral urbanization as a set of processes generating highly fragmented and heterogeneous urban spaces.

This introduction is followed by four sections. The first one introduces the evolution of the financing structure for social housing in Mexico and analyses how the privatization and financialization of housing policy materialized in the study region. The second presents the spatial patterns of the expansion of urban peripheries in the MCMZ and TMZ from a quantitative point of view. The third provides empirical evidence on the impacts of recent urban development trends in the MCMZ and the TMZ. In the conclusion, we offer observations on our findings.

Neoliberal Reforms of Housing Policy

The creation of social housing through the state became an objective of Mexican policy in the 1950s, when cities were rapidly growing, and the state's National Housing Institute (Instituto Nacional de la Vivienda) began to build dwellings for state employees. A system of subsidized credits for financing housing production was put in place with the creation of FOVI (Fondo de Operación y Financiamiento Bancario para la Vivienda) in 1963 (UN HABITAT, 2011, p. 13). In 1972, the main executors of housing policy for the low-income working class were founded: INFONAVIT (Instituto del Fondo Nacional de la Vivienda para los Trabajadores), and FOVISSSTE (Fondo de la Vivienda del Instituto de Seguridad y Servicios Sociales de los Trabajadores del Estado). In the 1970s and '80s, INFONAVIT and FOVISSSTE themselves managed social housing projects (Puebla, 2002).

In the context of the wider neoliberal reform agenda promoted by the World Bank, a 1992 policy reform abandoned the state's direct role as housing developer (Boils, 2004, p. 352). Besides the promotion of private financial intermediaries, INFONAVIT and FOVISSSTE were transformed into financially autonomous and profit-oriented mortgage banks. Credits ceased to be issued for the acquisition of land, thus largely abandoning the support of individual home construction (Boils, 2004, p. 352). Instead, housing promotion and construction was transferred to private developers, as also happened in other Latin American countries such as Brazil, Chile, Colombia, Peru, and Uruguay (Jha, 2007). The state's role changed from a project executor to a "facilitator" for private, profit-oriented investment into housing (Puebla, 2002; Schteingart & Patiño, 2006, pp. 171–172).

The financialization of the Mexican housing market was furthered after 2000 with the creation of a secondary mortgage market (Reis, 2017). The market took off with the World Bank financed transformation of FOVI into the Federal Mortgage Company (Sociedad Hipotecaria Federal, SHF) in 2001. The backing of credit risk by the SHF – that is, the Mexican state – attracted billions of investments from the global financial sector and led to a massive expansion of mortgage issuances by the national housing funds, especially by INFONAVIT. This created a boom in the social housing sector in Mexico, materializing in the mushrooming of large-scale housing projects (conjuntos urbanos) in the areas surrounding many of Mexico's cities, including the MCMZ and TMZ (see figure 4.1). The conjuntos urbanos usually contained several thousand houses, the largest ones having up to 13,000 dwellings. Data from the National Housing Commission (CONAVI) indicate that a total of 17.5 million housing loans were granted nationwide between 2000 and 2016, with 42% of these loans used for the acquisition of new housing in a large-scale housing project built by developers (see table 4.1). While this is less than half the number of loans, new housing accounted for almost 80% of the budget – that is, USD 183.5 billion were lent out as mortgages for the purpose of buying a dwelling (see table 4.1).

Figure 4.1 Large-scale housing project in Cuautitlan-Izcalli, State of Mexico. Photo by the authors.

Between 2000 and 2015, state and municipal authorities authorized about half a million housing units per year at the national level. In the MCMZ and the TMZ, the authorized projects were mainly targeted at borrowers with low purchasing power: 66.9% of housing units authorized in the MCMZ were of "social interest" housing and 16.9% of "progressive social type"[4] (Salazar, 2014a). Even though "social interest" housing is targeted at low-income groups, credits have only been available to workers with formal work contracts. However, around 60% of the Mexican population is engaged in informal work (ILO, 2014) and only 35% of the working population is part of the state social security system (Montoya, 2017: 239). This means that a large majority of the low-income population has from the outset been excluded from state housing policies (Monkkonen, 2011; UN HABITAT, 2011. p. 14; Soederberg, 2015, p. 483).

The large-scale housing boom came to an end in 2012–2013, when a series of factors led to millions of people abandoning their homes and a crash in the market (Reis, 2017). The lack of available land within cities and the high land prices there led real estate developers to acquire cheaper agrarian properties distant from consolidated urban areas

Table 4.1 Loans and budget for housing in Mexico, 2000–2016

	Mortgage		Budget (US Dollar)*	
Mortgage objective	Number*	%	Billion	%
New housing	7,445,050	42.4	183.5	77.9
Used housing	967,732	5.5	24.1	12.3
Self-construction	558,805	3.2	1.8	0.7
Housing improvement	8,010,138	45.7	8.1	3.6
Payment of liabilities	139,479	0.8	3.5	1.7
Other objectives	414,786	2.4	8.2	3.6
Unspecified	5,528	0.0	0.4	0.2
Total	17,541,518	100.0	229.6	100.0

Source: CONAVI (2017). In the original database, individual credits and housing co-financing are recorded separately. Given that an individual loan can be supported by co-financing, the latter were not considered in our calculation in order not to duplicate the information.

* To update the prices, the *Índice Nacional de Precios al Consumidor* was used at July 2016 prices. The July indices of each year were used.

that had not formally been designated as areas of urban development before and that consequently lacked any urban infrastructure. Nevertheless, the conjuntos urbanos were approved by state and municipal authorities[5] after the land had formally been accorded urban status in land-use plans and developers had made a commitment to providing the necessary infrastructure, including streets, lighting, electricity, water supply, public gardens, schools, and kindergartens. Conjuntos urbanos thus emerged as massive concrete islands in the peripheries of Mexican cities, often lacking public transport. Moreover, even though the projects were integrated into the formal planning process before building started, the construction sites lacked state oversight, resulting in extremely poor construction quality, the non-observance of safety regulations, and poor or inadequate infrastructure. According to interviews and media reports, urban developers often did not meet the legal requirements to which they had agreed, with many neighbourhoods lacking not only the schools, kindergartens, and public green spaces that had been promised to the inhabitants, but even a functioning water supply, sewage, and electricity networks. Some conjuntos urbanos have been regularly flooded as they were built in floodplains (interview, municipal urban development officer, 28.12.2015), while others have suffered regular fire outbreaks due to improperly installed electricity wires (Marosi, 2017). INFONAVIT contracts were based on high interest rates and a yearly increase of the mortgage principal in proportion to the inflation rate – an arrangement that delivered secure profits for international investors but turned out to be devastating for many debtors. Many borrowers were not aware of the usurious conditions when they signed the credit contracts (Interview INFONAVIT, State of Mexico, 13.4.2015; Marosi, 2017). When more and more people left their homes due to a lack of infrastructure, broken houses, and/or insolvency, abandoned houses were often occupied by organized crime, and insecurity aggravated the exodus from the conjuntos urbanos. According to the 2010 Census there were 35.2 million dwellings in Mexico, of which almost 5 million were declared uninhabited; 211,245 of these were in Mexico City and 538,220 in the state of Mexico (Sánchez & Salazar, 2011, p. 67). Especially in the biggest conjuntos urbanos to the north and northwest of the MCMZ, located in the municipalities of Tecámac, Huehuetoca, Zumpango, Nicolás Romero, Teoloyucan, Cuautitlán, Ecatepec, and Ixtapaluca, there was a high concentration of abandoned homes (*El Financiero*, 2014). At the end of 2017, 42% of homes built between 2000 and 2015 in the MCMZ were empty (Maya, forthcoming). Many people remain indebted for uninhabitable dwellings, while the nonperforming loans turned into public debt obligations of around USD 2 billion (Frente Mexiquense en Defensa para una Vivienda Digna, 2015).[6] Mexico's housing reform is thus an object lesson in how public policies have been transformed in order to redistribute wealth from the working class to globalized finance capital under neoliberalism (Reis, 2017).

In the next section we will explore these transformations of urban space production in the peripheries of the MCMZ and the TMZ.

Urban Sprawl and the Consolidation of a Polynuclear Megaregion

The Mexico City Metropolitan Zone (MCMZ) and the Toluca Metropolitan Zone (TMZ) are both located in Central Mexico (see figure 4.2). The MCMZ includes the 16 municipalities of Mexico City,[7] 59 of the State of Mexico, and one of the State of Hidalgo. The TMZ includes 15 municipalities of the State of Mexico. Since the 1970s, urban growth in the MCMZ has taken place mainly within the State of Mexico (*Estado de México*). Nowadays around 60% of the population of MCMZ lives in municipalities belonging to the State of Mexico.

Table 4.2 shows the growth of population, housing, and the built-up area between 2000 and 2015 for both zones. In 2015, the population in the MCMZ was ten times greater (20.9 million) than in the TMZ (2.1 million).

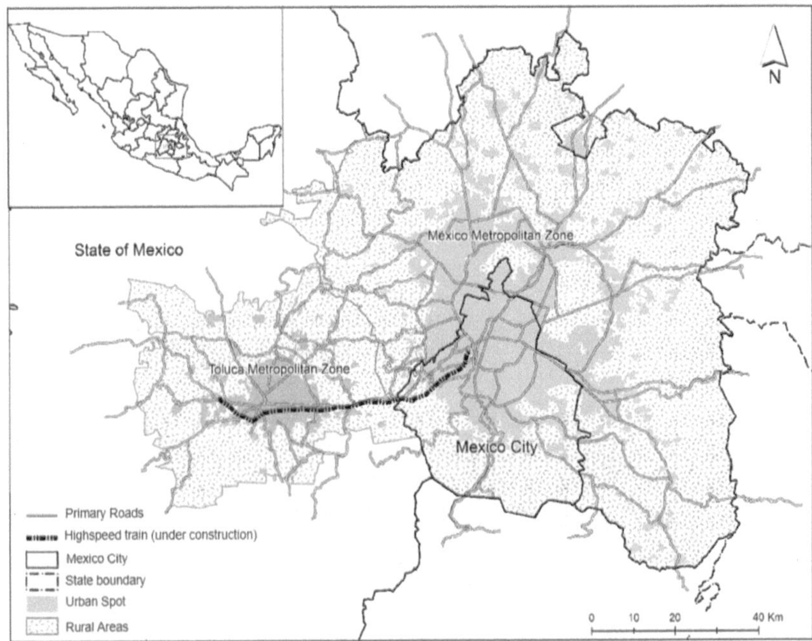

Figure 4.2 Mexico City Metropolitan Zone (MCMZ) and Toluca Metropolitan Zone (TMZ). Map produced by the authors.

However, the growth rate was significantly higher in the TMZ. In the same period, the TMZ grew faster than the MCMZ in terms of population, housing, and built-up area. While the MCMZ's population grew by 0.8%, the TMZ's population grew by 2.1%. The same trends can be observed for the increase in the number of housing units (2.1% in the MCMZ and 3.5% in the TMZ), and built-up area (0.9% and 1.9% respectively).

Second, population growth was not spread evenly across the two metropolitan zones: they grew faster in their peripheral municipalities than in their urban nucleus. For example, while the 14 municipalities of the TMZ grew at a rate of 2.4% per year, the municipality of Toluca grew only 1.8%. The situation was similar in the MCMZ: the 59 municipalities of Mexico City belonging to the State of Mexico grew by 1.3% per year, but the central nucleus (Mexico City) grew by just 0.2%. With around 9 million inhabitants, the population of Mexico City has remained almost constant since the 1980s.

Third, for both metropolitan zones, the housing growth rate was higher than the population growth rate: The housing growth rate

Table 4.2 Population, housing, and built-up area in the Mexico City Metropolitan Zone and the Toluca Metropolitan Zone, 2000–2015

Metropolitan zone and municipalities included	Population (million)		Housing (million)		Area (hectares)		Growth rate 2000–2015		
	2000	2015	2000	2015	2000	2015	Population	Housing	Area
MCMZ (76)	***18.4***	***20.9***	***4.2***	***5.8***	***212,115.30***	***242,695.60***	***0.8***	***2.1***	***0.9***
Mexico City (16)	8.6	8.9	2.1	2.6	79,656.80	87,494.40	0.2	1.4	0.6
State of Mexico (59)	9.7	11.9	2.1	3.1	130,163.60	155,201.30	1.3	2.7	1.2
State of Hidalgo (1)	0.0	0.1	0.0	0.0	2,294.90	3,369.40	6.5	7.7	2.6
TMZ (15)	***1.5***	***2.1***	***0.3***	***0.5***	***29,655.00***	***39,493.20***	***2.1***	***3.5***	***1.9***
Toluca (1)	0.7	0.9	0.1	0.2	10,167.60	13,422.00	1.8	3.1	1.9
State of Mexico (14)	0.9	1.2	0.2	0.3	19,487.40	26,071.20	2.4	3.8	1.9

Source: Authors' calculations from INEGI (2000; 2015).

Table 4.3 Growth in the built-up area in the municipalities of the Mexico City Metropolitan Zone and the Toluca Metropolitan Zone, 2000–2015

	Growth in urban area, 2000–2015				
	Total		Conjuntos urbanos authorised*		
Metropolitan zone and municipalities included	Hectares	% of total area*	Number	Hectares	% of growth in built-up area
MCMZ	***30,580.3***	***14.4***			
Mexico City (16)	7,837.6	9.8	na	na	na
State of Mexico (59)	**25,037.7**	**19.2**	**399**	**12,859.9**	**51.4**
State of Hidalgo (1)	1,074.5	46.8	na	na	na
TMZ	***9,838.2***	***33.2***	***142***	***2,880.8***	***29.3***
Toluca (1)	3,254.3	32.0	56	812.6	25.0
State of Mexico (14)	6,583.9	33.8	86	2,068.3	31.4
Greater Mexico City (MCMZ+TMZ)	***40,418.5***	***16.7***			

Source: Authors' calculations from INEGI (2000; 2015).

*Data based on Gacetas del Gobierno del Estado de México, various dates.

shows a significant increase between 2000 and 2015. In the TMZ, housing increased by 3.5%, in the MCMZ by 2.1%. This means that there has been a significant increase in the number of houses: 68% in the TMZ (210,000 units), and 37% (1.6 million units) in the MCMZ.

Fourth, both zones have seen excessive urban sprawl, particularly the TMZ (tables 4.2 and 4.3). Between 2000 and 2015, the built-up area of the MCMZ increased by 0.9% per year, which means that the MCMZ grew by 30,580 hectares and was 14.4% larger at the end of the period. Growth predominantly took place in the 60 peri-urban municipalities (19.2%), especially in the State of Hidalgo (46.8%), while the central nucleus grew moderately (9.8%). Surface growth of the TMZ was much stronger. Its built-up area grew at a rate of 1.9% (9,838.2 hectares) per year, while the central nucleus and the municipalities at the periphery grew at the same rate of 1.9% per year. This means that the built-up area of the TMZ was 33.2% larger in 2015 than in 2000.

Further, table 4.3 shows that conjuntos urbanos have significantly contributed to urban sprawl. The contribution of these large-scale housing projects to urban sprawl has increased over the years. Connolly (2012, p. 401) has pointed out that in the MCMZ, the share of new housing represented by the conjuntos urbanos increased from 17.6% to

Table 4.4 Large-scale housing projects built on ejido land and private property in the municipalities of the Mexico City Metropolitan Zone and the Toluca Metropolitan Zone, 2000–2015

	Area		Conjuntos urbanos	
Metropolitan zone	Hectares	%	Number	Housing units
MCMZ (59 State of Mexico municipalities)				
Ejido land*	261,117.8	100.0		
Ejido land privatized*	17,698.1	6.8		
Conjuntos urbanos authorized (total)**	12,859.9	100.0	399	881,964
Conjuntos urbanos on privatized ejido land **	3,425.4	26.6	83	233,240
Conjuntos urbanos on private property**	9,434.5	73.4	316	648,724
TMZ (15 municipalities)				
Ejido land*	118,273.4	100.0		
Ejido land privatized*	1,309.1	1.1		
Conjuntos urbanos authorized (total)**	2,880.8	100.0	142	131,112
Conjuntos urbanos on privatized ejido land **	41.4	1.4	3	3,240
Conjuntos urbanos on private property**	2,839.4	98.6	139	127,872

Source: Authors' calculations from *National Agrarian Register (*Padrón e Historial de Núcleos Agrarios, PHINA*), base-date, update, December 2016; and **Gacetas del Gobierno del Estado de México, several dates. Data for MCMZ from Varley and Salazar (2021), to end of February 2018.

35% between 1995–2000 and 2000–2005. Between 1990 and 2000, self-built neighbourhoods represented 72% of the growth in the housing stock of the MCMZ; but between 2000 and 2005, they contributed only 50% of the new housing stock (ibid., 399). Based on our data we can confirm that conjuntos urbanos have occupied a large amount of land in the urban peripheries.

How many of these large-scale housing projects have been built on privatized ejido land? In 2015, around 3% of ejido land had been privatized[8] at national level. However, in the 56 Metropolitan Areas of the country, this proportion reaches higher figures to those observed at national level, albeit showing a great heterogeneity. The data suggest that the privatization of ejido land has been an important phenomenon only in some urban peripheries. By 2015, 6.8% of ejido land in the MMZ had been privatized, while in the TMZ, privatization had been almost insignificant, at 1.1%.

Table 4.4 shows the extent to which privatized ejido land has been used for the construction of conjuntos urbanos. We found that for the State of Mexico periphery of MCMZ, privatized ejido land has contributed only 26.6% of the land where large-scale housing was constructed, and in the TMZ, it has contributed only 1.4% of the corresponding area. In the MCMZ, 3,425.4 hectares of privatized ejido land have been used for building 83 conjuntos urbanos with 233,240 housing units. In the TMZ, only 41.4 hectares of privatized ejido land have been used for building three conjuntos urbanos with 3,240 housing units. Hence, most of the conjuntos urbanos have been built on what was already private property (98.6% in the TMZ and 73.4% in the MCMZ), and not on ejido land. This means that the privatization of ejido land has played a much greater role for the expansion of large-scale housing in the MCMZ than in the TMZ, where its contribution has been negligible. We explore the reasons below.

Impacts of Large-Scale Housing Projects on the Production of Urban Space

Sharply Increasing Land Prices in Self-Built Neighbourhoods in the Mexico City Metropolitan Zone

This section shows that the presence of developers in the peri-urban area has had a significant impact on the informal land market and, hence, on the ability of the low-income population to acquire land and housing in the MCMZ. In order to track changes in land prices, a survey was conducted in the municipality of Cuautitlán Izcalli. Four of the ejidos in Cuautitlán Izcalli (San José Huilango, Santa María Tianguistengo, San Francisco Tepojaco and La Piedad) are located next to each other (see figure 4.3). These ejidos have some common characteristics. First, since the soil quality in most of these ejidos is poor, agricultural activities have usually been limited to production for subsistence and the local market. Second, as a result of population growth, these ejidos began to carry out a process of parcelling and transferring plots in the 1970s, but they did not always transfer these plots in an attempt to increase their capital. At that time, they gave away plots of around 600 square metres to siblings for free or sold plots to other relatives or to low-income groups who approached the ejidatarios when they needed a place to settle. According to the interviewees, the profits ejidatarios made from the land sales were used either to pay for their own living expenses or to help siblings or other relatives in financial difficulty. In this sense, ejidatarios selling land were acting according to a rationality that differed from that of economic agents. Since selling ejido

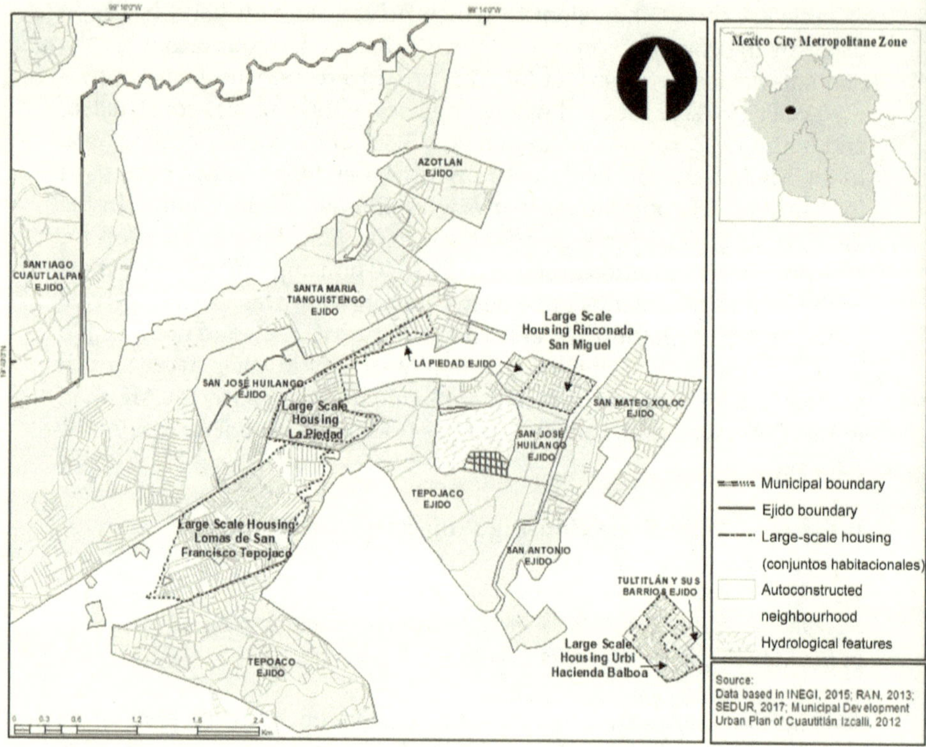

Figure 4.3 Self-built neighbourhoods and conjuntos urbanos authorized on ejido land. Map produced by the authors.

land was illegal prior to the 1992 reform, ejidatarios had no option but to transfer the plots without property titles. Consequently, buyers occupied their plots and built their dwellings themselves, despite the lack of urban infrastructure.

The low-income groups interested in acquiring ejido land usually do not have the cash to pay for the land in one go. The ejidatarios therefore accept payment in monthly installments. These agreements are based on trust and mutual support. When these conditions are met, ejidatarios normally sell their land to people to self-build their houses at a price that allows for the buyers' ability to pay. However, ejidatarios have been able to obtain higher prices for their land once self-built neighbourhoods have been legalized. This is because the state has compensated them for the expropriation of the land occupied by informal settlements, to give titles to low-income occupants.[9]

Over time, the self-built settlements have become consolidated urban areas, and demand for land has increased as well as its price. But this increase was gradual at the outset. In the mid-1990s, urban developers were authorized to acquire privatized ejido land, and government housing policy endorsed their decision to purchase the land to build large-scale housing. In the study area, housing production in the form of large-scale housing projects began in the late 1990s. In 1999, the government of the State of Mexico authorized the construction of the conjunto urbano Lomas de San Francisco Tepojaco, with 11,301 housing units on 220.1 hectares of the ejido Tepojaco. In 2001, it authorized the construction of 4,849 housing units on 44 hectares of the ejido La Piedad, and in 2005, it authorized another 2,849 dwellings on another 20.3 hectares of the latter (see figure 4.3).

At first, when the agrarian counter-reform had just passed, in 1992, ejidatarios sold land to developers at the same price they sold to low-income groups (Salazar, 2014b, p. 78). At that time, the difference lay in the size of the plot as well as the purpose of the purchase: while low-income groups purchased square metres of land to self-build their homes, urban developers acquired several hectares in order to do business by selling thousands of identical houses to low- and medium-low income groups, with mortgages from the state housing agencies.

In order to know how the land price increased in the study area when urban developers became buyers of ejido land, we conducted a survey between December 2015 and January 2016 among 120 low-income households that had purchased plots on the informal market between 1993 and 2015 and continued to live there.

Figure 4.4 shows that there was a great heterogeneity in prices registered before the conjuntos urbanos started to be built in 2000. The heterogeneity in prices can be explained by two circumstances. First, in the mid-1990s the siblings, relatives, or friends of ejidatarios were still the main applicants for ejido land. Transactions were made face to face, and the social proximity between sellers and buyers of land was the most important factor influencing the variation and the generally lower level of prices of the land. In addition, informal transactions did not include the transaction costs involved in formal economic transactions. For example, selling land to buyers without handing over legally binding property titles implies that the cost of paying taxes, bureaucratic procedures, and professional consulting can be avoided. Moreover, while a notary public can guarantee the validity of signatures on private agreements, the certified document recording private agreements is not a legal equivalent to a property title.

Second, we found that land prices increased around three times during the period analysed, from USD 35 to USD 90. Clearly, private

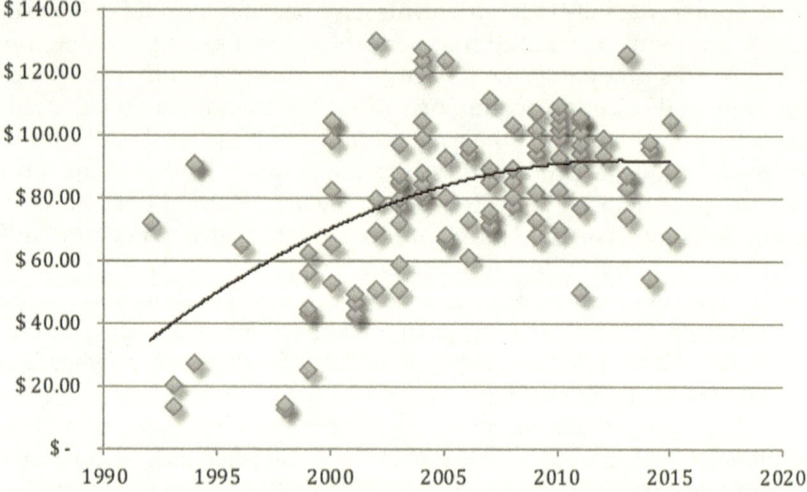

Figure 4.4 Land price in the study area by square metre and year of acquisition (in US dollars – constant prices as of January 2016).

investment from developers caused land value increments during the 2000s. The fact that urban developers were interested in buying ejido land created the expectation with ejidatarios that there would be a steadily increasing demand for ejido land from the housing industry. Consequently, they sought to obtain a higher price for the land they still possessed. Interviews with ejidatarios revealed that developers did not ask for more ejido land, but the large-scale investment they had already made created a speculative bubble. This led ejidatarios to start offering low-income buyers an elevated so-called expectation-price (*precio de expectativa*), as a result of their expectations about the future. Third, land prices tended to be homogenized upwards from 2010. Land prices stabilized because of the decreasing demand for land from developers in the course of the crisis in the housing market.

Taking into consideration the income gap between the two types of land buyers (developers and low-income groups), we can conclude that due to the elevated price it has become more difficult to access land for low-income housing in this area, especially considering that real wages have substantially decreased in Mexico since the 1980s (Soederberg, 2010, p. 82).

Rapid Expansion of Self-Built Neighbourhoods in the Toluca Metropolitan Zone

The TMZ has also seen the emergence of large-scale housing projects, but to a smaller extent than the MCMZ, and with qualitatively different impacts. As shown above, very few large-scale housing projects were built on privatized ejido land; instead, they were built almost exclusively on private property. According to interviewees, this is because historically more land was in the possession of ejidos in the MCMZ than in the TMZ. In the TMZ, developers therefore concentrated on acquiring agrarian land from farmers with private property, simply because there was no need to go through complex negotiation processes with a large number of ejidatarios and, later, the land privatization process, which involves many complicated and costly bureaucratic procedures (interview, land-use planner in the TMZ, 2.12.2015). Interviews with several officials and data from the National Agrarian Register (RAN) indicate that in contrast to the MCMZ, there have hardly been any cases of privatization of ejido land in the TMZ so far.

While the construction of large-scale housing has been a predominant factor shaping urban development in the MCMZ, the defining feature in the TMZ has been the massive expansion of autoconstruction since the 1990s. According to our interview data, there has been an increase in the activity on the informal market for ejido land, which has given rise to the much higher growth rates in terms of population, homes, and built-up area compared to the MCMZ. This growth has been spurred by two simultaneously occurring structural forces.

First, the valley has seen high population growth, due not only to natural growth but also to a positive migration balance. Between 2005 and 2015, the MCMZ had a net outward migration of around 270,000 persons, while the TMZ gained around 60,000 persons. Around 30% of the out-migrants from the MCMZ have moved within the megaregion, primarily to Toluca, Querétaro and Pachuca (CONAPO, 2010, pp. 40–41; Romo Viramontes & Velázquez Isidro, 2017, pp. 22–23).

Ejidatarios and municipal officials in the municipalities of Temoaya, Xonacatlán, and Otzolotepec, in the east of the TMZ (hence in proximity to the MCMZ, see figure 4.2) stated in interviews that many of the migrants acquiring land on the informal market for autoconstruction migrate from the MCMZ due to much lower land prices in the TMZ. According to interviewees in the three municipalities, the price for a square metre of ejido land is around MXN300–350 (USD 15–USD 18), and hence, only one sixth to one eighth of the cost of land in the MCMZ. With the construction of new highways substantially reducing travel time between the TMZ and the MCMZ, the substantial land price difference is likely to influence migration decisions. The completion of the Mexican government's recent prestige project, a highspeed train connecting the TMZ and the MCMZ, is likely to push population growth in the TMZ even further. Moreover, industrial and commercial zones in the TMZ itself have expanded greatly since the 1990s, providing new jobs in the formal economy and generating opportunities in the informal sector.

Second, the increasing demand for land in the TMZ has been met by an increasing supply, which, on the one hand, is related to the recent dynamics of agrarian change in Mexico, and, on the other, reinforced by the nature of the urbanization process. Since the 1990s, there has been a significant transformation of the agrarian sector in Mexico guided by the neoliberalization of policies and the liberalization of international trade, in particular since Mexico's entrance into the North American Free Trade Agreement (NAFTA). This has led to a far-reaching destruction of smallholder agriculture in ejidos (King, 2006; Fox & Haight, 2010; Appendini, 2014). Small-scale agriculture is still present in the periphery of the TMZ. However, in our qualitative empirical research conducted in 2015–2016, it was evident that ejido lands, if still used for agricultural production at all, were usually cultivated by the oldest generation in the households, while many households had given up cultivation. One reason is the lack of economic viability of production. Even if harvests turn out well and are not affected by more frequent climatic hazards – in which case the investment is completely lost – production is eventually a zero-sum game that does not result in financial gain. This finding agrees with the results of a survey conducted in the area by Lerner and Appendini (2012), which found that ejido households continuing maize cultivation do so out of socio-cultural preference for home-made tortillas or as an insurance strategy due to the insecurity and instability of wage-labour income. Another reason for the abandonment of cultivation is the drying up of soils due to groundwater overexploitation in the course of the expansion of industrial parks and urban areas (Reis, 2014). A phrase that is often used by locals is that "*la mancha urbana nos está comiendo* ("the city is eating us up"). The drying up of wells is specifically connected by some interviewees to the construction of large-scale housing.

Whether families can hold on to the land depends on the economic situation of the household. Many interviewees stated that they held on to their individual holdings (*parcelas*) so their own children or grandchildren could build their dwellings on the land. However, they also stated that many were forced to sell their parcels out of necessity, most often as a result of emergencies such as the illness or death of a family member. The informal land market is also stimulated by local intermediaries, who have the capital to buy parcels of agrarian land from people in need or want of cash, divide them into smaller plots and sell them on for autoconstruction. In this sense, the misfortune of some locals is the good luck of in-migrants, who may eventually find a better future in the new zones of autoconstruction.

Conclusions

Our findings reveal important insights regarding the modes of urban space production under neoliberal capitalism in Mexico. What can be concluded as regards the notion of peripheral urbanization, taking into account the prominent role of finance capital and private real estate in recent urbanization processes? Three conclusions can be drawn with respect to Caldeira's conceptualization of the term.

First, *finance capital and private real estate have become major actors in the production of urban space in Mexico*. This challenges Caldeira's conception of peripheral urbanization as a process of autoconstruction, producing "spaces that are always in the making" ("temporality and agency"). However, the substantial involvement of private capital has not solved the housing problem but exacerbated it. According to neoliberal discourse, the housing problem in the Global South has been caused by a lack of clearly established property rights and a lack of formal housing available to the poor, leading to the reproduction of "informal" urbanization and the associated problems, especially the lack of urban infrastructure. The policy solution for Mexico, promoted by the World Bank, was the inclusion of private real estate and international finance capital in the production of "formal" housing for the poor. However, these policy reforms have proved highly problematic for the poor. First, the housing that was made available has only been targeted towards workers with formal work contracts, omitting the majority of Mexico's poor, who depend on informal work. Moreover, the overall impacts of the privatization and financialization of housing policy have been devastating. Millions of people have incurred debt for poor quality homes that they often cannot even inhabit anymore due to infrastructural and security conditions that are often worse than those in many self-built neighbourhoods. As developers have failed to deliver necessary infrastructure, it is now up to the citizens and the state to turn the new neighbourhoods into liveable urban spaces. Moreover, in the course of the immense demand for cheap land by developers in the peripheries of the Mexico City Metropolitan Zone, which also involved ejido land, there has been a significant overall increase in land prices, undermining the ability of the poor to afford housing in these areas.

Second, a consequence of the first point is the *emergence of new spaces of autoconstruction in other areas*, as many have had to move farther out, to the peripheries of cities such as Toluca. The structural conditions of agrarian change has played a key role for the way in which autoconstruction is proliferating in these areas. The activities of private capital have thus hindered the ability of the poor to build their own homes in the MCMZ, while shifting autoconstruction to other areas. More research is required to investigate (a) to what extent the growth of medium-sized cities in Central Mexico, and the emergence of a polynucleated megaregion, has been taking place in the context of migration movements within the megaregion, and (b) to what extent these new spaces follow established processes of peripheral urbanization, such as the unsettling of official logics ("transversal logics") and the invention of new democratic practices ("new modes of politics"). However, we can unequivocally conclude that the poor remain the key actors in the production of urban space in Mexico ("temporality and agency").

Third, it is striking to note that the *mechanisms of the production of urban space by private real estate developers ultimately resemble those of autoconstruction: they operate with official logics in transversal ways* ("transversal logics"). Just like the poor in need of housing, developers have acquired cheap land without formal urban land use status and infrastructure in the urban periphery, albeit with a different aim: the realization of exorbitant profits through building on the cheapest land at the lowest quality standards and selling the homes on a monopolized market lacking state oversight. At the same time, the state has been complicit in authorizing land use changes and housing construction, creatively adapting existing regulations. While the new homeowners believed they acquired "formal" housing – that is, a properly built home in a pleasant and safe neighbourhood with appropriate infrastructure – many of those living in conjuntos urbanos have presumably been forced to return to political practices of negotiating with the state for basic infrastructure. The recent history of housing policy in Mexico shows that a key characteristic of peripheral urbanization is and continues to be that categories like "formal"/"informal" and "regulated"/"unregulated" are inherently unstable and shifting in the course of political negotiation processes, not only between residents and the state, but also between private developers and the state. Eventually, there are thus new ways of producing urban space in Mexico, but they have first and foremost contributed to the consolidation of peripheral urbanization as a set of processes that generates highly dynamic, fragmented, and heterogeneous urban spaces ("heterogeneity").

NOTES

1 For a comprehensive literature review on the Latin American *hábitat popular* see Connolly (2013).
2 Mexico's post-revolutionary constitution was the basis for a large-scale land reform. Between 1917 and 1992, more than 50% of the arable area was redistributed to the so-called social sector (Assies, 2008, p. 40ff.; World Bank, 2001), p. v). Peasant households were given land and organised into *ejidos* or "communities." The term "ejido" refers both to the agrarian community constituted by a group of *ejidatario* rights-holders and to their land. The land is the private property of the agrarian community, but it is worked individually (in *parcelas* transmitted across the generations) or collectively ("common use" land) (Azuela, 1989). The term *comunidades*, "communities," refers to communally held, mostly indigenous, property (Baitenmann, 2005, p. 174f.).
3 With the notable exception of debates on the regularization of informal settlements, see for example, Di Virgilio et al. (2014), Varley (2017), and Salazar (forthcoming).
4 Social Interest Housing (SIH) is a complete single-family house between 28 and 42 square metres in size. Social Progressive Housing (SPH) consists of a shell construction (between 21 and 25.1 square metres) with basic services, which must be finished by the buyer (Calderón, 2008, p. 17).
5 Article 115 of the Mexican Constitution empowers municipalities to formulate, approve, and manage zoning and urban development plans, but the government of the State of Mexico still regulates the height, density, and size of conjuntos urbanos in their municipalities.
6 In 2012, housing policy was reformulated, and it was announced that large-scale housing estates far from the consolidated urban area would no longer be funded. To tackle urban sprawl, the Ministry of Urban Development and Territorial Planning (*Secretaría de Desarrollo Urbano y Ordenamiento Territorial, SEDATU)* established that conjuntos urbanos located closer to consolidated areas were more likely to access credit than those located farther away.
7 The official name for Mexico City was changed from "Federal District" (*Distrito Federal*) to "Mexico City" (*Ciudad de México*) in 2016.
8 The privatization of ejido land is considered here as the change in regulation from an agrarian regime that does not allow land sales, to the civil regime that allows ejido land to be offered on the land market.

9 When the agrarian counter-reform was passed, the procedures to legalize "informal settlements" diversified, but most more recent settlements remain illegal (Salazar, 2011, p. 2018).

REFERENCES

Abramo, P. (2009). *Favela e mercado informal: A nova porta de entrada dos pobres nas cidades brasileiras*. Porto Alegre: Ed. Habitare-ANTAC. Caixa Econômica Federal do Brasil.

Appendini, K. (2001). *Land regularization and conflict resolution: The case of Mexico.* Document prepared for FAO, Rural Development Division, Land Tenure Service. http://www.fao.org/3/y3932t/y3932t01.pdf (Accessed: August 2021).

Appendini, K. (2014). Reconstructing the maize market in rural Mexico. *Journal of Agrarian Change, 14*(1), 1–25. https://doi.org/10.1111/joac.12013

Assies, W. (2008). Land tenure and tenure regimes in Mexico: An overview. *Journal of Agrarian Change, 8*(1), 33–63. https://doi.org/10.1111/j.1471-0366.2007.00162.x

Azuela, A. (1989). *La ciudad, la propiedad privada y el derecho.* México: El Colegio de México.

Baitenmann, H. (2005). Counting on state subjects: State formation and citizenship in twentieth-century Mexico. In C. Krohn-Hansen & K.G. Nustad (Eds.), *State formation: Anthropological perspectives* (pp. 171–194). London, Ann Arbor: Pluto Press.

Boils, G. (2004). El Banco Mundial y la política de vivienda en México. *Revista Mexicana de Sociología, 66*(2), 345–367. https://doi.org/10.2307/3541460

Calderón, J. (2004). Políticas de regularización y mejoramiento urbano en América Latina. *División de Desarrollo Sostenible y Asentamientos Humanos (CEPAL).* http://docplayer.es/3447707-Julio-calderon-cockburn.html (Accessed: 13 November 2018).

Calderón, J. (2008). *Vivienda progresiva en la Zona Metropolitana de Colima: Ivecol, aciertos y errores.* Colima, México: Universidad de Colima, Facultad de Arquitectura y Diseño.

Calva, J.L. (1993). La reforma neoliberal del régimen agrario: En el cuarto año de gobierno de Carlos Salinas de Gortari. *Problemas del Desarrollo, 24*(92), 31–39. https://www.probdes.iiec.unam.mx/index.php/pde/article/view/32730

Clichevsky, N. (2000). *Informalidad y segregación urbana en América Latina: Una aproximación*. Universidad de Buenos Aires, CONICET-UBA (pp. 1–24).

Clichevsky, N. (2006a). *Previniendo la informalidad urbana en América Latina y el Caribe.* Santiago de Chile: CEPAL, Serie Medio Ambiente y Desarrollo 24.

Clichevsky, N. (2006b). *Regularizando la informalidad del suelo en Amércia Latina y el Caribe.* Santiago de Chile: CEPAL, Serie Medio Ambiente y Desarrollo 50.

Comisión Nacional de Vivienda (CONAVI). (2017). *Base de datos a nivel municipal.* https://datos.gob.mx/busca/organization/conavi (Accessed: June 2017).

CONAPO. (2010). *Migración metropolitana.* http://www.conapo.gob.mx/work/models/CONAPO/Resource/2043/1/images/Migracion_metropolitana.pdf (Accessed: July 2018).

Connolly, P. (2012). La urbanización irregular y el orden urbano en la Zona Metropolitana del Valle de México, 1990 a 2005. In C. Salazar (Ed.), *Irregular: Suelo y mercado en América Latina* (pp. 379–425). México: El Colegio de México.

Connolly, P. (2013). La ciudad y el hábitat popular: Paradigma latinoamericano. In B.R. Ramírez Velázquez & E. Pradilla Cobos (Eds.), *Teorías sobre la ciudad en América Latina.* (pp. 505–562). México: Universidad Autónoma Metropolitana.

Cornelius, W.A., & Myhre, D. (1998). *The transformation of rural Mexico: Reforming the ejido sector.* San Diego, La Jolla: Center for US-Mexican Studies, University of California.

Di Virgilio, M.M. (2015). Urbanizaciones de origen informal en Buenos Aires: Lógicas de producción de suelo urbano y acceso a la vivienda. *Estudios Demográficos y Urbanos, 30*(3), 651–690. http://dx.doi.org/10.24201/edu.v30i3.1496

Di Virgilio, M.M., Guevara, T.A., & Arqueros Mejica, M.S. (2014). Políticas de regularización en barrios populares de origen informal en Argentina, Brasil y México. *Urbano, 17*(29), 57–65. https://www.redalyc.org/comocitar.oa?id=19836173008 (Accessed: August 2021).

Duhau, E. (1998). *Hábitat popular y política urbana.* México: Universidad Autónoma Metropolitana-A/Miguel Ángel Porrúa.

El Financiero. (2014). *Las 5 millones de casas abandonadas en el país* [The 5 million abandoned houses in the country]. http://www.elfinanciero.com.mx/archivo/las-millones-de-casas-abandonadas-en-el-pais.html (Accessed: 14 October 2015).

Fox, J., & Haight, L. (2010). Mexican agricultural policy: Multiple goals and conflicting interests. In J. Fox & L. Haight (Eds.), *Subsidizing inequality: Mexican corn policy since NAFTA* (pp. 9–50). Santa Cruz, CA: Woodrow Wilson International Center for Scholars, University of California Santa Cruz.

Frente Mexiquense en Defensa para una Vivienda Digna. (2015). *La bancarrota de Sociedad Hipotecaria Federal en la administración de Jesús Alberto Cano Vélez y de José Gabriel Fernández de la Concha* [The bankruptcy of the Federal

Mortgage Company under the administration of Jesús Alberto Cano Vélez and José Gabriel Fernández de la Concha]. http://www.frentemex.org.mx/component/content/article/7-comunicados/79-bancarrota-de-la-shf (Accessed: 5 December 2015).

Gacetas del Gobierno del Estado de México, various dates.

ILO. (2014). *Informal employment in Mexico: Current situation, policies, and challenges.* http://www.ilo.org/wcmsp5/groups/public/—americas/rolima/documents/publication/ wcms_245889.pdf

INEGI (Instituto Nacional de Estadística, Geografía e Informática). (2000). *Censo de Población y Vivienda,* 2000.

INEGI (Instituto Nacional de Estadística, Geografía e Informática). (2015). *Encuesta Intercensal,* 2015.

Jha, A.K. (2007, January). Low-income housing in Latin America and the Caribbean. *World Bank "en breve,"* No. 101. https://openknowledge.worldbank.org/handle/10986/10301

King, A. (2006). *Ten years with NAFTA: A review of the literature and an analysis of farmer responses in Sonora and Veracruz, Mexico.* CIMMYT Special Report 06–01. Mexico, CIMMYT/Congressional Hunger Center.

Lerner, A., & Appendini, K. (2012). Dimensions of peri-urban maize production in the Toluca-Atlacomulco Valley, Mexico. *Journal of Latin American Geography, 10*(2), 10–106. https://doi.org/10.1353/lag.2011.0033

Lomnitz, L. (1973). *Como sobreviven los marginados.* México: Siglo XXI.

Marosi, R. (2017). Mexico's housing debacle: Mexico promised affordable housing for all. Instead, it created many rapidly decaying slums. *Los Angeles Times.* http://www.latimes.com/projects/la-me-mexico-housing/ (Accessed: March 2018).

Maya, L. (forthcoming). *Efectos de la política habitacional en la economía de los hogares.* Tesis de Doctorado en Estudios Urbanos y Ambientales. México: El Colegio de México.

Monkkonen, P. (2011). The housing transition in Mexico: Expanding access to housing finance. *Urban Affairs Review, 47*(5), 672–695. https://doi.org/10.1177%2F1078087411400381

Montoya, M. (2017). Condiciones de vida de los hogares trabajadores en las zonas urbanas de México durante la crisis 2008–2010. In J. Nájera, B. García, & E. Pacheco (Eds.), *Hogares y trabajadores en México en el siglo XXI* (pp. 227–275). México: El Colegio de México.

National Agrarian Register, Padrón e Historial de Núcleos Agrarios (PHINA). Update, December 2016. http://www.ran.gob.mx/ran/index.php/sistemas-de-consulta/phina

Navarro-Olmedo, S., Haenn, N., Schmook, B., & Radel, C. (2016). The legacy of Mexico's agrarian counter-reforms: Reinforcing social hierarchies in

Calakmul, Campeche. *Journal of Agrarian Change, 16*(1), 145–167. https://doi.org/10.1111/joac.12095

Pírez, P. (2016). Las heterogeneidades en la producción de la urbanización y los servicios urbanos en América Latina. *Territorios, 34,* 87–112. http://dx.doi.org/10.12804/territ34.2016.04.

Puebla, C. (2002). *Del intervencionismo estatal a las estrategias facilitadoras: Los cambios en la política de vivienda en México (1972–1994).* México: El Colegio de México.

Reis, N. (2014). Coyotes, concessions and construction companies: Illegal water markets and legally constructed water scarcity in Central Mexico. *Water Alternatives, 7*(3), 542–560. https://www.water-alternatives.org/index.php/alldoc/articles/vol7/v7issue2/263-a7-3-6/file

Reis, N. (2017). Finance capital and the water crisis: Insights from Mexico. *Globalizations, 14*(6), 976–990. https://doi.org/10.1080/14747731.2017.1315118

Romo Viramontes, R., & Velázquez Isidro, M. (2017). La dinámica inter e intrametropolitana de migración y movilidad entre la Zona Metropolitana del Valle de México y zonas metropolitanas vecinas. In N. Baca Tavira, Z. Ronzón Hernández, R. Romo, R.P. Román Reyes, & M. Padrón Innamorato (Eds.), *Migraciones y movilidades en el centro de México* (pp. 15–34). CONAPO, UAEM.

Salazar, C. (2011). La privatisation des terres collectives agraires dans l'agglomération de Mexico: L'impact des réformes de 1992 sur l'expansion urbaine et la régularisation des lots urbains. *Revue Tiers Monde, 206,* 95–114. http://doi.org/10.3917/rtm.206.0095

Salazar, C. (2012). Los ejidatarios en el control de la regularización en la periferia urbana. In C. Salazar (Ed.), *Irregular: Suelo y mercado en América Latina* (pp. 301–305). México: El Colegio de México.

Salazar, C. (2014a). Suelo y política de vivienda en el contexto neoliberal mexicano. In S. Guiorguli Saucedo & V. Ugalde (Eds.), *Gobierno, territorio y población: Las políticas públicas en la mira* (pp. 343–371). México: El Colegio de México.

Salazar, C. (2014b). El puño invisible de la privatización. Territorio. Estudios Territoriales, *Universidad del Rosario, Colombia, 30,* 69–90. http://dx.doi.org/10.12804/territ30.2014.03

Sánchez, L., & Salazar, C. (2011). Lo que dicen las viviendas deshabitadas sobre el censo de población 2010. *Coyuntura Demográfica: Revista sobre los procesos demográficos en México Hoy, 1,* 67–72. coyunturademografica.somede.org/wp-content/plugins/coyuntura_demografica/DEMOGRAFICA/ARTICULOS/PUB-2011-01-015.pdf

Schteingart, M., & Patiño, L. (2006). El marco legislativo, programático e institucional de los programas habitacionales. In R. Coulomb & M.

Schteingart (Eds.), *Entre el estado y el mercado: La vivienda en el México de hoy* (pp. 153–191). México: Universidad Autónoma Metropolitana/Porrúa.

Soederberg, S. (2010). The Mexican competition state and the paradoxes of managed neo-liberal development. *Policy Studies, 31*(1), 77–94. https://doi.org/10.1080/01442870903368181

Soederberg, S. (2015). Subprime housing goes south: Constructing securitized mortgages for the poor in Mexico. *Antipode, 47*(2), 481–499. https://doi.org/10.1111/anti.12110

UN HABITAT. (2011). *Housing finance mechanisms in Mexico.* http://unhabitat.org/books/housing-finance-mechanisms-in-mexico/ (Accessed: 5 May 2015).

Valtonen, P. (2000). *The politics of agrarian transformation in Mexico.* Helsinki: Academia Scientiarum Fennica.

Varley, A. (1985). La zona urbana ejidal y la urbanización de la Ciudad de México. *A: Revista de Ciencias Sociales y Humanidades, 6*, 71–95.

Varley, A. (1987). The relationship between tenure legalization and housing improvements: Evidence from Mexico City. *Development and Change, 18*(3), 463–481. https://doi.org/10.1111/j.1467-7660.1987.tb00281.x

Varley, A. (2017). Property titles and the urban poor: From informality to displacement? *Planning Theory and Practice, 18*(3), 385–404. https://doi.org/10.1080/14649357.2016.1235223

Varley, A., & Salazar, C. (2021). The impact of Mexico's land reform on periurban housing production: Neoliberal or neocorporatist? *International Journal of Urban and Regional Research.* doi:/10.1111/1468-2427.12999

World Bank. (2001). *Mexico land policy: A decade after the ejido reform. World Bank Report No. 22187-ME.* http://www-wds.worldbank.org/external/default/WDSContentServer/WDSP/IB/2002/02/02/000094946_02011904004560/Rendered/PDF/multi0page.pdf (Accessed: January 2015).

5 Peri-urban Megaprojects in Santiago de Chile: The Urbanization by Holdings and the Paradoxical Happiness of Middle-Class Dwellers

CÉSAR CÁCERES SEGUEL

This chapter focuses on the residential experience of emerging middle-income groups in peri-urban megaprojects of Santiago de Chile. Although the process of peri-urbanization of Santiago has been widely studied, the impact of private housing projects on the quality of life of inhabitants remains unclear. Drawing on the findings of sixty interviews conducted with inhabitants of Larapinta and Valle Grande in the peri-urban commune of Lampa, this paper confirms the paradoxical happiness of middle-class peri-urbanites of Santiago de Chile. Such a contradiction emerges from the satisfaction related to the achievement of higher residential standards, the financial vulnerability derived from a lifestyle based on mobility, and the availability of private neighbourhood facilities. These projects show the multiple roles played by real estate holdings in the urbanization process of Santiago de Chile, and social vulnerability of middle-income groups with precarious tools to inhabit settlements based on the privatization of basic urban services.

Peri-urbanization in Santiago de Chile and the Urbanization Phase Led by Holding Companies

During the last two decades, Chile experienced an intensive urbanization process (44% of houses are less than twenty years old) that revealed a change in the actors who lead the transformation of Chilean cities. At the turn of the 1990s, the role of the State changed from being the main agent of urban development – via infrastructure investment and housing programs – into an actor coexisting with a predominant urbanization process led by private agents. In 1980, the private sector was responsible for 23% of the whole of national urban development, while in 1997 this had increased to 81%. A particular feature of this urban phase is the participation of national and international holdings.

The pro-market reforms introduced during the 1970s and '80s were the ideological and legal basis for the subsequent consolidation, since the '90s, of a process of urbanization led by real estate, retail, and highway holding-companies with an interest in capitalizing through the Chilean cities. A holding is a special type of business that owns investments for other companies: stocks, bonds, mutual funds, real estate, or virtually anything else that has value. A holding usually controls a higher percentage of bonds of companies and allows a major control on strategic businesses decisions. The term usually refers to a company that does not produce goods or services itself; instead, its purpose is to own shares of other companies. A Real Estate Holding operates in two forms: owning real estate properties (in New York, for example) or executing new urban megaprojects (Shanghai, Melbourne, Lima, Santiago de Chile).

Here, it is argued that the urbanization led by holding companies, or *holdingbanism* (Cáceres, 2016a; Cáceres, 2016b), shall be interpreted as one of the main milestones for the process of urbanization of capital initiated in Chile during the '70s. Since the '90s, holding companies increased and sophisticated their participation in the urban development creating a new generation of entrepreneurial landscapes: malls, businesses parks, leisure boulevards, satellite towns, and private highways.[1] An important aspect is that the urbanization process led by holding companies does not only relate to investment in sophisticated residential spaces to high-income families, but more importantly, to the production of urban spaces for the daily use of emerging middle-income groups. These companies are aware that their power lies not only in the capacity to act on the cityscape, but also to transform the culture and modes of life in the city. Today, the presence of holdings in different Chilean cities, and their social ubiquity, installs the notion that holding companies are the architects of a new urban phase in Chile. Megaprojects built by holdings are already installed in the mindscape of Chilean´s urbanites and regarded as liveable and modern urban spaces.

Although the political and territorial effect of urban megaprojects in Chile has been widely studied (Hidalgo & Borsdorf, 2005; Poduje, 2006; Lizama, 2007; De Mattos, 1999), the impact of residential megaprojects on the quality of life for emerging middle-income groups remained unclear. The study of the residential experience of middle-income groups is relevant considering that the expansion of this segment is one of the most significant social processes of the last thirty years in Chile (Barozet & Fierro, 2011; Espinoza & Barozet, 2009). Between 1990 and 2007, middle-income groups in Chile increased from 1.2 million in 1990

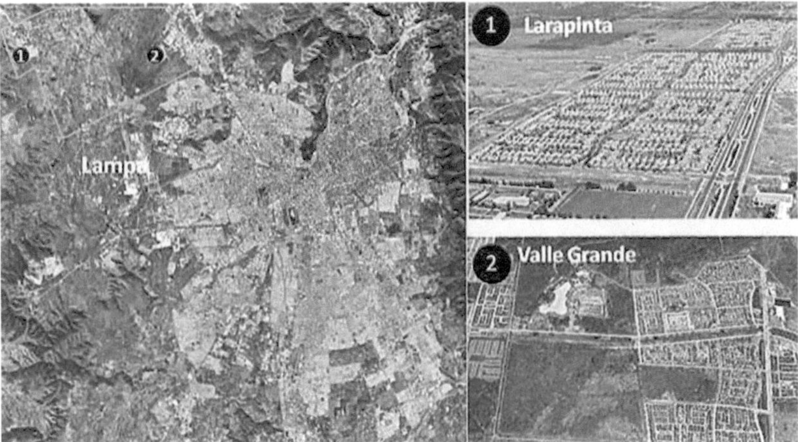

Figure 5.1 Location of satellite towns in Lampa, Santiago de Chile.
Source: Google Earth.

to 2.3 million in 2007. Nowadays, upper-middle-income groups make up 18% of national population, with an average income of USD 1,300, meanwhile, the middle-low income segment represents 29.1% with an income average of USD 806. The sustained economic growth and the effectiveness of the social policies of the last three decades allowed to reduce the poverty by income from 38.6% in 1990 to 11.7% in 2015. Chilean middle-income groups are a heterogeneous segment defined by a highly social fragility (Contreras et al., 2005). This high fragility would be explained by its dependence on consumer credits and by the limitations of the social security system in Chile, which, in the face of illness, unemployment, or retirement, can lead people to episodes of downward mobility.

This chapter investigates the residential experience of middle-income groups in private megaprojects built in the peri-urban commune of Lampa, Santiago de Chile.[2] Sixty inhabitants (30 each) of the projects of Larapinta and Valle Grande were interviewed. The main results reveal (1) unprecedented capitalization strategies from real estate companies; (2) new urban roles of real estate holdings; and (3) unknown forms of social exclusion of middle-income groups in Chilean cities.

The following section analyses the urban normative planning approaches concerning to satellite towns. Later, the main results of the interview process in two private megaprojects of Lampa are stated, and finally, a general discussion and conclusions are presented.

Satellite Town: A Neighbourhood Concept 30 Kilometres Away from the City

Satellite towns such as Larapinta and Valle Grande in Lampa are part of the urban megaprojects built by real estate holdings in Chile. These towns have been built since the early 2000s in peripheral and peri-urban communes located in the metropolitan region of Santiago. Satellite towns were born out of a normative that permitted the creation of new settlements in conditioned urbanization zones located in peri-urban communes of this metropolis. The main goal of the "urbanization under conditions" normative was that private agents would bear the costs of the ensuing urban sprawl by imposing neighbourhood equipment standards. The Ministry of Housing and Urban Planning presented the reasons for the benefits of this norm as follows: "To allow the creation of new towns in areas with aptitudes to the location of new components of the urban regional system, differentiated and separated from the current cities and metropolis, avoiding the conurbation and the urban sprawl" (Ministerio de Vivienda y Urbanismo, 1996, p. 30). The control of the plan rested with local governments that negotiated on a project-by-project basis within an adaptive concept of the norm. This normative reform can be read as a political legitimation given by the central State to the creation of private middle-size settlements by real estate holdings.

The satellite town of Larapinta was built (USD 150 million of investment), and is managed, by the real estate holding Socovesa. It is located in a conditioned urban area (ZODUC) in Lampa. The company projects an urbanization of 389 hectares on a macro-plot of 1,200 hectares which is the property of the company. The projected total number of homes is 10,376 that will house approximately 51,880 inhabitants. Currently, 10,000 residents are accommodated in 2,500 houses. The projected densities are expected to range from a minimum of 70 inhabitants/hectare to a maximum of 150 inhabitants/hectare. Moreover, according to the norm, the project envisages that 5% of all housing shall be high-density housing (i.e., social housing) with a 300 inhabitants/hectare density at least. The project is focused on a socioeconomic segment that includes inhabitants of both, low- and high middle-class income. Prices range between USD 40,000 and USD 90,000 respectively. The project anticipates the development of 44.5 hectares of parks and green areas, of which 21.5 hectares will be freely accessible.

> We project 15,749 houses. Currently, there are 2,800 houses between Valle Grande and Santo Tomas. We are just beginning. The project is designed

> for the next thirty years. Valle Grande and Santo Tomas are the property of an insurance company. Valle Grande is envisaged to have approximately 25,000 houses, or 100,000 inhabitants. (Employee of Valle Grande, name withheld, September 2010)

The satellite town of Valle Grande is owned by the Valle Grande Consortium which is the property of three insurance companies: Holding BiceCorp, Holding FFV-Fernandez León, and Holding CorpGroup. Beginning in 2002, the company projected the urbanization of 437.9 hectares, where 180 hectares were designated for residential use. This project has been designed to include two macro projects: Santo Tomas and Valle Grande-Orient. The project will provide 25,000 homes that can house 100,000 inhabitants. The project's horizon is 30 years. Currently, the private settlement accommodates approximately 20,000 people. Similar to the project of Larapinta, this is focused on a middle-class socioeconomic segment. The price of the houses range between USD 51,000 and USD 140,000, respectively. The projected densities vary from a minimum of 70 inhabitants/hectare to a maximum of 100 inhabitants/hectare.

According to the conditioned urban development norm, the new settlements shall accomplish the following requirements: (1) a minimum of 85 inhabitants/hectare; (2) urbanization of at least 300 hectares; (3) 30% of the dwellings built shall be acquired through the housing subsidy program, with prices between 300 and 1500 UF (Unidad de Fomento, an indexed Chilean currency, and of these, 30% to 40% shall be high-density housing between 401 and 500 inhabitants/hectare (social housing); and (4) facilities such as health care, education, public safety, green areas, parks, sports, and services shall be included.

An interesting aspect relates to the urban planning approach used by companies: in satellite towns, the real estate holding shifts from a classical model of mono-functional suburbia towards multi-functional private cities where the predominant value is a life experience in multi-functional neighbourhoods. The success of these projects depends on the skill of the promoter to produce symbolic distances from the "traditional city." The characteristics and dimension of these projects are described below by the urban administrator of Valle Grande:

> The satellite town is designed to allow walking or cycling to supermarkets, schools, or nursery schools. We aim to re-invent the neighborhood as a key urban unit ... Children play in squares. Young people cycle in parks.

> Families meet in public squares to socialize with other neighbor families. (Socovesa, 2003)

> The opportunity for creating a new city allows [us] to plan and generate high-level life conditions, regardless of the socioeconomic status, with good highway standards, and offering green areas and open spaces, supply of services, and recreational facilities. (Sociedad Inmobiliaria Valle Grande, 2002)

The promoters of satellite towns carried out marketing campaigns via the comparison with traditional residential areas of Santiago. The pursuit of superior residential standards for middle-income inhabitants of Santiago was seen as a "market opportunity" to build liveable peri-urban settlements. Companies examined classical urban planning principles such as the garden city of Howard and the neighbourhood unity of Perry to transform the residential typologies from mono-functional suburban residences to settlements based on a mixture of neighbourhood facilities. Paradoxically, although promoters invite Santiago´s inhabitants to project their urban life into the countryside, their planning proposal is originated from the most basic urban concept: *the neighbourhood.* The argument behind the business concept of neighbourhood life, 30 kilometres outside the city, is based on an opportunity to gain access to a superior urbanity denied by former neighbourhoods and communes.

In summary, satellite towns are interesting case studies since they portray the multiple roles entrepreneurs currently play in the urban development of Santiago. These can be characterized as follows: (1) They are *financiers;* peri-urban megaprojects are recognized as part of the financialization strategies developed by holding companies in Santiago de Chile. These operations were accompanied by parallel lobby operations with the regional government of Santiago in order to approve the normative which allows the urbanization of rural-peri-urban soil. (2) They are *landowners;* a central aspect concerning the strategy of capitalization through urban landscapes refers to the acquisition of land plot "banks" in expansion areas of Chilean cities, aimed to develop urban megaprojects. (3) They are *administrators;* entrepreneurs also realized that their businesses opportunities include the administration of an entire settlement (100,000 inhabitants estimated) and offering and charging for neighbourhood facilities during a period of twenty years. This feature promotes the provision of local services by the companies (schools, public transportation, or sport centres) and creates particular modes of co-government between the municipalities and the holding companies.

Examining the Residential Experience of Middle-Income Groups in the Private City

Asked about their motivation for middle-income groups to move into satellite towns, the interviewees reported that their decision for leaving their former homes at inner communes of Santiago (Conchalí, Renca, Recoleta, San Miguel, La Cisterna) and moving into private settlements was driven by the pursuit of superior residential standards. This superiority was characterized by a variety of subjective factors such as quietness, family security, new "levels" of neighbours, higher abundance of green areas, bigger houses, and the escape from air pollution. Lampa was also selected since it appeared as a "realistic" option considering the housing market of Santiago.

Improvements in the living conditions of inhabitants in satellite towns were described as follows:

> Yes, we chose this place because we did not want to live somewhere like Maipu or Puente Alto, where all the houses are crowded and semi-detached. There is more space here; we have a yard that is twice what it would be in a house in Santiago. Here, we have space for two cars. Similar houses in Santiago would cost twice as much … (Carolina, Larapinta)

> I think that (regarding the quality of life perception), we got more in terms of public spaces … We have a bigger house, a bigger yard. For the same amount of money, a house bought in Santiago would have been much smaller. It would have a much smaller yard … Here, children can play. We have space for a pool. In San Miguel, our yard was only one meter wide. (Daniela, Larapinta)

The interviews showed that middle-income groups are more ambitious in the definition of their habitat. Urban sceneries with large green areas, rows of pastel houses, and private security represent urban symbols of social mobility. They choose the isolation from the traditional Santiago, since in their exile they have access to urban experiences that make sense in their pre-fixed notion of urban quality of life. Satellite town companies appear ingenious in fulfilling the expectations of superior residential standards of Santiago´s

Figure 5.2 Recreational facilities in satellite towns. Photo by the author.

middle-income groups (71% of interviewees in the urban "quality of life survey,"[3] stated that the lack of green areas was a serious problem in their neighbourhoods). Asked about an evaluation of squares, sports facilities, and parks, inhabitants expressed a high satisfaction with the landscape design and the quality of the recreational facilities. These results give rise to the notion of satellite towns as settlements able to encourage new modes of recreational practices and public life. As portrayed by a young mother:

> Children went back to play on the streets.
>
> We go out two or three times a week with the girls [to parks and squares]. We did not do that in the commune where we lived before because there were no squares nearby. To go to a square, we had to take the car and go to the "9 de Gran Avenida (bus stop)" where there was a park. But, you could go out for a walk. This is something I like here, we can go out with the kids on bikes, to walk quietly. It feels safer. We did not have that quality of life where we lived. (Daniela, Larapinta)

A second feature of satellite towns mentioned by inhabitants that affects life quality is the value of the quiet. Exploring this value during

Figure 5.3 Educational and sports facilities owned by the company Socovesa in Larapinta. Photos by the author.

the interviews, this seems to be primarily related to the settlement design, characterized by the larger space of homes and design. The conquest of more space appears as a precious value when compared to life conditions in the former communities where small houses and overcrowded neighbourhoods affect mental health and social relationships of their inhabitants. A next feature related to the quiet is the way of life in semi-rural settings. The peri-urban areas appear as an urban lifestyle freed from the air pollution and stress of the inner city. A third form of the expressed quiet is associated with an escape from crime such as thieving or burglary (a sensitive issue in Santiago´s population).

In order to evaluate the routines of access to basic services and goods, satellite town inhabitants were asked how they satisfy their needs relating to education, work, health services, and consumption. The studied projects exhibit a significant ability to satisfy the educational needs of families at a neighbourhood scale. Even though most of the interviewees mentioned the high costs of monthly fees for schools in private projects, they rely on nearby schools. Regarding the access to health care centres, all the interviewees revealed the use of medical services located in inner communes of Santiago. The use of the health care centres located in satellite towns was cited by just one person referring to an emergency case, but interviewees stated that nearby health care centres are expensive and lacked specialists, which means that these had no influence on their health attention routines. Regarding their preferred shopping practices, interviewees stated that they shop in supermarkets and markets located in inner communes of Santiago.

> The supermarket of Larapinta has less variety of products, I buy only small things there. I shop our groceries in the LIDER (supermarket) of Plaza Norte. Actually, most of families from Larapinta do their shopping in the Mall Plaza Norte because of the quality and price. We also attend the Integramedica medical center in the shopping mall; we do not attend the medical centre from Larapinta. Normally, I go to the mall a lot with my children, for shopping, medical care and also for a stroll … (Johanna, Larapinta)

This data allows three observations: (1) Although satellite towns have been advertised as self-contained settlements, the ability to provide health care and to satisfy consumption needs is still marginal. (2) It is often argued (Frey, 1999; Calthorpe, 1995) that a neighbourhood composed of a variety of nearby services improves the quality of life for the inhabitants by reducing the need for mobility; however, the evidence above shows that in the case of peri-urban megaprojects in Santiago, the inhabitants' priorities lie in quality (health care) and costs (supermarket) and not in proximity. (3) The dwellers of satellite towns constitute a cohort of metropolitan inhabitants who organize their urban life based on the use of diverse and distant urban territories (mall, mega-supermarket, satellite towns, and clinics or jobs in inner communes).

Although settlements accommodate between 12,000 and 24,000 inhabitants and this number will increase to between 52,000 and 100,000, they are not included in the metropolitan public transport system of the Greater Santiago area ("Transantiago"). This causes an absence of nightly bus services and higher costs to families since they pay twice per trip. The lack of a private car and an efficient public transport system creates dramatic situations for families in need of medical assistance from hospitals located 30 kilometres away from their homes. The cost of transport can also exclude people from jobs simply because salaries do not provide minimum earning margins to cover the transport on top of their living expenses. All of this represents a sensitive factor damaging the quality of life.

> (Public transport) the quality is poor, there are no services at night; from 10 p.m. there is no public transportation … I think people do not come (to the satellite town), not because it is far away from Santiago, but since it is inaccessible. (Luisa, Larapinta)

A second variable damaging the quality of life in the private peri-urbia is related to exclusive locations of private neighbourhood services (private clinics, private schools, private sport centres). As the conditioned urban development norm does not regulate whether services like schools, health care centres, sports infrastructure or nursery schools shall be private or public, this current legal window allows the development of captive markets for the company. The supply of these services constitutes an opportunity for the companies to provide basic daily services (schools, transportation to Santiago, sport centres, etc.). The use of "good clubs" as a business strategy is pointed out by Glasze (Glasze, 2003): "Developers may profit from the fact that the establishment of a neighborhood governance structure, with the powers to exclude free riders, as well as the power to regulate the use of common spaces and facilities, reduces the risk of an economic decline of the neighborhood." With respect to Valle Grande, the schools and public transportation that connect to Santiago are owned by the company. Being allowed to manage these towns for up to twenty years represents a market opportunity where profit strategies are not only aimed at house sales, but also to the provision of private daily services. Although the cited normative is regarded as demanding (in terms of the standards of the accomplishments) by the companies, it constitutes a friendly normative for entrepreneurs due to the absence of a specification regarding a private or public character of recreational, health care or educational facilities, allowing the creation of "good-clubs."

> They only thought of providing PROCLUB (private sports club), which is a club where you have to pay if your children want to play football. They did not provide additional municipal multipurpose courts ... only some sectors have ones ... They (the company) say if your children want to play football take them to PROCLUB and pay. (Catalina, Valle Grande)

Peri-urban Megaprojects in Lampa: An Undisputed Conquest of Quality of Life for Middle-Income Groups?

The exodus of middle-income groups to private towns can be understood as spatial testimony to their social mobility. In these projects, middle-income groups reclaim the right to redefine their habitat, exalting new identities and differences. Asked about the wish to stay in the satellite town, 85% of the interviewed residents projected their future life in these projects. Peri-urban satellite towns of Santiago de Chile show that middle-income groups are more ambitious in the definition of their habitat. They move to these projects searching for a life experience like a "city a la carte," that is, by selecting their valued urban attributes, such as green spaces, neighbourhood facilities, and pedestrian-oriented town design, while excluding the negative aspects of the traditional city of Santiago like stress, pollution, lack of safety, and lack of green areas. Rather than an escape from the city, the exodus to satellite towns is expressed as a search for the most basic of urban concepts: the neighbourhood. For these inhabitants, satellite towns are an option for not abandoning urban life completely, but leaving behind unwanted experiences.

> I really won. The place design invites you to go out. My daughter told me, "We can bring a tablecloth and have a picnic." The grass seems like a carpet. This is a place to go out with the family. There's no comparison – see these open skies! There, I had the neighbour's wall, I had the cables, the apartments with the clothes hanging out. (Paulina, Larapinta)

> I bought a house to stay here ... I prefer living here than in Santiago, even if I shall stay all the time in Santiago. Here we got pure air...here it is agreeable; it is just a better quality of life. Here you can sleep quietly. I never had a problem I go for a walk at night and there isn't problem ... and also with the technology – I can solve many things by internet. (Victor, Larapinta)

Nonetheless, do satellite towns in every case present an improvement in life quality? Even when a majority wish to stay in satellite towns, the questions about access to basic services revealed that a significant part of inhabitants did not agree with an urban life mode based exclusively on private facilities and with an inefficient public transport system. Therefore, when asked about their financial tools to handle the cost of the peri-urban life, 42% of residents reported that maintaining the suburban dream posed serious economic problems.

> Stay here? I am applying for another school for my son, the National Institute, which is very far from here, so I am thinking, a few more years and then I´ll move. The supermarket is far and transport is very expensive. I don´t have a car and it is very expensive. Also, the housing payment; it is complicated. Many people have returned their homes to the banks because they cannot fund the housing payments. (Johanna, Larapinta)

How to interpret this apparent contradiction between residential satisfaction and social precarity? Inhabitants of the studied satellite towns experience what Lipovetsky (Lipovetsky, 2010) describes as a "paradoxical happiness" characterized by an apparent contradiction in societies between higher consumption ability, but also higher levels of scarcity. Interviewees consider suburban life as a precarious combination between neighbourhood satisfaction and economic scarcity. These periurbanites experience high satisfaction with the residential standards proposed by real estate holdings, but the change to private peri-urbia also implied an unprecedented cost associated with expenses for private neighbourhood facilities and daily mobility. The transition to a city administered by holdings came as an economic shock to emerging middle-class groups, since Santiago´s inner communes are still a source of welfare services and consumer opportunities (30% of interviewees are users of the public health care system).

> Yes, the family budget is limited. My son wanted to buy a car when we arrived here, but as our life cost is higher, he couldn´t afford it. The life cost is higher because of the transport ... If you look around, there are many houses for sale. Here, there are many people selling their homes. People come, buy a house, and then they realize that the family budget is not enough. (Aida, Valle Grande)

Conclusions

Residents of peri-urban megaprojects experience high satisfaction with the entrepreneurial urbanity, but the change to the private housing market also implies an unprecedented economic scarcity associated with higher expenses for private neighbourhood facilities and daily mobility. The urbanization of holding companies, or *holdingbanism*, diversified the residential options for middle-income groups; however, in the case of peri-urban satellite towns, the evidence shows a conflict associated with the creation of settlements based on the privatization of basic urban services. The privatization of basic urban facilities is exemplified by the satellite town of Lampa as a residential space configured under the logic of what Harvey (Harvey, 2004) defines as the concept of "accumulation by dispossession" – a neoliberal strategy based on the suppression of rights to common goods and privatization of social rights; a phase characterized by a process of privatization and commodification of public urban assets, natural resources, and the emergence of urban megaprojects under private regime of administration.

As argued above, the neoliberal reforms during the '70s and '80s in Chile were the foundation of the ideological and legal basis for the subsequent consolidation since '90s, a process of urbanization led by real estate, retail, and highway holding-companies. In this framework, satellite towns should be understood as residential typologies designed to maximize profit margins of real estate holdings and to expand the financialization networks in Chilean cities. Finally, the social and political impact of the process of urbanization by holdings in Santiago de

Chile can be summarized the rise of three outcomes: (1) new types of entrepreneurial urban landscapes; (2) the appearance of new urban life modes and exclusion dynamics in middle-income groups; (3) and the emergence of private-public governance arrangements at local level.

The pursuit of higher liveability standards (neighbourhood facilities, green areas, security) was the main reason behind the suburban exodus. Although people want to live in a "private" city, this is a fragile urban lifestyle that can be only kept as long as they are able to fund the cost of an entrepreneurial peri-urban life. People decide to maintain a superior level of urban life despite the higher degree of economic uncertainty; however, as soon as some family member loses their job or an unforeseen expense emerges, for sale signs appear, and they return to the city. With the Chilean economic development of the last decades, Santiago´s emerging middle-income groups enjoy more personal freedom than ever. They can decide between several alternatives of urban habitat typologies and move freely to an extended and multi-centred metropolis. In contrast to the notion of peri-urbia as a boring, homogeneous, non-political space, the commune of Lampa in Santiago de Chile appears as a complex space opening new forms of social vulnerability and capitalization strategies from real estate holdings.

NOTES

1 Among the main holding companies linked to Chilean urban development Socovesa, Cencosud, Mall Plaza, Fallabella, Holding D&S, Salfacorp, Highroad Vespucio Sur S.A. Highway Costanera Norte S.A. stand out.

2 This commune had a population of 79,421 inhabitants in 2012, which represents a population increase of 98% (39,323 new inhabitants) in comparison to 2002. This is the commune with the fifth highest population increment in the region of Santiago. Considering the houses prices, Lampa can be identified as the principal peri-urban residential pole of middle-class groups.

3 National Urban Quality Survey 2010, Ministry of Housing and Urban Development.

REFERENCES

Barozet, E., & Fierro, J. (2011). Clase media en Chile, 1990–2011: Algunas implicancias sociales y políticas. *Santiago, Serie de Estudios Nro. 4. Fundación Konrad Adenauer.* http://www.kas.de/wf/doc/kas_29603-1522-4-30.pdf?111202200649

Cáceres, C. (2016a). Ciudades satélites en Lampa, Santiago: Un caso de co-gobierno urbano entre el municipio y holdings inmobiliarios. *Cuadernos Geográficos, 55*(2), 265–281.

Cáceres, C. (2016b). La urbanización de holdings empresariales en Chile 1990–2015: Una Industria de Paisajes en Serie. *Biblio 3w: Revista Bibliográfica de Geografía y Ciencias Sociales, XXI*(1.171), 1–15.

Calthorpe, P. (1995). *The next American metropolis*. New York: Princeton Architectural Press.

Contreras, D., Cooper, R., Herman, J., & Neilson, C. (2005). Movilidad y vulnerabilidad en Chile. *Expansiva en foco, 56*(1), 1–16. http://repositorio.uchile.cl/bitstream/handle/2250/152184/Movilidad-y-vulnerabilidad.pdf?sequence=1&isAllowed=y

De Mattos, C.A. (1999). Santiago de Chile, globalización y expansión metropolitana: Lo que existía sigue existiendo. *EURE, 25*(76), 29–56. http://dx.doi.org/10.4067/S0250-71611999007600002

Espinoza, V., & Barozet, E. (2009). De qué hablamos cuando decimos "Clase media"? Perspectivas sobre el caso chileno. In A. Joignant & P. y Güell (Eds.), *El arte de clasificar a los chilenos: Enfoque sobre los modelos de estratificación en Chile.* (pp. 103–130). Expansiva. http://www2.facso.uchile.cl/sociologia/1060225/docs/clase_media_ex.pdf

Frey, H. (1999). *Designing the city: Towards a more sustainable urban form.* London: Taylor & Francis Group.

Glasze, G. (2003). Private neighbourhoods as club economies and shareholder democracies. *BelGeo. Revue belge de géographie, 1*, 87–98. https://doi.org/10.4000/belgeo.15317

Harvey, D. (2004). The new imperialism: Accumulation by dispossession. *Social Register 40*, 63–87.

Hidalgo, R., & Borsdorf, A. (2005). Los megaproyectos residenciales vallados en la periferia ¿Barrios cerrados autosuficientes o nuevas ciudades? *Revista Urbano, 8*(12), 5–12.

Lipovetsky, G. (2010). *La felicidad paradójica*. Barcelona: Anagrama.

Lizama, J. (2007). *La ciudad fragmentada: Espacio público, errancia y vida cotidiana*. Santiago de Chile, Universidad Diego Portales.

Ministerio de Vivienda y Urbanismo. (1996). *Memoria explicativa, modificación Plan Regulador Metropolitano de Santiago, Provincia de Chacabuco*. Santiago, MINVU.

Poduje, I. (2006). El globo y el acordeón: Planificación urbana en Santiago, 1960–2004. In A. Galetovic (Ed.), *Santiago: Dónde y vamos* (pp. 231–276). Santiago de Chile: Centro de Estudios Públicos.

Sociedad inmobiliaria Valle Grande. (2002). *Memoria Explicativa proyecto de desarrollo urbano Valle Grande*. Santiago de Chile: Sociedad Inmobiliaria Valle Grande.

Socovesa. (2003). *Estudio de impacto urbano proyecto Larapinta*. Santiago de Chile: Socovesa.

6 Financialization and Social Reproduction in the Buenos Aires Urban Periphery

LIZ MASON-DEESE

Introduction

From large-scale land takeovers and informal settlements in the 1980s to the factory occupations, roadblocks, and alternative economies of the unemployed in the 1990s and early 2000s, Buenos Aires's urban periphery has long found itself in the centre of struggles over neoliberal transformations (Stratta & Barrera, 2009; Zibechi, 2012). In recent years, the urban periphery has undergone another transformation: from the vital, popular, and solidarity economies developed by the poor and unemployed during the economic crisis of 2001 to the increased operation of financial mechanisms that seek to "include" these territories in the dominant circuits of capital accumulation. This chapter argues that as consumption has dramatically increased, it becomes impossible to speak of these sectors of the population as purely "excluded" or "marginalized"; yet this "inclusion" often puts the reproduction of life itself at risk due to the everyday violence of financial capitalism.

Drawing on research conducted between 2003 and 2015 with the Unemployed Workers' Movement (*Movimiento de Trabajadores Desocupados* or MTD) of La Matanza, in this chapter, I follow the organization's trajectory from a movement dedicated to creating alternatives to neoliberalism to its increasing incorporation into the financial system, most notably through a partnership with Banco Santander. La Matanza, a predominately low-income and working class zone of Buenos Aire's urban periphery, now finds itself at the heart of struggles over the expansion of finance and new forms of consumption. These new forms of finance seek to incorporate the popular sectors into the dominant economic system by expanding consumer credit to the poor and informal workers, and financializing movements' alternative economic practices in a variety of ways. While the case of the MTD La Matanza is in some sense unique because of their extensive ties to the business and financial sector (Vommaro, 2016), it also exemplifies broader trends in the urban periphery.

As I will show in what follows, the processes of financialization in low-income areas of Buenos Aires's urban periphery are closely linked to the everyday alternative practices of production and social reproduction developed by the poor and unemployed during the crisis. These practices were what allowed the poor and the unemployed to survive the crisis when they were simultaneously denied access to formal employment and state welfare (Colectivo Situaciones, 2012, 2014). The unemployed workers' organizations, such as the MTD La Matanza, were vital in the process, creating forms of "work with dignity" through worker-managed cooperatives, as well as autonomous forms of social reproduction, such as community-managed schools, health clinics, and soup kitchens (Mason-Deese, 2017).

Various authors have identified financialization as a key element of the contemporary neo-extractivist economy that emerged following the country's economic crisis (Gago, 2015; Gago & Mezzadra, 2017; IIEP, 2014). The Kirchner government defended this economic model, under the concept of neo-developmentalism, in which taxes on the extraction of natural resources and agriculture would be reinvested in the country through social programs and infrastructure and technology development (Gago et al., 2014). Thus, much of the analysis of this model sets the rural and the urban in opposition to one another with the rural as the site of natural resource extraction, while residents of the urban periphery benefit from this extraction through social programs. However, citing the importance of finance, Gago calls for us to "broaden the concept of extractivism beyond the reference to the reprimarization of Latin American economies as exporters of raw materials in order to understand the particular role played by territories in the urban peripheries in this new moment of accumulation" (2015, p. 22).

Building on this call that recognizes financialization as the key element of the neo-extractivist economy, I argue that the recent drive toward financialization in the urban periphery can be seen as an attempt to incorporate these popular and solidarity economies and extract value from their forms of cooperation and vitality. Financialization seeks to directly capture value from activities of social reproduction as well as popular economy activities on the margin of dominant capitalist practice, by extracting value without going through the wage relation (Brighenti & Mezzadra, 2013). This occurred, first, through a new generation of welfare programs aimed to strengthen the popular economy by providing loans to cooperative enterprises and incorporating the poor into the financial system by distributing welfare benefits through banking institutions. Later, a proliferation of other forms of finance, including different types of legal and illegal, formal and informal consumer credit, extended this process.

I start the chapter by exploring the alternative practices of production and reproduction that were created by social movements and the poor during Argentina's economic crisis. The following section explores the proliferation of different forms of finance in the urban periphery, focusing on their relationship to the popular and solidarity economy created by social movements. Particularly, I look at the role of microcredit, financialized welfare programs, and consumer credit to show how these different forms of finance have transformed the urban periphery through incorporating more and more areas of everyday life into financial flows and subsequently increasing precarity and vulnerability.

Argentina in Crisis

In 2001, Argentina experienced a devastating economic crisis following years of growing unemployment and inflation, forcing the country to default on much of its public debt. More than a financial crisis, Argentina's 2001 crisis was also a crisis of legitimacy of the entire political and economic system. Referring to the high inflation, the end of convertibility between the peso and dollar, the collapse of the banking system and confiscation of savings (the *corralito*), and the simultaneous existence of multiple currencies in the country, Gago describes the crisis as "the crisis of money as sovereign authority" (2014, p. 206). In other words, the crisis not only led to a loss of trust in the banking system and financial institutions, it also revealed the social relation and power behind money itself. This crisis also opened up space for experimentation for new forms of economic and financial relations, outside of the dominant economy.

The economic crisis can also be understood as a crisis of social reproduction. As the state withdrew from its role in ensuring social reproduction during the period of neoliberal structural adjustment in the 1990s, cutting spending on health care, education, unemployment benefits, and aid to the poor in general, women were forced to pick up the extra costs in order to protect their families (Dalla Costa, 2008). It was in response to this crisis of social reproduction that the unemployed workers' movements, such as the MTD La Matanza, emerged. La Matanza is the most densely populated area of the Buenos Aires region with a mixture of working class neighbourhoods and self-constructed, informal housing, and access to safe and affordable housing is constantly an issue. In the neighbourhood where the MTD formed, schools are often dilapidated and suffer structural damage, health clinics are understaffed and under-resourced, many streets are unpaved, and public transportation does not reach them. These are some of the many issues that led people to begin organizing in the neighbourhood.

The MTD La Matanza began in 1995 as a group of neighbours concerned about the rising price of gas, electricity, and food in their neighbourhood, as well as the problems mentioned above. These neighbours participated in several actions with other neighbourhood organizations, including *ollas populares* (popular meals) in a main square and taking over a municipal government building to demand aid and price controls. Eventually, they realized that the root of their problems lay in unemployment and began to organize more directly around that question by blocking roads to demand jobs and unemployment benefits (Flores, 2005). The MTD brought together individuals with diverse experiences and expectations of work. The original members included a combination of older men who had been recently laid off from factory employment, women who had worked as domestic workers or caretakers in houses in the city centre or ran small businesses out of their homes, and younger people with much less experience in formal work. Despite these diverse backgrounds, all these members found themselves suffering from the economic crisis.

After a few years of organizing roadblocks, the MTD La Matanza took another turn when they occupied an abandoned school building in order to establish a social centre, preschool, and cooperative. With support from both national and international NGOs, the MTD was able to start a cooperative textile workshop and cooperative bakery. Toty Flores, one of the group's leaders, speaking about these cooperative enterprises, states: "Cooperation not only represented an economic response to the needs of life, but was also the organizational form that we found to break with isolation and to counteract the politics of neoliberal individualism predominant in our society" (2005, p. 36). In other words, the cooperatives were attempts not only to meet members' basic needs through providing a small income, but also to create different social relations and ways of organizing society. These cooperatives and others like them formed an essential node in the burgeoning popular and solidarity economy that emerged in Argentina during this period (Giarracca, 2008).

At the same time, barter clubs and alternative currencies also proliferated throughout the country. By 2002, hundreds of thousands people across Argentina participated in one of the two principal alternative currency networks (North, 2005). In order for these currencies to function, new relationships of trust had to be built between the market-goers, trust not backed by the state force as in the case of a national currency, but in face-to-face relationships and solidarity between neighbours and backed by the political power of social movements. The MTD La Matanza hosted a barter club in the courtyard of their social centre, open to any neighbourhood resident who agreed to the club's principles. While initially the club worked through direct barter and a system of local alternative currency notes, by 2005, the club was mainly using the official Argentine peso. Items for sale ranged from homemade goods, such as food and clothing, to resell items, such as cheap clothing bought in bulk from a larger informal market or food items from government aid baskets. The MTD also sold goods from its cooperatives, such as bread and pastries and screen-printed T-shirts.

Their flexible production strategies, producing small batches and cultivating a wide-ranging network, drawing on the skills and relationships that members brought from their different experiences of paid and unpaid labour, as well as neighbourhood organizing, allowed them to provide stable employment to thousands of neighbourhood residents. Other cooperative enterprises developed by similar movements in other areas of the urban periphery had similar effects (Zibechi, 2012). Although developed on the margins of the state and capital, these practices were more than mere survival strategies, they were also highly productive and innovative value-creating enterprises, based on diverse forms of social cooperation and know-how (Flores, 2005, Zibechi, 2008). And these activities were what made the urban periphery such an attractive zone for the expansion of financial capital.

Proliferation of Finance in the Urban Periphery

Following these intense years of crisis, Argentina slowly began to experience an economic recovery. However, this recovery was not based on a return to formal, stable employment for much of the population. Instead, it has been based on an increase in informal and precarious work, state subsidies, and extractive industries that create few permanent jobs. As Maristella Svampa notes: "A key feature of neoextractivism is the immense scale of the projects, which says something about the size of the investment: they are capital-intensive, not labor-intensive, activities. For example, in the case of large-scale mining, for every $1 million invested, between 0.5 and 2 jobs are directly created" (2015, p. 66). However, there has been a noted increase in consumption, especially by popular sectors, that is largely funded by different credit mechanisms, as I will discuss in what follows.

In order to understand this financialization, the *Instituto de Investigación y Experimentación Política* (IIEP) describes how financial capital as operates both "from above" and "from below":

> This dynamic of financialized capitalism branches off, in conditions of neo-developmentalist hegemony, according to a double line: from below, it feeds dynamics of popular consumption and indebtedness that make possible a type of inclusion beyond the world of employment; from above, it permits a fund of resources that at this point become irreplaceable for the functioning of state agencies. (2014, pp. 6–7)

In one sense, this increase in the capacity for popular consumption is the positive outcome of struggles against austerity measures in the 1990s: the consumption is, in large part, funded by social programs and subsidies demanded by the movements. It is a refusal, on the part of the poor, to have their needs and *enjoyment* subordinated to capital, a refusal of austerity and poverty in a society filled with so much wealth, a refusal that takes its ultimate form in recurrent outbreaks of looting (as occurred, for example, in December 2012 and December 2013). On the other hand, this consumption includes the poor in an unequal and subordinate way through mechanisms of debt and relationships of dependency that ultimately threaten their very ability to reproduce themselves. This expansion of credit is part of what Gago (2015) terms *popular neoliberalism* or *neoliberalism from below:* the extension of practices of consumption and a type of neoliberal subjectivity into peripheral areas.

The territorial expansion of finance also entails a proliferation of *forms* of finance and financial bodies in the urban periphery, from direct cash loans to credit cards, from transnational banking institutions to local or regional non-banking financial institutions, informal and shadow banking systems, as well as mechanisms of debt and credit that operate within circles of families and friends (rotating credit circles, lending credit cards, etc.). Different relationships of credit allow people to pay for goods in instalments or offer credit to other neighbourhood residents when selling items in kiosks. Communal or family practices of sharing credit cards or creating rotating credit circles allow people to share their access to formal credit with others without that access (Wilkis, 2015). Other forms of financing, such as the *pasanaku* or *anticrético* system, are rooted in Indigenous practices, brought to Buenos Aires through successive waves of migration and transformed in the process (Gago, 2017).

However, a report from the Attorney's Office for Economic Crimes and Laundering of Assets (Procelac) shows that despite the proliferation of different forms of financing, the large banks are increasingly the major players (Feldman, 2013). Large banks, under different names, slogans, and advertising campaigns, are also responsible for offering consumer credit to low-income sectors, as well as financing cards for supermarkets and domestic appliances. The Procelac report highlights how non-banking financial institutions, which are more prevalent in low-income neighbourhoods, not only face less regulation but also tend to charge substantially higher interest rates. This means that the poor pay significantly more for the same products compared to middle- and high-income sectors who have access to different types of credit. This "vulnerable indebtedness" to use Procelac's term, "places people in a critical situation in regards to their subsistence" (Feldman, 2013, p. 13) and puts their material reproduction and livelihoods in question. This vulnerability arises when people depend on credit to meet their basic needs, meaning if they default or are otherwise unable to access that credit, they risk not being able to meet these needs.

Returning to the experience of the MTD La Matanza, their trajectory clearly illuminates this financialization of the urban periphery and the role that social movements played in it. As the country entered into economic recovery, the organization saw their membership decline as many people were able to find some sort of paid employment and therefore did not need to rely on the MTD for an income or other forms of support. Activists from another unemployed workers' organization, in a different zone of the urban periphery, saw a similar decline in membership, which they link to an increase in consumption, as well as drug use and violence in the urban periphery (interview 8 September 2012). They argue that the growth in consumption and violence corresponds to an increase in individualism and a disintegration of the social ties that made the unemployed workers' movements possible. This is similar to processes described by Teresa Caldeira (this volume) in which the autoconstruction of peripheral urbanization does not necessarily contest the logics of the state and capital, but rather interacts with them in *transversal ways*, and in many ways contributes to the privatization of common land.

The MTD La Matanza responded to this decline in membership by redoubling their efforts in their cooperative enterprises. In an effort to fund these projects, they began a partnership with the Banco Santander in 2012. They started by using their social centre to sign up neighbourhood residents for credit cards and then worked to open the bank's first branch in the neighbourhood. They also offered help with filling out forms and other banking activities to neighbourhood residents, and in 2016 they opened a Banco Santander call centre employing members of the organization. The bank branch is important because it means "avoiding walking kilometers to receive a payment, it offers banking culture and services to people with low literacy rates, thanks to employees that come from the neighborhood itself … It has fostered the formalization of small economic initiatives and also employment," Toty Flores stated. "Although we might not have formal employment, we can have access to credit cards, and this makes us equal to everyone else. It is a concrete tool for inclusion," (quoted in Vommaro, 2016) Flores stated on another occasion.

While this alliance with Banco Santander has helped to create jobs in the neighbourhood and allowed for an increase in consumption by making credit more available, there have also been unintended consequences. "They don't care about the neighborhood anymore. They don't care about dignity, they just want easy money," states Carlos, a former member of the organization (interview, March 23, La Matanza). Another organizer in a popular school in the neighbourhood describes how the new participants in the MTD, drawn in by the promises of access to credit, don't have a sense of solidarity or community, and cites this as the cause of the increase in robberies and other violence in the neighbourhood. Or as Vommaro (2016) states, describing one member of the MTD who is in charge of the organizations new financial ties and managing international donors: "This distances her from neighborhood social and political life. She isn't, according to her neighbors, a leader, like the social mediators of territorial movements. She doesn't mobilize people nor politicize collective demands." In other words, as the MTD becomes more involved in financial partnerships, its ability (or interest) both to politically intervene in the issues that most matter to neighbourhood residents and to serve as a site for the production of alternative practices and values diminishes.

While the case of the MTD La Matanza might be the most blatant example of social movement involvement in the proliferation of finance in the urban periphery, it does allow us to draw out some broader trends and processes. First, financial institutions (and state institutions) recognize the economic potential of those spaces, an untapped resource from which to extract value. Second, those institutions rely on the territorial knowledge and know-how of social movements in order to coordinate and prepare their insertion into those spaces. I will now examine three forms of the proliferation of finance in the urban periphery in more detail: microcredit, financialized welfare programs, and consumer finance.

Microcredit

Now I want to turn specifically to microcredit programs as another form of the financialization of popular economies. While not as widespread in Argentina as in other countries in Latin America, due to a pervasive "culture of work" tied to Peronism (Gago, 2017), a number of microcredit projects targeting the urban periphery were started during the crisis. No-interest or low-interest micro-credit first became available on a small-scale in the neighborhoods of the urban periphery in the late 1990s, originally funded by international NGOs and international financial institutions. Now, not only are international bodies involved in providing funding for micro-credit programs, but Argentinean NGOs and the national government do so as well. With the economic recovery these programs have multiplied as a way of implicating the poor in financial capitalism and controlling and capturing value from the myriad of alternative economic activities that emerged in the wake of the crisis.

The unemployed workers' organizations, besides often taking on micro-credit loans themselves to fund their own collective projects and small enterprises and cooperatives, have, in some cases, become involved in managing microcredit programs for the government and NGOs. The MTD La Matanza, for example, managed the microcredit project called the *Banco Popular de la Buena Fe*, funded by the Ministry of Social Development, in the neighbourhood of La Juanita. This program provides small loans to individuals (at first only to women but then it was expanded to include men) to start up or expand small business projects. The initial loans charge no interest but can be renewed with a 6% interest rate. The projects funded through the MTD La Matanza were often based in the participants' homes and included making and selling food or clothing or, even more frequently, reselling clothing or other products purchased cheaply in a central market. Participants were placed into groups of five people (segregated by gender) that then would be held collectively responsible for ensuring that everyone in their group repaid their loans. MTD members served as "promoters," who were trained by the government to facilitate the small groups and manage distributing the loans and collecting repayments in the neighbourhood. While some members complained of the administrative work involved in these tasks, the leadership of the MTD lauded the program as a way to promote the self-sufficiency of neighbourhood residents without relying on the state (despite the fact that the loans came from the state to begin with).

When some members left the MTD La Matanza to form the school Yo Sí Puedo (YSP), the people largely responsible for managing that micro-credit program went over to the school as well, abandoning their duties with the Banco Popular de la Buena Fe. YSP decided not to continue involvement in micro-credit programs, as they were very critical of the way the program had developed and the overall effect it had on the MTD. One of the women in charge of managing the program for the MTD La Matanza, who later joined YSP, explained her critique of the micro-credit program in this way: "It makes everything be about money, everyone's relationships are centered around money" (interview 23 September 2011, La Matanza). Another former MTD member critiqued the program for promoting individual solutions to unemployment as opposed to collective ones and not actually aiding in building the movement, since movement members were required to spend considerable time managing the programs. This further demonstrates how the introduction of finance disrupted the movements' neighbourhood organizing and its capacity to institute new social relations and values.

Additionally, by putting movements in the position of managers of these programs, it incorporates the movements themselves into the regime of governmentality and finance (Lazzarato, 2015), responsible for monitoring people's behaviour and producing compliant subjects who repay their debts. In this way, the programs rely on the relationships of trust and the communitarian knowledge developed by the social movements in order to be sustainable. Thus, the microcredit projects are capitalizing on that trust that the movements spent years building. One man, a former member of the MTD La Matanza, opposed that movement's use of microcredit for promoting financialization in the neighbourhood: "Even if these [micro-credit loans] don't charge high interest and put people in debt, they promote a culture of using credit for everything. Now [the MTD La Matanza] even distributes real credit cards" (interview, Sept. 23, 2011 La Matanza).

Micro-credit has been critiqued by various authors and social movements for functioning to incorporate the poor into relations and subjectivities of debt. For example, the Bolivian feminist collective Mujeres Creando is critical of the entire enterprise of micro-credit, arguing that it serves to transfer money from the informal sector to the formal sector with no benefit for the participants themselves. They see micro-credit as a method of capturing the value created by the social networks and practices that the poor, mostly women, create as survival mechanisms (Toro Ibáñez, 2010). Micro-credit programs aim to integrate these informal networks and survival mechanisms into a capitalist system. In this way, micro-credit serves as a mechanism to privatize and constrain women's reproductive work, capitalizing on women's informal relationships and introducing competition into otherwise cooperative mechanisms. Mezzadra and Neilson also see micro-credit as a form of capture: "The arrangements of micro-credit are one means by which the entire life of these masses is coded as "human capital" that should not be wasted (although it is often wasted) but rather compelled to generate value according to the logic of abstract labor" (2013, p. 93). The experiences of the MTDs with micro-credit programs validate these critiques and demonstrate the intimate relationship between these low-interest micro-loans and formal and informal forms of consumer credit, which I will discuss in more detail below.

Social Programs and Finance

If, during the period of neoliberalism and crisis of the 1990s and early 2000s, the territories of the urban periphery could be seen to be abandoned by the state as the government cut back on funding on welfare and other social programs, the economic recovery has seen the emergence of the state as a key actor in these territories (Gago et al., 2014). First responding to the demands of the unemployed workers' movements for unemployment benefits and other forms of aid to the poor and unemployed, the national government created a series of new social benefits packages, starting with the *Plan Jefes y Jefas de Hogar* (Heads of Household Plan) that provided a small monthly benefit to unemployed heads of households, which was complemented by provincial and municipal level programs providing similar types of aid. There have been intense debates in Argentina about the intentions and effects of these social benefits packages, with some movements maintaining that the benefits increase dependence on the state and foster clientelist relationships (c.f., Flores, 2005, Marcioni, 2010). However, the benefits also demonstrated a clear victory for the movements who demanded them; winning control over the distribution of benefits allowed movements to build their political power in specific neighbourhoods through determining what counted as "work," and also allowed organizations to collectivize the benefits and use them to fund cooperative endeavours.

In recent years, these social programs have undergone significant transformations. On one hand, they have become ever more massive, as well as seemingly permanent. If the Jefes y Jefas Program was proposed as a temporary response to a crisis of unemployment, the current national program Argentina Trabaja, is proposed more as a long-term jobs program, providing subsidies to cooperatives, recognizing the impossibility of a return to the full employment of previous decades (Gago, 2017). Other programs, such as the Asignación Universal Por Hijo, also aim to provide universal benefits to poor families due to what is now understood to be a permanent crisis of formal work. On the other hand, these new benefits increasingly operate through distinct forms of financialization: either distributing individual benefits through bank cards or providing loans to cooperatives and other productive enterprises that interpellate the activities of the popular and solidarity economy into the logic of debt, as I will discuss below.

Many of the social programs enacted during the Kirchner government directly attempted bring social movements and their productive activities into the financial system. Under the slogan "Your productive dream is the motor of all of that we do," one of the programs of Argentina Trabaja provides small, low-interest loans to cooperative businesses. These loans allow cooperatives to purchase supplies or equipment and are designed for medium or small-scale enterprises, including family groups, associations of producers, and recuperated enterprises, providing an important source of funding and financing for these enterprises. However, some have critiqued this program for serving as a new form of clientelism where only government supporters could receive funds (Marcioni, 2010). As other sources of funding dry up or are incorporated into these programs, the organizations that cannot or chose not to receive these funds are often forced to turn to commercial banks or other financial agencies for loans, as in the case of the MTD La Matanza. In turn, members of the MTD La Matanza thus argue that these benefit packages based on loans are only capable of creating precarious jobs and don't allow for the creation of sustainable cooperative enterprises (interview 12 September 2012).

These programs are also critiqued for incorporating cooperative enterprises into a logic of profit-seeking and expansion in order to pay back loans, rather than privileging solidarity and the well-being of the community. Others have complained about the rigidity of loans, saying that cooperatives are forced to adjust to the rhythm of repayments and lose much of their flexibility to respond to issues within the organization (for example, they cannot withhold a loan payment in order to help a movement member pay medical bills) or that to receive these benefits, cooperatives must institutionalize some form of hierarchical leadership roles. However, a member of a different social organization in La Matanza, emphasized that these programs provide access to important resources and can be still be used creatively by the movements to fund their collective projects.

Additionally, Gago et al. (2014) argue that these subsidies and social programs are a form of capturing popular knowledge:

> The social programs allow for the development of an intelligibility of the popular world profoundly disrupted by the mutations that have taken place since the 1990s and the 2001 crisis. It is a way of recording and classifying the modes of life that cannot be considered within the formal salaried world nor with the classic cannons with which the state operates. Consequently, it was necessary for the state to incorporate into its roster many officials coming from the movements and the social sciences. Their knowledge of their groups and their operative, territorial, and organizational knowledge are the foundation of a new dialogue (but also a system of exclusion).

It was the social movements, namely the movements of the unemployed, that made the heterogeneity of labour and diversity of economic and social practices in the urban periphery legible to the state through collectivizing them and translating them into the language of demands. Thus, the social programs, as well as forms of consumer credit, seek to capitalize on that labour of social cooperation and the knowledges in order to expand finance into the urban periphery.

Consumer Credit

Microcredit and financialized welfare are not the only forms of the financialization of the urban periphery. State assistance serves as a guarantee for other forms of credit, where proving you are a recipient of state benefits is enough to allow you access to other forms of credit. Gago and Mezzadra (2015) describe how this works in La Salada, the largest informal market in South America, in the urban periphery of Buenos Aires and a site where an immense array of goods and services are bought and sold. In the market itself, there are stands offering cash loans to purchase goods in the same market with the state subsidies serving as backing of those loans for people who do not have formal employment. Gago and Mezzadra describe this in more detail:

> In the belts of the periphery of Buenos Aires, there are the financiers that are established on the same premises where sports clothes and domestic appliances are sold. Only a stairway away, they offer loans for consumption meant to be spent in that same physical space. In turn, those immediate cash loans are obtained through a very precise accreditation: the beneficiary number that a person has for receiving a social benefits package or state subsidy, such that financial extraction is organized around sectors that do not have a capacity for solvency given by the traditional labor

> market and that, however, on being recognized as a subsidized population, the state accredits their inscription into the banking system. Thus, the financiers literally extract value from a set of activities, forms of cooperation and obligations to labor in the future, guaranteed by the state.

Thus, the social benefits programs are connected to financialization not only through their direct connection with banking institutions in order to distribute benefits, but they are also linked to an entire "shadow" economy of informal lending practices. As the previously cited Procelac report made clear, these cash loans take place outside of banking regulations and are often extremely predatory in nature, preying on people who, in many cases, would not qualify for more formal loans, and who often face dire consequences from this indebtedness, such as losing their homes or even the ability to purchase basic food items. Therefore, it is impossible to speak of completely separate spheres of the formal and informal economy or legal and illegal financial practices (Gago, 2017).

Recognizing the role of the social benefits packages as guarantees for other forms of credit opens the door for analysing other forms of finance in the urban periphery as well – the most important of these being the different forms of consumer credit available to low-income residents. As mentioned above, these range from the informal to the formal, the illegal to the legal, often without clear distinctions between categories. These forms of consumer credit include the credit cards offered by commercial banks and the option to purchase goods in quotas from large retail chains, as well as other forms of informal credit used to buy goods.

As Ariel Wilkis (2015) has demonstrated, nearly 40% of residents in the Buenos Aires slums have access to credit cards, even though a significantly smaller percentage are actual banking customers. In explaining this, Wilkis argues: "Lending of credit cards between individuals has become a common practice of reciprocity between those in the lower classes. The family members who have the documentation required to apply for a loan and can help other family members play an important role; credit cards circulate among relatives, friends, and acquaintances" (772). Thus, relationships between friends and family members take on an added role as they are what allow credit cards to circulate, as well as access to other forms of credit. Additionally, people often borrow directly from family members, friends, and neighbours in order to pay back other loans, leading to a complicated web of debt and credit in these neighbourhoods. This creates tension and often explicit violence when people neglect to pay back the debt owed to their family members or neighbours, or refuse to share their cards.

The increasing availability of different forms of credit is largely behind the increase in consumption in the low-income urban periphery of Buenos Aires and other Argentine cities. Without a parallel growth in formal wage labour, and with increasing inflation rates, individuals turn to credit in order to make purchases. Yet media narratives and government campaigns encourage spending and consumption. This consumption has led to a marked transformation in the quality of life in many of these neighbourhoods, for example, through increased access to time-saving household appliances and private transportation.

It is important to recognize that much of this consumer credit goes to purchase products and goods that are directly related to social reproduction: from domestic appliances, such as refrigerators and washing machines, to cars to the goods and motorcycles for transportation, clothing and toys for children or other goods involved in childcare, and sometimes even basic food or household items. Thus, in a sense, this credit is compensating for the state's withdrawal from services and activities of reproduction, and, in theory, alleviating women's burden. However, women also usually take on the burden of managing and repaying loans, and maintaining the social relations that allow them to access credit cards or rely on friends and family for additional loans. This opens these relationships up to distrust and, at times, violence.

This consumption has changed the quality of life in the urban periphery. Residents consistently comment that debt is necessary in order to live how they want to live. Yet, it has also changed social relations An organizer at a popular school describes an increase in robberies, which are sometimes violent: "Before there were places that were off limits – the schools, the social centers – or you would know they would leave the widows alone; now there are no codes, nowhere is off limits" (interview, October, 2015, La Matanza). The IIEP, formed after drug dealers attacked and burned the homes of various members of the MTD Solano, investigated the situation in the neighbourhood as well as its links to broader processes affecting the country. Linking that instance of violence against the MTD Solano and the indebted neighbours' unwillingness to intervene with similar instances of violence across the country, they discovered a complex web linking different forms of finance and violence, that ultimately directly attack the social fabric:

> The extractive character of the aforementioned businesses is a mark of up to what point the process of valorization underway requires subordinating and exploiting or directly destroying the existing social fabrics. Accumulation by dispossession is another way of referring to this business pattern

> that seizes common goods in exchange for astronomical monetary gains. The result is a social and communitarian disarticulation that, however, benefits with a general increase in incomes. This increase in the capacity for popular consumption does not mean, however, a horizon of collective enrichment or of social equality. (IIEP, 2014, pp. 5–6)

In other words, this financialization, by seeking to capture value from more and more areas of everyday life and social reproduction, also directly attacks the social fabric itself.

Conclusions

Thus, the microcredit programs and financialized government social plans served as entryways for an increased financial penetration into the urban periphery. A whole set of other both legal and illegal, formal and informal, forms of consumer credit also emerged. Often these are tied to immediate consumption – purchasing household appliances or electronic goods – yet other times they are for buying land and houses. Sometimes credit is used to start or expand a small business, again often tied to informal markets and the other forms of credit that those entail. While some of these forms of credit are tied to major banking and financial institutions, others are supported by money from illegal trade, and many of them escape regulation and safeguards for the indebted. The use of credit to resolve many everyday life issues means that people are less reliant on social movements to resolve those same needs, in some cases leading to less participation in those movements. Additionally, the increasing incorporation of movements themselves into financial operations have limited much of their political power, forcing them to instead focus on making loan payments on time, managing microcredit programs, making business decisions, and producing obedient financialized subjects instead of focusing their attention on other aspects of movement building.

The experience of the MTD La Matanza shows how financialization becomes embedded in the everyday lives of residents of the urban periphery through the popular economy practices organized and fostered by social movements themselves. If the social and solidarity economy activities, and especially the alternative practices of social reproduction, were what allowed the poor and unemployed to survive the economic crisis, they were also what fuelled the nation's economic recovery. Both government programs and financial institutions have sought to incorporate these initiatives in order to be able to extract profit from them. Targeting activities of social reproduction as the objects of credit allows value to be extracted from those activities directly without the mediation of a wage.

The growth in consumption, fuelled by credit, means that those parts of the urban periphery, previously considered marginal, as the dumping ground for surplus populations, or only occasionally considered useful as the source of cheap labour or service providers, have now become the central battleground for a new expansion of capitalist relations. Examining the increase in consumption fuelled by credit points to an increase in exploitation and dispossession, forms of extracting value from more and more spaces and times of everyday life. This spread of finance is intimately linked to the alternative economic practices and autonomous forms of social reproduction created by social movements of the poor and unemployed during the crisis. It is an attempt to incorporate these activities that exist on the margins of capital, that are organized externally to capital, into the dominant system in order to be able to more efficiently extract value from them. In this regard, financialization can be seen as a form of *extraction* paralleling natural resource extraction (Gago & Mezzadra, 2015).

However, despite these attempts to incorporate the everyday practices of social reproduction and popular economies into the dominant system of capitalist extraction and exploitation, resistance to the financialization of everyday life continues. It will be necessary to conduct further research on this resistance as it rarely takes the form of the large, organized social movements or protests, as it did in early 2000s. Despite all the attempts to produce an obedient financialized subject, one who repays his or her debts, still much resistance exists to this subjectivation on the part of Argentines who still remember the 2001 collapse. For example, the Grupo Elektra, a credit agency and domestic appliance companies, the subsidiary of the Mexican Banco Azteca, which targets loans toward low-income communities and informal workers who would not otherwise have access to credit, opened its first branch in La Matanza in 2007. In 2013, they decided to stop doing business in Argentina, denouncing certain government macroeconomic policies as well as the existence of "a culture of not paying debts that makes the credit business unfeasible" in Argentina.[1] This "culture of not paying debts" is still apparent in much of Argentine society, especially in the urban periphery, where significant distrust of the financial system remains. It is these forms of resistance to becoming "indebted subjects" and the tactical subversion of finance that require more investigation in order to understand current transformations in Buenos Aires's urban periphery.

NOTE

1 http://rpp.pe/economia/economia/grupo-elektra-abandona-argentina-por-entorno-inviable-para-negocios-noticia-635962.

REFERENCES

Brighenti, M., & Mezzadra, S. (2013, December 9). Extractivismo y política de Lo Común: Entrevista a Sandro Mezzadra. *Lobo Suelto!* http://anarquiacoronada.blogspot.com/2013/12/serie-nuevo-conflicto-social.html (Accessed: 01 August 2018)

Colectivo Situaciones. (2012). *19 & 20: Notes for a new social protagonism.* New York: Minor Compositions.

Colectivo Situaciones. (2014). Crisis, governmentality and new social conflict: Argentina as a laboratory. *ephemera: Theory and Politics in Organization, 14*(3), 395–409.

Dalla Costa, G.F. (2008). *The work of love: Unpaid housework and sexual violence at the dawn of the 21st century.* Brooklyn, NY: Autonomedia.

Feldman, G. (2013). *Créditos para el consume: Análisis del fenómeno socioeconómico y su impacto en los sectores populares.* Buenos Aires: Procuraduría de Criminalidad Económica y Lavado de Activos.

Flores, T. (2005). De la culpa a la autogestión, Aclaraciones preliminares. In T. Flores (Ed.), *De la culpa a la autogestión: Un recorrido del Movimiento de Trabajadores Desocupados de La Matanza* (pp. 13–46). Buenos Aires: MTD Editora.

Gago, V. (2015). Financialization of popular life and the extractive operations of capital: A perspective from Argentina. (Liz Mason-Deese, Trans.). *South Atlantic Quarterly, 114* (1), 11–28. https://doi.org/10.1215/00382876-2831257

Gago, V. (2017). *Neoliberalism from below: Baroque economies and popular pragmatics.* (Liz Mason-Deese, Trans.). Durham: Duke University Press.

Gago, V., & Mezzadra, S. (2017). A critique of the extractive operations of capital: Toward an expanded concept of extractivism. (Liz Mason-Deese, Trans.). *Rethinking Marxism, 29*(4): 574–591. https://doi.org/10.1080/08935696.2017.1417087

Gago, V., Mezzadra, S., Scolnik, S., & Sztulwark, D. (2014). ¿Hay Una Nueva Forma-Estado? Apuntes Latinoamericanos. *Utopía Y Praxis Latinoamericana: Revista Internacional de Filosofía Iberoamericana Y Teoría Social, 19*(66), 177–84.

Giarracca, N. (2008). Producción y mercados para la vida: Una posibilidad emancipadora para el siglo XXI. In N. Giarracca & G. Massuh (Eds.), *El trabajo por venir: Autogestión y emancipación social* (pp. 36–41). Buenos Aires: GEMSAL: Editorial Antropofagia.

Instituto de Investigación y Experimentación Política. (2014). *Apuntes Del Nuevo Conflicto Social.* Buenos Aires: Tinta Limón.

Lazzarato, M. (2015). *Governing by debt.* South Pasadena, CA: Semiotext(e).

Marcioni, N. (2010). La lucha por lo público de las políticas: El caso del Programa Argentina Trabaja. In Centro de Estudios para el Cambio Social (Eds.), *Pensamiento critico, organizacion y cambio social: De la critica de la economia politica a la economia politica de los trabajadores y las trabajadores* (pp. 159–166). La Plata, Argentina: Centro de Estudios para el Cambio Social; Editorial El Colectivo.

Mason-Deese, L. (2017). Unemployed workers' movements and the territory of social reproduction. *Journal of Resistance Studies, 2,* 65–99.

North, P. (2005). Scaling alternative economic practices? Some lessons from alternative currencies. *Transactions of the Institute of British Geographers, 30*(2), 221–33. https://doi.org/10.1111/j.1475-5661.2005.00162.x

Stratta, F., & Barrera, M. (2009). *El tizón encendido: Protesta social, conflicto y territorio en la Argentina de la posdictadura.* Buenos Aires: El Colectivo.

Svampa, M. (2015). Commodities consensus: Neoextractivism and enclosure of the commons in Latin America. (Liz Mason-Deese, Trans.). *South Atlantic Quarterly, 114*(1), 65–82. https://doi.org/10.1215/00382876-2831290

Toro Ibáñez, G. (2010). *La pobreza, un gran negocio: Un análisis crítico sobre oeneges, microfinancieras y banca.* La Paz, Bolivia: Oficina contra la Usura Bancaria; Mujeres Creando.

Vommaro, G. (2016, November 29). *Pro-mediadores de lo sensible.* Revista Crisis. http://www.revistacrisis.com.ar/notas/pro-mediadores-de-lo-sensible (Accessed: 1 August 2018)

Wilkis, A. (2015). The moral performativity of credit and debt in the slums of Buenos Aires. *Cultural Studies, 29*(5–6), 760–80. https://doi.org/10.1080/09502386.2015.1017143

Zibechi, R. (2008). El trabajo libre contra la economía política. In N. Giarracca & G. Massuh (Eds.), *El trabajo por venir: Autogestión y emancipación social* (pp. 173–179). Buenos Aires: GEMSAL: Editorial Antropofagia.

Zibechi, R. (2012). *Territories in resistance: A cartography of Latin American social movements.* (Ramor Ryan, Trans.). Oakland, CA: AK Press.

PART 3

Community, Commoning, and Political Agency on the Urban Margins

7 The Self-Built City as Palimpsest: (Re)Constructing Urban Memory in Lima's Hybrid Peripheries

KATHRIN GOLDA-PONGRATZ

In memory of Juan Tokeshi (1955–2013), "the barefoot architect."

I am in Lima, that immense town, head of the false Wiraqucha.
In Pampa de Comas, on the sand, with my tears,
with my strength, with my blood, singing, I built a house.
The river of my town, its shadow, its great wooden cross,
the grasslands and flowering shrubs surrounding it,
are there, alive inside that house;
a golden hummingbird plays in the air, above the roof.
– José María Arguedas, *Katatay y otros poemas huc jayllicunapas*[1]

Introduction

Contemporary Lima gives the impression of an endless city spreading along the Pacific coast: low-density residential areas sprawl beyond a new skyline of skyscrapers, form nodes in the arid hills, and finally fade into the desert. At present, more than 10 million of Peru's total population of 32 million live in greater Lima, 60% of them in the more or less consolidated self-built or non-formal city made up by the cone-shaped extensions Cono Norte, Cono Sur, and Cono Este.[2] Close examination of the metropolitan expansion along the milestones on the Pan-American highway, bearing in mind the predictions of some local urban planners, would suggest that greater Lima as a macro-region and magnet of migration will gradually become an urban agglomeration that will grow to some 400 kilometres in length over the next few decades. The metropolitan area would then stretch from the northern city of Barranca to Ica in the south, and extend way beyond the so-called Cono Norte and Cono Sur that now form the metropolitan area (Golda-Pongratz, 2015, pp. 32f.).

The recent impact of the COVID-19 pandemic, though, visibilized the extreme fragility in which a major part of Lima's population lives, working in the informal sector and living from hand to mouth, and showed first signs of a possible inversion of that phenomenon: people have massively started to return to their villages of origin in rural areas – a trend that might persist and retard the growth curve in the long run.

The city of Lima has perhaps a more conflictive relationship with its memory and definition of identity than other cities do. On the one hand, it cultivates its colonial legacy, without establishing institutions really capable of protecting its physical heritage and of overcoming social structures that are still defined by a strong rejection of the population with Indigenous roots by the dominating white and mestizo population (Golda-Pongratz, 2016, p. 4). At the same time, its urban development and discourse – be it in the debate about the historic city centre, in the development of modern suburbs and the creation of non-formal self-built settlements since the 1940s, and in current vertical growth and infrastructural expansions – has always turned its back on its pre-Hispanic heritage. Contemporary planning entities and cultural institutions fail to assume their responsibility for protecting it as value for mankind and neglect its integration and urban-landscaping potential and cultural value for the city. The denial and the widely tolerated destruction of pre-Hispanic heritage is, in fact, a form of violence against people's roots, which makes it difficult for them to identify with the territory they inhabit. Recent valuable initiatives and debates about the millennial past of the city[3] and the rediscovery of *huacas*[4] as tourist attractions still need to look much more at the integrative potential of this largely neglected urban heritage, especially in the self-built peripheries.

At the same time, Lima, as most Latin-American cities, embodies social inequality and conflicts alongside hybrid or unfinished urban identities[5] and multiple forms of non-formal urbanism. For more than half a century, the so-called self-built city has been the predominant form of housing production in Peru. Having emerged without planning, through occupation and collective building processes, only later on did it acquire legitimacy and infrastructure (cf. Caldeira, 2016). While such urban development from below is recognized in its potential for place-making and as an effective response to failed housing policies, its unwritten history and memory is still unexplored and underestimated. Yet, collective remembrance and the enunciation of collective production have the power to drive neighbourhood renewal, sustainable consolidation, and resilient development in a fundamental way. This is developed as the key thesis of the text.[6] Here, also a hypothesis comes into place: In spite of the hard layers of commercialization and modernization that cover the current urban agglomeration of Lima, its inhabitants will sooner or later lay claim to the memory of their origins, especially in the self-built neighbourhoods.

The chapter opens its argument by presenting Lima and its peripheries as a macro-region with new migrational pressure but identifies a decreased spirit of the collective identity that the generation of "invaders" used to have, a loss of interest in its origins and struggles and territorial attachment. It reviews the genesis of the formation of the *barriadas*[7] and related policies until the present, based on the revision of the legacy of the British architect John F.C. Turner, an important theorist of the self-built, to which I have been committed for over a decade. Two major issues challenge the process of consolidation and integration of the self-built settlements within the current context of a globalized hybridity and of new urban economies: a massive increase in land speculation[8] through the illegal subdivision of remaining empty lands and land trafficking, especially in the urban fringes, natural reserves, and fragile territories; and a simultaneous rise in crime levels and a dismantling of civic solidarity and traditionally strong community ties.[9] This is especially worrying in a context where no metropolitan development plan for the Peruvian capital has normative validity at present.[10] Therefore, there are no adequate planning responses and tools to deal with the spatial and social challenges of the extended macro-region that Lima is today.

The argument is developed using the case of Lima's Cono Norte, its district of Independencia and two specific, originally self-built areas within it: El Ermitaño and Pampa de Cueva, where I have been developing projects, since 2013, that specifically deal with the reactivation of local memory.[11] A longitudinal study there is based on, and acts a kind of follow-up to, John F.C. Turner's research and documentation of these very same places in the early 1960s. Instead of producing a new sense of identity, the area's current hybrid character is increasing a loss of the feeling of permanence and belonging. This weakened sense of community comes along in a place where valuable pre-Hispanic heritage is present, but not recognized as such. However, I argue that this heritage has the potential to serve as a motor for urban integration when it is understood as a palimpsest: just as a manuscript or glassine in ancient Greece on which inscriptions were erased in order to give space to new ones on the same surface, nevertheless the traces of the previous ones

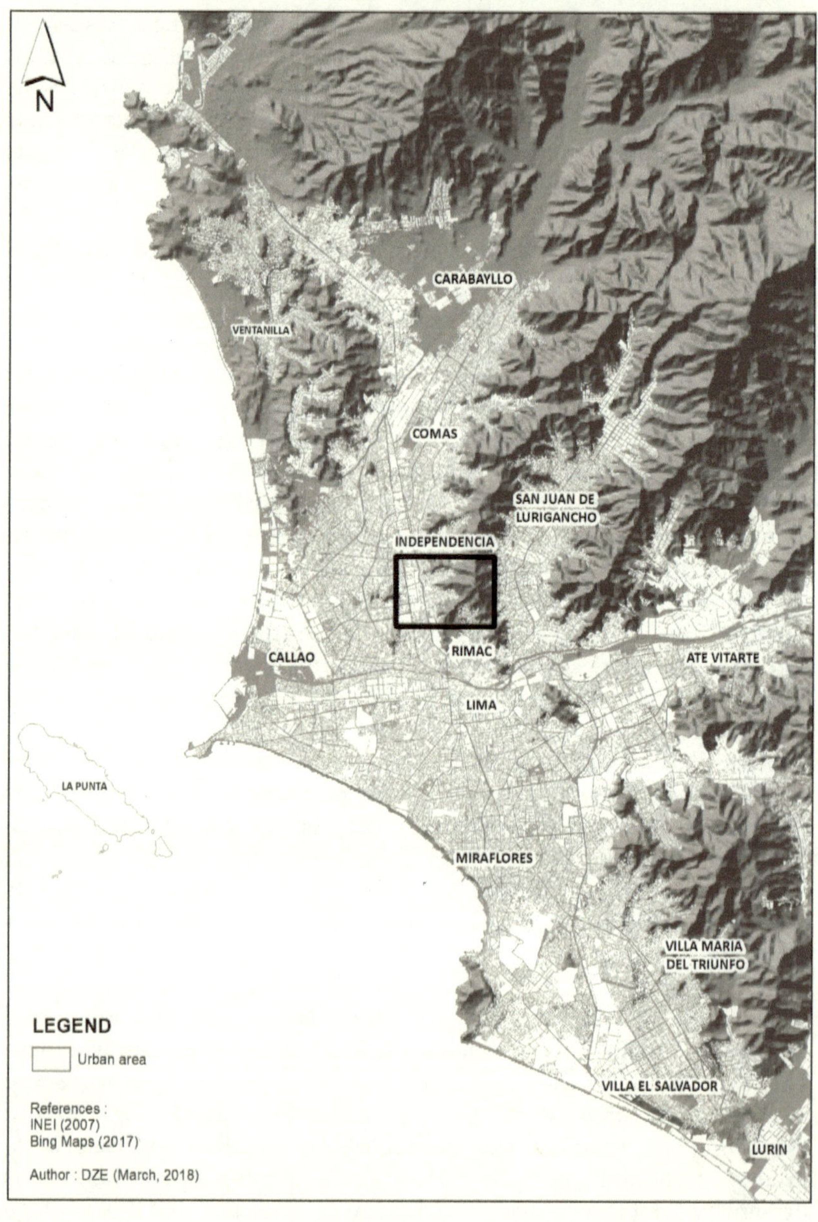

Figure 7.1 Location of the Pampa de Cueva and El Ermitaño neighbourhoods within the metropolitan area of Lima, Peru. Map by Dayan Zussner, 2018.

remain, and the document acquires more meanings and density as more layers are added to it. In a similar way, territories and urban contexts hold such multi-layered traces and inscriptions.[12] Strengthening these layers within a broader concept of community and public space renewal, giving vernacular layers of the self-built similar importance as traces of historic inscriptions, and connecting and reactivating them is a primary concept developed here and described in the case study.

The chapter argues, furthermore, that the identity of this macro region has to be understood from a territorial perspective: It is possible to reconstruct a new community spirit through the activation and reconstruction of urban memory, the connection of vernacular memory with pre-Hispanic sites, and systems of pathways and territorial linkages, and to engage a population with primarily Indigenous roots in a palimpsestic cohabitation that creates new forms of cultural hybridity and a renewed sense for place-making.[13]

Which identities inhabit this macro-region today? By acknowledging that they are multiple and multi-layered, what can we learn? How can we contribute to strengthen these layers, help develop new emancipatory urban platforms, and enhance interventions in the territory capable of creating spaces of identification? With a population composed primarily of originally rural migrants and their descendants of a second and third generation, what kinds of rural urbanities, or maybe urban ruralities, do we find? Do we have to speak of a loss of territorial knowledge, both in the rapidly urbanized desert landscape of Lima and in the abandoned places of origin? Can we reconstruct this knowledge out of the notion of the urban space as a palimpsest? And how are the origins of the barriadas, the non-formally built young towns, remembered in the contemporary inhabitants' perspectives? Which collective memories accompany the processes of consolidation beyond the stigmata which still persist in their perception?

In order to answer these questions, we will look at the already mentioned case study in the northern periphery of Lima, the so-called Cono Norte, originally mainly self-built since the 1960s. This is where the British architect John F.C. Turner studied informal housing production and strategies of self-organization; he finally theorized about the potential of these non-formal settlements as "dwelling resources in South America" (Turner, 1963, p. 375) and wrote the influential *Housing as a verb* (Turner & Fichter, 1972). Turner at the time describes the making of housing as a process and an activity that corresponds to the real needs of its inhabitants in each step of their personal biography and development. The documentary film *A Roof of My Own*,[14] made in 1964, which portrays the formation of the neighbourhoods Pampa de Cueva and El

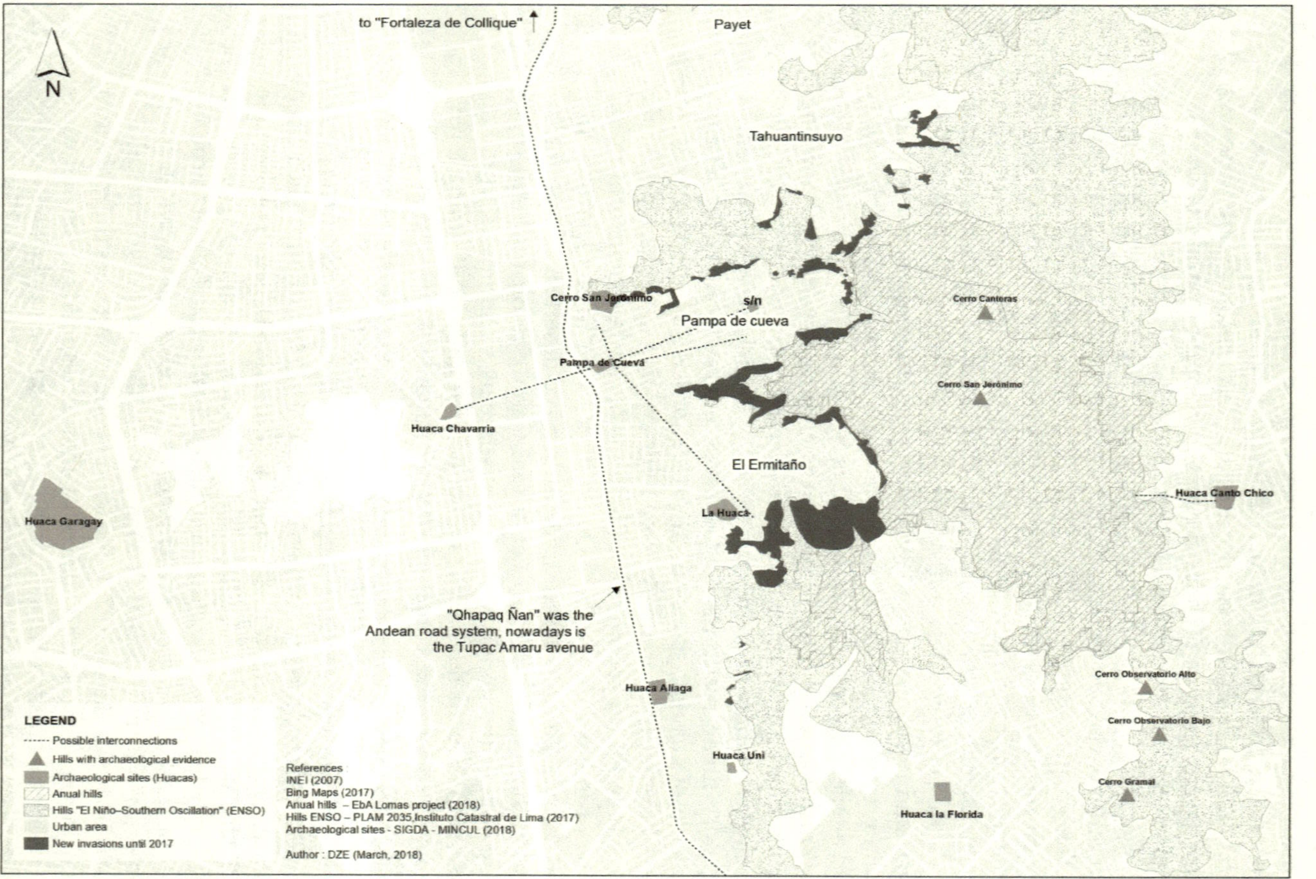

Figure 7.2 Pre-Hispanic sites, their possible connections, and the Lomas natural heritage in the area, endangered by the current ongoing urban expansion through recent invasions. Map by Dayan Zussner, 2018.

Ermitaño, is the basis for a research and audiovisual follow-up project I have recently been directing.[15] By activating memories and unwritten city-making knowledge through the projection of the historic film document and related workshops and interviews, the local response and perception are outlined and analysed, in order to find urbanistic, artistic, and civil formulas to depict their traces in the urban space, protect heritage from below, and determine new forms of citizenship and identity beyond the periphery. The making of a new documentary film acts as a tool of memory activation and will hopefully become a powerful means to communicate the contemporary needs and challenges of those historic barriadas to the authorities and to a broader audience.[16]

The making of the documentary film and related research aims at the following four crucial steps: first, to formulate frameworks towards a territorial identity and place-making in emerging neighbourhoods and in the self-built city; second, to understand the value of the unwritten urban history and its potential for collective identification and culturally rooted upgrading processes; third, to activate and set into value of urban memory and development of tools and methods for its participatory translation into public urban space; and finally, to support residents as responsible actors and protectors of the cultural heritage and of the ecological equilibrium of their habitat and landscape they live in.

Lima: The Genesis of a "Mega-Barriada-Metropolis"[17]

Lima's dominance and the roots of centralism in Peru can be traced back to colonial times, when the city was the seat of political and ecclesiastical power and a centre of culture and knowledge. The promise of progress, health, education, work, and better living conditions attracted and continues to attract many people from rural areas to the capital. A failed attempt at agrarian reform from the late '60s onwards, increasing inequality between urban and rural areas, the terrorism of the Maoist Shining Path movement (*Sendero Luminoso*), and the violent counterattacks by the armed forces and the police in the '80s and '90s led to the displacement of tens of thousands of people from the Peruvian Andes, particularly the Ayacucho area and the Central Highlands (Golda-Pongratz, 2004, p. 40). Twentieth-century and early twenty-first-century migration was and is also driven by push factors such as the social and economic marginalization in Amazonian and Andean areas, the damage caused by natural disasters, earthquakes, flash floods, and the consequences of the El Niño phenomenon in the '90s. The floods in 2017 in the north of Peru, the on-going exploitation of mining territories in the Andes, and the progressive destruction of livelihoods in the rainforest might enhance other migrational waves towards the capital in the near future.

Since the '40s, the migration of rural populations has been the most decisive factor of life in Peru. From 1940 to 2000, Lima's population has increased from 645,000 to 7.5 million inhabitants, and its urbanized land area has multiplied by a factor of almost sixteen, from 5,000 to approximately 78,000 hectares. "The barriada that was scorned in the fifties by an elite that called for its violent expulsion, and was later accepted and grudgingly integrated, has now, less than fifty years later, led to a total blurring of the boundaries between the formal and informal city" (Ludeña Urquizo, 2012, p. 4).

Peru´s politics towards the barriada have changed radically over the last 50 years, from providing a solid legal framework and recognition in the sixties to a lack of political attention and the favouring of speculation and land grabbing based on neoliberal politics since the 1990s. The creation of the historic *Barriadas Law* (Law 13517) in 1961 recognized the unplanned occupation of public and private land in urban fringe areas, provided a legal framework, and brought infrastructures and technical assistance into the self-construction process (Calderón Cockburn, 2001). The political step suggested by architect John F.C. Turner who worked as consultant in Peru at the time – "the idea was that everything should be a political program"[18] – was a kind of "mixed formula with state support for the self-construction sector:"[19]

> The moral [...] is very simple: it is that the proper role of government is to ensure that those who are best able to build, either for themselves or for their neighbours, have access to the tools or basic resources for the job. By "tools" I mean the supports and resources of the kind that the Peruvian government provided for the people of El Ermitaño: land, or titles to land illegally but rationally occupied, technical assistance for the economic design of subdivisions and dwellings. (Turner, 1976, p. 33)

In 1971, with Juan Velasco Alvarado's military government in power, the first case of an urban scheme for a future self-build city was strategically planned after a violent and massive occupation of land south of Lima. Villa El Salvador, an urban settlement for 50,000 inhabitants, was the government's way of showing its concern for poverty and its active involvement in social reform and modernization. Civil servants working for SINAMOS[20] provided water, health care, and public transport links to Lima. Since then, this urban prototype has been more or less promoted as a model by the state. In 1979, the name of the barriadas was officially changed to *pueblos jóvenes* (young towns) and the areas with titled lots were recognized as ordinary districts with municipal rights and the responsibilities that went with them.

Coinciding with the rise of neoliberal ideas, the '80s were a decade of unprecedented ethical, social, and economic decline in Peru. The inequality and persistent poverty of rural areas paved the way for the emergence of the Maoist terrorist movement Shining Path (*Sendero Luminoso*), which brought the country to a standstill with extreme and almost archaic violence. This triggered state and paramilitary counter-violence that left almost 70,000 dead and sowed mistrust and trauma that still persist to this day, particularly among the rural population.[21]

The increasing informalization of the Peruvian economy during those years influenced the ideas of the economist Hernando de Soto, who described informality as "the third way" in his book *El otro sendero* ("The Other Path," in reference to Shining Path). He maintained that the self-built young towns would be the key actor in a new, emerging, popular capitalism and the engine for the creation of urban employment, as well as a place of enormous potential due to its invisible, dormant, property capital. His main argument was to formalize the slum housing and bring it into the market so that the poor would be able to access credit (and thus become part of an interest-based financial economy) (De Soto, 1986).

Alberto Fujimori's dictatorial, neoliberal government (1990–2000), which De Soto advised on economic matters, promoted a new stage of informal, self-built suburban growth. The first stage was to get rid of the whole state apparatus linked to housing that had been created in 1960. The 1993 constitution removed the right to housing, along with the Ministry of Housing, leaving the sector completely in the hands of the market and the private sector. COFOPRI, the commission for the formalization of state property, was set up and run with a loan from the World Bank. As a populist instrument for the centralization of power, it took over the issuing of property titles, which was previously carried out by the municipalities. "The Ministry of the Presidency issued over one million property titles and indiscriminate credit through the Bank of Materials, and set up water and power networks financed by the Fondo Nacional de la Vivienda (FONAVI)" (Calderón Cockburn, 2013, p. 52). The government itself incited social uprisings and occupations in order to then resettle people in state-owned desert areas and give them property titles, thus demonstrating its social commitment and at the same time win the votes of the poor. The paternalistic dependency cultivated during this political era certainly left the most destructive traces on community building and civic engagement until today.

The lack of political attention, corruption within municipalities, and a generalized absence of urban planning and regulatory instruments

Figure 7.3 Illegal subdivision of vacant land in Lima Norte mainly enhanced by land trafficking. Photo by Kathrin Golda-Pongratz, 2016.

is persistently favouring speculation in an indiscriminate way. The criticisms of the "COFOPRI phenomenon," as it is widely named, and which had been taken up as a model by other developing countries, are diverse: that it is primarily a cheap and easy strategy that avoids any serious attempt at a redistribution policy; that access to land is too easy and favours speculation; that more titled land pushes up prices and thus restricts the possibilities of acquiring titled property and decreases the means of accessing land on the market.[22] What seems most objectionable in these polices is the fact that by favouring speculation and land grabbing in peripheral areas, they undermine consolidation processes, the chance to develop a common vision and to build community and identity. This will be shown in the specific case of the neighbourhoods of El Ermitaño and Pampa de Cueva in Lima Norte, both having their origin in collectively organized land occupations of privately-owned land in the early 1960s and facing massive new invasions by mafia-like land trafficking organizations in 2017.[23]

Lima Norte in the Context of Metropolitan and Cultural Restructuring

The Rise of a New Urban Economy

Lima Norte began out of self-built settlements set up by workers' and family associations that John F.C. Turner initially portrayed in a special edition of Architectural Design in 1963 (Turner, 1963). It gradually consolidated and grew in density, and now has a population of 2 million. Since the late 1990s, Lima's Cono Norte has developed into a fast-growing commercial district, generating a new centrality. At the start of the twenty-first century, it is a new metropolitan centre in its own right, with its own motivations and aspirations and its own economy. It is thus a major decentralization force and a sign that Lima is gradually moving away from the typical centralized city model and becoming a multi-centre metropolitan zone. The values of the people who live in the Cono Norte at the start of the twenty-first century are shaped by modern elements such as Internet access and technological equipment and by a strong shift towards mass consumption and leisure activities.

In the Independencia district, where El Ermitaño and Pampa de Cueva are located, 93% of the 3,500 companies (surveyed in 1996) were registered as a micro-enterprise employing between one and four people, and generated intensive activity at the local and regional levels. Many of them are family-owned companies that provide work for members of the immediate or extended family and, in most cases, are directly linked to housing (Williams, 2005, p. 52), in the sense that these small businesses are mainly located in the self-built and properly adapted houses themselves. Homes have been converted into productive spaces and in that sense provide all the flexibility and adaptability that distinguishes them from formal housing schemes.

Along the Tupac Amaru Avenue, which connects the Independencia district with the core city – since 2013 with the BRT (Bus Rapid Transit) system *Metropolitano* – the colourful façades of the new shopping malls, building supply stores, tailors, hairdressers, restaurants and nightclubs, private clinics, cyber cafes, gas stations and garages make it hard to believe that life there started from scratch only forty years ago with an invasion of desert dunes. It seems to prove the thesis formulated by John F.C. Turner and William Mangin, according to which progress-driven migrants would significantly determine the urban economy of future generations (Turner & Mangin, 1969).

Lima Norte is the place where most cash turns over in the entire country.[24] An initial consolidation of night spots, clubs, casinos, card rooms, and amusement arcades like Bulevard El Retablo was followed in the late '90s by the advent of large shopping malls such as Mega Plaza Norte in 2003, which, 15 years on, is the most successful centre of its kind in greater Lima and a point of reference for all citizens. Plaza Lima Norte, another mall that opened in 2009, covers a surface area of 200,000 m^2, making it the largest in the country to date. There is a largely positive response to this phenomenon. Citizens, especially urban middle classes, use the malls extensively in an accelerated pace

Figure 7.4 The Mega Plaza Norte and other commercial centres have completely changed the face of the area and indicate its important economic activity. Photo by Kathrin Golda-Pongratz, 2010.

of consumption. The low-income population from the immediate surroundings sees them as a major sign of economic improvement of their neighbourhood: even if their own income does not allow them to spend money there, people are given the possibility of spending time in such spaces without being discriminated against, as often happens in wealthier neighbourhoods. Also, urban scholars acknowledge the "power of the malls" and their capacity to reaffirm identities:

"San Juan de Lurigancho, Independencia, Comas, and other areas where the Lima's poor masses live, proudly boast enormous supermarkets that have changed the nature of the city. They are not just shopping centres, malls ... On the contrary, they are a new kind of dynamic based on reaffirming values, creating relationships, and reconsidering habits and lifestyles, which will shape the average inhabitant of Lima over the next few years" (Tokeshi & Zolezzi, 2003, p. 82).

Deciphering Identities and Globalized Urban Imaginaries

The fusion of the global and the local is blurring the boundaries between the formal and informal economy and has clearly generated a new urban conglomerate that is no longer based on the traditional distinction between centre and periphery, which gives the periphery the role of marginal and of constant dependency from the consolidated centre. Undoubtedly, the "counter power from below" that Raul Zibechi attributes to the urban peripheries from a social point of view (Zibechi, 2007, p. 21), has also translated into economic terms. Urban dreams and opportunities are manifold, and they are fuelled by the snares and possibilities of free trade treaties, of the challenges of greater access to the world, of global opportunities, transnational credit, and migrant remittances sent from abroad.

At the same time, new forms of segregation have appeared in the residential areas of the emerging city in response to the increase in social diversification and in the population's purchasing power. The creation of gated communities or compounds in Lima's North, South and East *conos* reflects rising levels of crime and has increased the already high fragmentation and segregation of urban space in Lima in general. At the same time, in recently invaded territories on the slopes of the hills, poverty is structural, and social decay is a major issue in the educational centres nearby. Such is the case in Lima Norte and specifically in El Ermitaño and Pampa de Cueva.

The demographics of Lima have kept pace with its rate of growth: while at the turn of the century 75% of residents had not been born in the city, the figure is now down to around 50%. A third and fourth generation born in the capital to immigrant parents and grandparents are now forming a new urban middle class, with migrant backgrounds struggling with their own identity of being, or not being, citizens of Lima. And as families outgrow parental homes, there is a tendency to occupy and gradually urbanize vacant land in existing urbanizations, resulting in deterioration of public space and of quality of life in the neighbourhoods.

Over the last decade, in conjunction with a real estate boom, the growing shortage of available land, an increasing housing demand, and the rise of land prices all over greater Lima have led to a significant process of vertical densification in the districts of higher income, partly also in North and South Lima. There is a new shift towards tenancy in self-built settlements as a way of adding to the family income. This shift is based on living experiences of family members abroad and on the remittances sent from abroad. In a study carried out by the Observatorio Socioeconómico Laboral (OSEL) in Lima Norte, 86.2% of households surveyed reported having one family member abroad, 13.5% said they had two family members abroad, and almost 0.3% of households are known to have three family members abroad.[25] The remittances sent home by Peruvians living in Argentina, Chile, Europe, and the United States have significantly driven the completion of semi-finished houses and the vertical extension of existing houses for the rental market or to start up small businesses. At the same time, these multi-locational and transnational families have expanded the frame of cultural references.

The concept of a "new adaptive culture" is used by José Matos Mar already in the 1980s, when describing the activities of cultural associations of migrants and especially their "transferred cultural system" with regards to music and dancing in the new context of Lima (Matos Mar, 1984, p. 83). A "positive adaptation," defined as the capacity to build community in a material and a spiritual sense, the feeling of being able to improve the living quality in comparison to the abandoned life and place and the feeling of being able to construct a future (Lobo, 1984, pp. 17f.) is followed by being forged into a kind of "hybrid culture" (Garcia Canclini, 1995) with many connections and places of reference. But even so, the definition of identity still refers to old traditions, which shape the moods and situations of life, although it is getting increasingly difficult to decipher the points of reference within a merge of values that defines the self-built city or *ciudad popular*. Today's third- and fourth-generation migrants often get together and create associations,[26] which effectively maintain links with their homeland. Andean traditions of a shared economy and mutual aid endure, and often lead to the creation of self-employment in the new urban context. In fact, a mix of rural customs and habits, the assimilation of urban life, and global cultural influences have created a new culture of the *chicha*[27] city, as a hybrid of styles and influences that permeates all social classes and all parts of the metropolis.

Against this background of hybrid or merged values, a booming modernization, and an apparent increase of wealth is determining the self-built city today, even while it is accompanied by persistent marginality and inequality, and a growing crisis of identity, belonging, and citizenry. The reasons for and expressions of it are manifold: first, the coverage of the basic needs requires less community engagement; the years of terrorism under *Sendero Luminoso* and posterior politics of paternalism have weakened social networks and ties decisively; people mistrust each other as crime rates have gone up and the public sphere has become very unsafe; newcomers from outside who settle in the new extensions produced by land trafficking are considered as intruders; a large number of young people without higher education and with limited employment prospects show a weakened sense of belonging and search for opportunities elsewhere.[28] A generation of teachers and social workers in Lima Norte is now looking for means by which to strengthen identity. In that context, Acuña et al observe a lack of identity, that people do not feel part, and that citizens need to have a stronger identification and commitment to the development of their district.

And they ask how such an identification could be achieved (Acuña Damiano et al., 2010, p. 9).

In the consolidated informal city, it is clearly time now to establish cultural references, moorings that reconnect citizens to their cultures of origin, and to create tools for conviviality and mutual respect that go beyond what Ivan Illich calls "survival conditions" (Illich, 1973, p. 13) and that respond to environmental fears and the increasingly superficial knowledge of the land.

And while urban consolidation and the economic boom also strengthen the rise of a new middle class of Andean background, current research and related interviews have shown that there is a strong desire and, in fact, even a need to remember the collective strength of the origins of these originally self-built neighbourhoods, in order to strengthen contemporary community ties and to foster place attachment, a sense of belonging and self-understanding that might guarantee a more sustainable urban development as it is controlled and actively influenced by the dwellers themselves.

The Role of Urban Memory in Lima Norte – A Longitudinal Study of Pampa de Cueva, El Ermitaño, and "La Bella Durmiente"

Oftentimes collective memory is lost when it is not looked after, especially in the case of non-monumental urban spaces, or when we talk about spaces that are subjected to urban transformations. This is the case both for the memory of the achievements of former generations to build up the self-built city and to consolidate it, as well as for ancient territorial connections and a collective knowledge of the living environments. The connection of both – the fundamental approach of the work in progress – can engender new dynamics of its reactivation. Scattered local initiatives to rescue local heritage are in need of a reinforcement in that sense.

In Lima, the denial and destruction of pre-Hispanic heritage could in fact be considered as a form of violence against people's own roots which, in the end, makes it difficult for them to identify with the territory. As a consequence, people do not relate to it in an affective way, do not cultivate it and do not care for it. The reading of some of Lima's pre-Hispanic sites, such as the Huaca of Pampa de Cueva or other partly or fully dismantled former sacred places, is therefore a reading of scars: through ad hoc or forced decisions, ephemeral events or historical occurrences, sometimes because of a large urban plan, sometimes out of caprice or a short-term dream of some mayor, that the territory and the collective memory are inscribed and, in turn, erased or replaced by other inscriptions successively producing visible or invisible layers within the territory. There are thousands of such palimpsests in urban space, which in itself is an example of a collective memory palimpsest: it is composed of inscriptions that preserve traces of other former inscriptions on the same surface, but which are expressly erased to give way to what exists, both physically and as memories.

This palimpsestic nature of such places also holds a strong potential: The traces of the various pasts can be recovered in the process of creating a solid future for citizens of the emerging city aside from economic growth, which is after all subject to global dynamics and disconnected from citizens and their relationship to their living environment. The connection of vernacular memory with pre-Hispanic sites and systems of connections in the territory might engage a population with primarily Indigenous roots in a palimpsestic cohabitation and herewith create an alternative form of cultural hybridity, a sense for place-making and an enlarged sense of responsibility for the territory itself.

Based on these convictions, two projects were put into practice since 2013 in two related sites in the Cono Norte: We explored the possibility of building up a new community spirit through the reconstruction of urban memory. Work started with an urban memory workshop[29] that looked in depth into the Pampa de Cueva settlement, which, since its founding in December 1960 through a land invasion, has generated various layers of memory: the spatial and organizational ideas of its original migrant settlers, the changes in the course of three generations, and the current pressure on land in the context of a booming Cono Norte. Stone construction techniques from Cajamarca, the existence of small roadside gardens cultivated by settlers, and solidly built and beautifully painted multi-story buildings characterize this area, with origins as single-family modules with small courtyards built on the dunes that were recognized at the time as valuable determining factors of bottom-up city making. It is an unwritten "history from below" that is key for the construction of a contemporary identity, and key also for redefining the territorial strategies of the *pueblos jóvenes* (young towns) with a view to the future.

There are traces of the memory of the founding of the settlement, even though not an awareness of them. Some of the street names in the area catch the eye: "17 de Noviembre," "Kilómetro 4," "Niños Mártires," "37 días," "23 de Diciembre" are clear references to the key dates, the suffering, and the milestones of the invasion and foundation: 17 November 1960 was the date of the initial occupation by 1,800 families; then the original landowners, the Nicolini family, claimed ownership

Figure 7.5 The 17th of November Avenue that holds the name of the date of the initial invasion of the area; its recent renewal by the municipality does not take into account people's memories and place-making strategies. Photo by Kathrin Golda-Pongratz, 2013.

Figure 7.6 The mainly dismantled pre-Hispanic site, used as a base for two high-voltage towers and currently accessible only through a school building, should be integrated into a system of public spaces. Photo by Kathrin Golda-Pongratz, 2013.

and forced all the settlers to retreat to "kilometre 4" of what is now Tupac Amaru avenue, where they remained for 37 days, and where several children – the *Niños Mártires* (the Martyr Children) – died in a clash with the police. During this time the two parties reached an agreement for a renewable ten-year lease on the land, so on 23 December the families were able to return to Pampa de Cueva and start building their city.[30]

There is also a hidden memory layer that is bound up with the origins of human presence in the area: Pampa de Cueva takes its name from an agricultural-archaeological complex and a U-shaped *huaca,*[31] built over 3,000 years ago, during the so-called Early Horizon, around 1200 BC – an important archaeological site that is part of a larger civic-religious temple complex (Agurto Calvo, 1984, p. 34).

Thirty years after its occupation by a settlement, in 1995, a team of archaeologists from Universidad Nacional Mayor de San Marcos carried out the first excavations at the site to confirm the prior findings. Interestingly, unlike other temples that are built facing rivers, this complex in the foothills faces the mountains, probably the end of the gorge where the water converged. Originally, there was a building over the second platform at the Pampa de Cueva ceremonial centre, which has been completely destroyed, leaving only imprints.[32]

Now almost totally dismantled, it is surrounded by informal housing and Alberto Hurtado Abadía high school (now the entrance to the site), which was built in 1963 and is also known as "Colegio El Morro." For the past few decades, the area is also home to enormous electrical towers that are part of the power supply grid for the whole of Lima Norte. The area is consequently known as *"las dos torres"* (the two towers) and in the memory of the locals it is not linked to its past and its pre-Hispanic value, but to the fact that it was "dynamited" in the not-too-distant past: in the '90s, one of the tactics used by the Shining Path was to blow up the city's electricity towers with dynamite, in order to bring about greater short-circuits (and incite terror). The locals are thus marked by the threat and live beneath the high-power line, with no links to signs of the importance of the territory in the past.

Towards a New Self-Understanding of Peripheral Landscapes of Memory

Until today, there have been no attempts to consider the potential of the place itself as part of a more holistic or integral point of view, as part of a network of urban spaces with pre-Hispanic origin which could transform into a place of reference both on a neighbourhood and a metropolitan scale. However, it could come to play a part in important initiatives such as the attempts by teachers and social workers to redefine the identity of Lima Norte, which is described as follows: "Family property or assets: material, like a house; immaterial, like the mores and customs transmitted from parents to children. Local heritage: Centro Ceremonial de Pampa de Cueva in Independencia; immaterial: the social organizations in the district."[33] In that sense, the workshop introduced a holistic approach, in a Geddesian[34] spirit, to encourage Pampa de Cueva to discover its own roots, define its own centre, and design its own public spaces as the essence of the self-built neighbourhood. The fact that the civic-religious archaeological remains politically unified the various human groups that lived in the region historically could be an element of the redefinition of the place as a public space, a common space with a power arising from the overlapping of archaeological and vernacular memory.

It also stretched the vision to the neighbouring district, El Ermitaño, where the making of the documentary film *Ciudad infinita – Voces de El Ermitaño* (City Unfinished – Voices of El Ermitaño) is a second project of memory activation: it follows up the historic document *A Roof of My Own*, which at the time had been censured by the Peruvian president Fernando Belaúnde, as he considered his own short congratulatory speech in the film, in which he praised mutual aid as a production method, as too positively oriented towards the barriada process.[35] The follow-up project of that historic documentary started with a simple but complex gesture: the re-edited and subtitled UNTV document was brought back to the place where it was filmed. After the first screening on 2 October 2016 in the Octavio Sánchez Medina community centre, which had an overwhelming reception and sparked a lively debate, several further screenings were requested and – in combination with mapping workshops – played in other sectors of the neighbourhood during the following months. The screenings brought together several generations that normally do not participate in public activities; a young generation in particular showed much interest in the origins of the neighbourhood and said that they had had no idea under what difficulties the first settlers and their own grandparents had to fight for their right to stay and build homes of their own. The screenings helped bring together several local residents from the first generation as well as families of the musicians who had composed and played the music in the 1964 documentary, which evoked a lot of emotions. A series of interviews were held with several original invaders and dwellers of the first generation, a community leader's son,[36] a family of tenants, and a young female community leader who struggles against land trafficking and on-going new invasions in the remaining hills, which affects

Figure 7.7 Original invasion of the agricultural lands flanked by the "Bella Durmiente" hill range in the early 1960s. Photo by H. Romani, 1960, © Caretas.

Figure 7.8 The fully urbanized area that today makes up the Pampa de Cueva and El Ermitaño neighbourhoods in the Independencia district, Lima, Peru. Photo by Kathrin Golda-Pongratz, 2017.

the fragile microclimate and Lomas landscape[37] of La Bella Durmiente, a ridge with a very specific shape that many dwellers want to defend from urban pressure and rescue as a natural space and even a sacred territory.[38]

From those efforts and the interests developed through several interviews and field visits, the film project derived a first working title, *Migrantes: Hijos de la Bella Durmiente*, dedicated to those generations of migrants who are now sons and daughters of that hill ridge which determines their livelihood and might be the essence for the creation of a new collective identity. It asks the following questions: What is El Ermitaño today? What are its dynamics, its problems and challenges? What are its inhabitants' aspirations and needs? And what is its urban identity? The film serves as a tool for tracing these urban identities of the progressive city and for the reconstruction of the place's collective memory. In practical terms, it has created new networks of solidarity among dwellers themselves and among those who, through collaborating in the project and through seeing the film, have established emotional ties and a sense of commitment with the neighbourhood.

Conclusion

New territorial practices, such as collective walks, ceremonies, and the planting of trees reconnect the inhabitants with this natural, historic, and sacred landscape. Such practices, other pedagogical activities, and finally the development of urban policies[39] can provide the population with cultural tools to confront the aggressive land trafficking and invasion practices that have nothing in common with the way people occupied land in the 1960s, when John Turner and his fellow researcher documented their dynamics. The collective reading of layers, the transmission of the possibility of a palimpsestic cohabitation of different cultural practices, the reinterpretation of the sacredness of the territory, and the setting-into-value of the historic traces in a network of contemporary public spaces is suggested as a most promising framework for an integral upgrading of places like El Ermitaño and Pampa de Cueva, but also of other comparable peri-urban areas.

It will be difficult to recover or recreate the kind of profound relationship with nature suggested in the words by José María Arguedas, quoted at the start of this text. Nonetheless, there is nothing romantic about arguing that the principal task in the expanded city should be to develop a sensibility towards place, and an identification with the territory, precisely to ensure that the existing hegemonies and environmental threats produced by further land occupation no longer escape local control in the consolidated barriada. Moreover, there is a need to strengthen the conviction that the future of the metropolis and its hinterland depends on the periphery and on what were once marginal neighbourhoods. Hence the need to materialize the awareness of its memory, the sense of responsibility, and the prospect of a holistic vision of the macro-region city.

NOTES

1 Arguedas (1972, p. 23).
2 In 1956, 120,000 people lived in peripheral settlements; by 1983, the figure was 2 million, and according to the census it had grown to 3.1 million in 2003. The last national census was held on 22 October 2017. Source: http://www.inei.gob.pe (Accessed: 18 February 2018).
3 An important step and platform, therefore, is a blog created in 2010 by the Peruvian journalist Javier Lizarzaburu: http://limamilenaria.blogspot.com. The Peruvian contribution to the Venice Biennial 2018, co-curated by Lizarzaburu and entitled "En Reserva" dealt with the pre-Hispanic heritage in Lima.
4 *Huacas* (from the term *Wak'a* in Quechua) are sacred places where pre-Hispanic deities where venerated through various rituals and where the people communicated with them. On the Peruvian coast, they are mainly mud pyramids or elevations in the landscape. Many of them have been destroyed by urban development.
5 The Argentinian anthropologist Nestor García Canclini has centred his studies and observations of the globalizing Latin American nation states and their processes of construction of democracy in a certain conflict between tradition and modernity around the concept of "hybridization" (García Canclini, 1995). The Brazilian anthropologist Joao Biehl, together with Peter Locke, theorize about "the transformative potentials of 'becoming' (in a Deleuzian sense) and of 'the unfinished'" (Biehl & Locke, 2017).
6 First formulated and published in Golda-Pongratz (2014). Mainly two other city contexts serve as references to sustain this thesis and approach: first, the successful integral neighbourhood upgrading program in Medellin, Colombia, and second, the neighbourhood initiatives in Barcelona, Spain, and the city's current memory programs. See: http://ajuntament.barcelona.cat/programesmemoria/en/.
7 The *barriadas*, in the Peruvian context first described and systematically analysed since the 1950s in the studies of José Matos Mar (1966), are non-formal settlements originated by group-organized land occupation, mainly in suburban public land, which are subsequently tolerated by the state. The flat desert land around Lima and the dry climate favour invasion and

make life in a precarious shed tolerable over months, until savings suffice to build a solid and more permanent home.

8 Speculation was already an issue in the early years of the barriada formation, where, by far, not all land invasions were spontaneous but partly organized by land promoters. The role of *líderes populares* (peoples' leaders) has always been fundamental with regards to territorial control. Interview with the American-Peruvian anthropologist Marcia Koth de Paredes shortly before her death, Lima, 8.10.2016.

9 In the context of the Fujimori regime, to which dismantling processes are contributed, the report of the Truth and Reconciliation Commission (CVR) mentions this phenomenon, see Arroyo (2003).

10 The reasons for the failure of the current PLAM 2035 are described in Puente Frantzen (2017).

11 Local residents and school children were interviewed as part of the urban memory public space workshop directed by Kathrin Golda-Pongratz at the Architecture Postgraduate Section of Universidad Nacional de Ingeniería de Lima in March 2013. The results of the workshop were presented at PROLIMA on 1 April 2013, and were exhibited at the Colegio de Arquitectos de Perú (Peruvian Architect's Association, CAP) on 4 April 2013.

12 Within the memory discourse and with regards to politics of commemoration, the term is coined by Huyssen (2003).

13 The concept of place-making is understood and interpreted from the current work of John F.C. Turner. A recent summary of this research framework is published in Turner (2016).

14 *A Roof of My Own, International Zone 41* (1964), uncensored version, Producer: G. Movshon, Cinematographer: D. Myers, Commentator: A. Cooke, Consultant: J.F.C. Turner. [DVD]. New York. UNTV. After its rediscovery, the re-edition of the historic film was done by Chris Berry, with the advice of John F.C. Turner and Kathrin Golda-Pongratz and with the assistance of Amarun Turner. The re-edited and subtitled film was, together with the related follow-up work, presented at the Habitat III Conference in Quito in October 2016 in the Urban Library event *Towards an Autonomy of Housing: Legacy and Topicality of John F.C. Turner's Work in Latin America.*

15 The documentary *Ciudad Infinita – Voces de El Ermitaño* (City Unfinished-Voices of El Ermitaño) was premiered at the Goethe-Institute in Lima and in the neighbourhood in October 2018. It has been selected for several festivals since then. The team is composed as follows: Kathrin Golda-Pongratz (concept and project direction); Dayan Zussner and Rosa Paredes (assistance and research); Rodrigo Flores (direction); Rodrigo Flores, Noelia Crispin and Kathrin Golda-Pongratz (script); Claudia

Chávez (production); Totó Flores (director assistant); Ian Ilbert (director of photography), Audrey Córdova Rampant (additional photography); Miguel Reyes (editing); José Carlos Valencia (boom operator); Rafael Benavides (sound designer); Jorge Pickman (audio post-production); Oswaldo Montúfar (digital colourist and post-production). See http://www.communityplanning.net/JohnFCTurnerArchive/index.php and https://www.facebook.com/barrioautoconstruido. The project received support from the Building and Social Housing Foundation (BSHF – now: World Habitat) and from a Crowdfunding campaign. The English subtitling was made possible by the Barcelona Knowledge Hub of the Academia Europaea. The trailer is available at https://vimeo.com/345314122

16 Parts of this article are part of a larger research and collaboration with John F.C. Turner and are published in Golda-Pongratz (2018). The book *John F.C. Turner – Autoconstrucción: Por una autonomía del habitar* (Golda-Pongratz, Oyon, & Zimmermann, 2018) translates fundamental texts by John F.C. Turner into Spanish and revises his legacy as well as contemporary work and thought.

17 The term "mega barriada-metropolís" is used in Ludeña Urquizo (2012), manuscript p. 3.

18 Personal interview with John F.C. Turner in Hastings, 18.7.2011, published in Golda-Pongratz, Oyon, and Zimmermann (2018), p. 268.

19 Oyon (2018), p. 244.

20 SINAMOS (Sistema Nacional de Apoyo a la Movilización Social), a state organism to mobilise the population during the military government of General Juan Velasco Alvarado (1968–75), which ran several government development programs.

21 In order to overcome the wounds of the internal war in Peru, the transition president Valentín Paniagua (2000–2001) set up the *Comisión de la Verdad y Reconciliación* (Committee of Truth and Reconciliation, CVR). In the conclusions to its final report, released in August 2003, the CVR stated that the internal armed conflict in Peru between 1980 and 2000 caused 69,280 deaths, a number that virtually doubles previous estimates. The CVR showed a clear relationship between levels of poverty and social marginalisation and the probability of being a victim of violence, and confirmed that the Maoist terrorist movement Shining Path (*Sendero Luminoso*, SL) was the main body responsible for the crimes and human rights violations, attributing 54% of the deaths to the SL. The CVR also found that the democratic parties Acción Popular (AP), which governed with Fernando Belaúnde Terry between 1980 and 1985, and the Aprista Party (APRA), in government with Alán García Pérez from 1985 to 1990, shared political responsibility regarding the events, and that Alberto

Fujimori, president from 1990 to 2000, was guilty of criminal responsibility. The left-wing opposition was also considered responsible due to its initial ambivalence to the insurgents. According to the CVR, during the urban attack, large swathes of all levels of society where willing to sacrifice elements of democracy in exchange for more security, and tolerated human rights violations as a necessary price to guarantee the end of terrorism. See Arroyo (2003), pp. 1f., quoted in Golda-Pongratz (2014), p. 10.

22 See Rush (2014), p. 72, and Williams (2005), p. 84.

23 The phenomenon is widely known, but very difficult to grasp, as the leaders of land trafficking networks and organizations remain largely unidentified within a system of threats, corruption, and connections to people within the public sector who provide them with information. Their influence has caused "a total control over the territory" while the creation of "annexes" to urbanized areas "turns out to be a more and more common mode of illicit land accumulation." Interview with the Peruvian anthropologist Eduardo Arroyo, Lima, 4 April 2017.

24 The handling of large amounts of cash in this area has given rise to phenomena such as the safety deposit box service available to customers at various fast-food restaurants on Tupac Amaru avenue. The massive cash flow is also attributed to major activities of money laundering related to the mining businesses in the north of the country and real estate and land trafficking activities in the capital.

25 Observatorio Socio Económico Laboral de Lima Norte (2008), p. 59. In the wake of the economic crisis in Europe, this figure was revised downwards in 2013 with the return of many Peruvians migrants from countries like Spain and Italy.

26 This was already described by the anthropologist William Mangin in the 1950s: see Mangin (1959).

27 The word "*chicha*," originally a traditional maize drink, has come to mean informal, popular, or cheap; "*cultura chicha*" refers to a mix of concepts created by immigrants.

28 The beginning of this crisis of identity, belonging, and citizenry is clearly attributed to the 1990s and the neoliberal and paternalistic politics of the Fujimori era. "There is a before and an after," says the community leader Dante Sánchez. Interview with D. Sánchez, Lima (El Ermitaño), 27 September 2017. Market and state dynamics of the 1990s are analysed in Riofrío (1991).

29 See note 11.

30 These events are recounted in the film *A Roof of My Own* (1964), which also documents the invasion and early consolidation process in the El Ermitaño neighbourhood in close proximity. See note 14.

31 *Huacas* (from the Quechuan term *Wak'a*); see note 4.

32 Chumpitaz Llerena (1995). In April 2017, the municipal worker José Alberto Chacón commented to the author about some further findings in the hills which might give evidence of a far larger extension of the pre-Hispanic site than expected until now.
33 Cépeda García (2010), p. 18.
34 In allusion to the biologist and urban thinker Patrick Geddes (1854–1932), whose ideas have deeply influenced the architect John F.C. Turner.
35 See notes 14 and 15.
36 See note 28.
37 The Lomas are hills where coastal fog gathers, and between July and October it causes the desert to turn green.
38 The existence of pre-Hispanic trails and the shape of the hills inspires dwellers of Andean origin to impose their memories from their place of origin onto the desert landscape of Lima Norte and produces a very interesting and fruitful identity shift towards the acceptance and appropriation of the territory. Interview with Victor Quispe, Lima (El Ermitaño), 6 October 2016.
39 Just to mention one example, the Peruvian NGO PREDES, in collaboration with the environment department of the Municipality of Independencia and the support of the PNUD of the United Nations, is developing a reforestation program that engages the population and its bottom-up initiatives within the so-called Lomas.

REFERENCES

Acuña Damiano, N., Almonte Flores, J.E., Arizaga Villalba, G., Ataucure Condo, F., & Mauricio Vivar, A.L. (2010). *Fortaleciendo nuestra identidad: Aportes para conocer la historia del distrito de independencia en el contexto de Lima Norte.* Lima: Tarea.

Agurto Calvo, S. (1984). *Lima prehispánica.* Lima: Municipalidad de Lima Metropolitana.

Arguedas, J.M. (1972). *Katatay y otros poemas huc jayllicunapas.* Lima: Instituto Nacional de Cultura.

Arroyo, P. (2003). *Resumen de las conclusiones de la comisión de la verdad y reconciliación.* Lima.

Biehl, J., & Locke, P. (2017). *Unfinished: The anthropology of becoming.* Durham: Duke University Press.

Caldeira, T. (2016). Peripheral urbanization: Autoconstruction, transversal logics, and politics in cities of the global south. *Environment and Planning D: Society and Space, 35*(1), 3–20. https://doi.org/10.1177%2F0263775816658479

Calderón Cockburn, J. (2001, May). *Official registration (formalization) of property in Peru (1996–2000)* [Paper presentation]. The ESF/N-AERUS International Workshop, Brussels, Belgium.

Calderón Cockburn, J. (2013). La ciudad ilegal en el Perú. In W. Jungbluth Melgar (Ed.), *Peru Hoy – El Perú Subterráneo* (pp 39–56). Lima: DESCO.

Cépeda García, N. (2010). Algunos conceptos involucrados en el desarrollo de la identidad. In N. Acuña Damiano, J.E. Almonte Flores, G. Arizaga Villalba, F. Ataucure Condo, & A.L Mauricio Vivar (Eds.), *Fortaleciendo nuestra identidad: Aportes para conocer la historia del distrito de independencia en el contexto de Lima Norte*. Lima: Tarea.

Chumpitaz Llerena, D. (1995). *El centro ceremonial pampa de Cueva: Un sitio formativo en forma de "U."* Lima: Universidad Nacional Mayor de San Marcos.

De Soto, H. (1986). *El otro sendero*. Lima: ILD.

Garcia Canclini, N. (1995). *Hybrid cultures: Strategies for entering and leaving modernity*. Minnesota: University of Minnesota Press.

Golda-Pongratz, K. (2004). The *barriadas* of Lima – Utopian city of self-organization? In A. Ruby & I. Ruby (Eds.), *The challenge of suburbia, architectural design AD 74* (pp. 38–45). London: Wiley.

Golda-Pongratz, K. (2014). Memoria urbana – Palimpsestos, huellas y trazados en Lima Metropolitana. In V. Marzal (Ed.), *EST: Revista de Estudios sobre Espacio, Sociedad y Territorio* (Vol. 1. No. 1, pp. 10–22). Lima: UNI.

Golda-Pongratz, K. (2015). Transformaciones espaciales, identidades urbanas emergentes y conceptos de ciudadanía en el Cono Norte, Lima/Perú. In A. Sethman & E. Zenteno (Eds.), *Continuidades, rupturas y emergencias: Las desigualdades urbanas en América Latin* (pp. 31–44). Mexico City: Fondo Editorial UNAM.

Golda-Pongratz, K. (2016). Urban memory: Palimpsests, traces and demarcations in metropolitan Lima. In K. Golda-Pongratz & K. Teschner (Eds.), *Spaces of memory. Trialog 118/119*, 4–17. https://www.trialog-journal.de/en/journals/trialog-118119-spaces-of-memory/

Golda-Pongratz, K. (2018). Lecturas contemporáneas de las barriadas turnerianas: Nuevas identidades y nuevos retos de la Lima emergente. In K. Golda-Pongratz, J.L. Oyon, and V. Zimmermann (Eds.), *John F.C. Turner – Autoconstrucción: Por una autonomía del habitar*. Logroño: Pepitas de Calabaza Editores.

Golda-Pongratz, K., Oyón, J.L., & Zimmermann, V. (Eds.). (2018). *John F.C. Turner – Autoconstrucción: Por una autonomía del habitar*. Logroño: Pepitas de Calabaza Editores.

Huyssen, A. (2003). *Present pasts: Urban palimpsests and the politics of memory*. Stanford: Stanford University Press.

Illich, I. (1973). *Tools for conviviality*. New York: Harper & Row.

Lobo, S.B. (1984). *Tengo casa propia*. Lima: IEP.

Ludeña Urquizo, W. (2012). *Barriadas y ciudad: Critica de la razón urbana*. [Conference paper delivered]. *Escuela del Hábitat 30 años*. Escuela del Hábitat, Faculty of Architecture, Universidad Nacional de Colombia,

Medellín. https://de.scribd.com/document/263994630/Barriada-y-Ciudad-Critica-de-la-Razon-Urbana (Accessed: 27 July 2021).

Mangin, W. (1959). The role of regional associations in the adaptation of rural population in Peru. *Sociologus, 9*(1), 23–36.

Matos Mar, J. (1966). *Las barriadas de Lima 1957.* Lima: IEP.

Matos Mar, J. (1984). *Desborde popular y crisis del Estado.* Lima: IEP.

Observatorio Socio Económico Laboral de Lima Norte (Ed.). (2008). *Remesas y desarrollo económico local en Lima Norte: Un enfoque territorial para políticas generales.* Lima.

Oyon, J.L. (2018). John Turner, el arquitecto geddesiano. In K. Golda-Pongratz, J.L. Oyon, and V. Zimmermann (Eds.), *John F.C. Turner – Autoconstrucción: Por una autonomía del habitar.* Logroño: Pepitas de Calabaza Editores.

Puente Frantzen, K. (2017). Plan Metropolitano de Desarrollo Urbano de Lima y Callao 2035: Análisis de un intento fallido. *Revista Iberoamericana de Urbanismo, año 8,* nr. 3, 111–134. http://hdl.handle.net/2117/108619

Riofrío, G. (1991). *Producir la ciudad (popular) de los 90.* Lima: DESCO.

Rush, E. (2014). Die selbstgebauten Städte. Limas Randbezirke – vorbildliche Siedlungen oder Spekulationsobjekte? In *Moloch, Kiez & Boulevard, le Monde Diplomatique* (Edition No. 14, pp. 69–73). Berlin: TAZ Verlags–& Vertriebsg.

Tokeshi, J., & Zolezzi, M. (2003). LIMA PARAquién/*LIMA PARAdojas.* In *QueHacer* No. *141.* Lima: DESCO.

Turner, J.F.C. (1963). Dwelling resources in South America. *Architectural Design, 33*(8). London.

Turner, J.F.C. (1976). A new universe of squatter builders. *The UNESCO Courier: Habitat: More than a roof overhead. XXIX*(6) 12–14.

Turner, J.F.C. (2016). An emerging placemaking paradigm. In K. Golda-Pongratz & K. Teschner (Eds.), *Habitat III – Quito 2016. Trialog 124/125, 1–2,* 20–23.

Turner, J.F.C., & Fichter, R. (1972). *Freedom to build: Dweller control of the housing process.* New York: Macmillan.

Turner, J.F.C., & Mangin, W. (1969). Benavides and the barriada movement. In P. Oliver (Ed.), *Shelter and society.* Barrie & Jenkins.

Williams, S.M. (2005). *"Young Town" growing up: Four decades later: Self-help housing and upgrading lessons from a squatter neighborhood in Lima* [Unpublished Master's thesis]. Department of Urban Studies and Planning, MIT.

Zibechi, R. (2007). *Territorios en resistencia: Cartografía politica de las periferias urbanas latinoamericanas.* Buenos Aires: Lavaca.

8 Occupy the Periphery: Housing Occupations and the Production of Urban Commons in Belo Horizonte

JOÃO TONUCCI AND RODRIGO CASTRIOTA

Introduction[1]

One of the most disruptive phenomena in Brazilian urbanization in the last decade has been the upsurge of housing occupations, a process linked to a vicious context combining economic growth, income and credit expansion, rising land prices, and housing policies that, instead of tackling the housing needs of the urban poor, favour the building industry and raise land rents (Magalhães et al., 2011). Although occupation for housing is not a new phenomenon in Brazil – on the contrary, this has been one of the most important means by which the poor secure their place in the city – the present moment is marked by a new scale of mobilization and by different socio-political dynamics, forms of organization, and possibilities.

Usually organized by social movements, occupations take place in a very short period through the occupation of vacant (public or private) land, mainly located in the metropolitan peripheries, and the building of housing and basic infrastructure by the residents themselves or by community joint efforts. In the Metropolitan Region of Belo Horizonte (MRBH) – the third largest in Brazil, inhabited by more than 5 million people – there are nowadays more than 15.000 families living in housing occupations (Campos, 2020). Most of them were organized over the last fifteen years through the agglutination of diverse networks of new social movements, homeless families, and varied supporters.

In this chapter, we argue that housing occupations are – despite their peripheral, subaltern and precarious condition – platforms for cooperative work and for production of the common, whether as a condition of survival or fruit of political and social experimentation. Occupations usually have different forms of collective spaces, such as community centres, communal kitchens, community daycare, urban gardens, and so on. Therefore, they may be better seen as *common spaces*: neither exclusively private nor public, but instead urban spaces that are collectively produced and appropriated, to some extent, by their own dwellers.

However, occupations also experience contradictions and ambivalences, lying between the potentialities of autonomy and collective construction of commons, and the harsh realities of an extreme condition of exclusion, deprivation, segregation, and violence. Hence, in this chapter we explicitly address this ambivalence and analyse new occupations as a privileged laboratory for our understanding of how urban commons are entangled in peripheral conditions, therefore hoping to reveal some of the shortcomings, potentialities, and contradictions of thinking of urban commons in the periphery of a Global South metropolis.

To do so, in section 2 we begin by presenting some brief theoretical and historical underpinnings of the commons, especially in Global South metropolises. In section 3 we provide historical and geographical contextualization of housing occupations in Brazil. Section 4 then, the main empirical part of this chapter, focuses on Belo Horizonte and two of its most emblematic occupations, Dandara and Eliana Silva. First some antecedents on the processes of metropolitanization and rising inequality in Belo Horizonte are presented, to then explore how common spaces and resources are produced and maintained in Dandara and Eliana Silva, focusing on four interrelated dimensions and elements of commoning: public space, women and social reproduction, agriculture and urban nature, and the question of land.

Most of our empirical data presented in those topics are drawn from personal involvement and fieldwork carried out on some occupations in Belo Horizonte over the last few years.[2] The conclusion tries to summarize some potentialities of thinking of Brazilian metropolises through the lenses of the urban commons.

Beyond Public or Private: Urban Commons in the Periphery

Generally, the commons refer to goods, spaces, and resources (material and immaterial) that are collectively used and managed by a given community through *commoning*, that is, a series of practices and relations of sharing and reciprocity, beyond the scope of the state and the market, of public and private property (Tonucci & Cruz, 2019). In this sense, before being a thing or a good, the commons is best understood as a social relation, a practice of collective production and social sharing between a community and some resource (Bollier, 2014; Linebaugh, 2014).

As a contemporary political discourse of resistance countering neoliberal capitalism, the common stands against the growing wave of enclosures (of nature, cultures, knowledge etc.), and against the extension of the commodity logic of competition and private property to all spheres of social life. Furthermore, it helps to sustain and guide the construction of multiple experiences of production and reproduction that are intended to be autonomous, democratic and self-managed (Dardot & Laval, 2015; De Angelis, 2007; Federici, 2010).

Commons have sustained human societies for a long time, but the formation – and the ongoing reproduction – of capitalism, an economic system based on private property and market relations, was (and still is) achieved through their enclosure, expropriation, and commodification (Marx, 2013; Polanyi, 2012; Wall, 2014). But commons are not only those pre-capitalist relics: they are today at the centre of capitalist production. Hardt and Negri (2009) distinguish between the commonwealth of the material world and the commons produced by social interaction, expressed in the immaterial results of human labour and creativity, such as ideas, images, languages, affections, codes, and the like. These new commons are mostly produced and embedded in the metropolis, which is not only a vast commons produced by collective work but also a generative space for the production of many other commons (material or immaterial).

Most recently, the idea of urban commons has been invoked by movements, demonstrators, collectives, researchers, activists, and even policymakers, to claim that resources and urban spaces could be more widely shared among city dwellers (Foster & Iaione, 2016). Many authors have been identifying the potential for a wide range of commonalities at different scales of urban space, such as urban land itself (particularly vacant and non-constructed land), a variety of open spaces and infrastructures (such as streets), and claims to the use and occupation of abandoned or underused public or private buildings and structures (Dellenbaugh et al., 2015). But, according to Simone (2014), the urban commons is something more than specific shared resources or infrastructures, common land property or bounded communal spaces: it also refers broadly to the multiple sociospatial relations and practices of commonality and sharing that take place in and are sustained through everyday urban life.

Blomley (2004) asserts that, regarding urban property, a variety of claims are made to urban space that are more collective in orientation than those offered by the public-private dualism. To him, the struggle for property should not always be understood as a struggle for alienation rights (rights to freely buy and sell land), as it can also manifest itself as rights of use and access against the enclosure of the commons. These community claims usually are not made on behalf of an abstract "public," as they are backed by the sense of belonging to a given community, and sustained by acts of occupation, use, and representation. Dardot and Laval (2015) also insist that the common is based on the democratic establishment of rights of use and social appropriation, in opposition to private (and public) exclusive appropriation.

In addition, practices and spaces considered as pre-modern, archaic and informal – such as slums and working-class peripheries – are beginning to be recognized by their powers of commonality and cooperation. There´s no doubt that the unequally divided metropolitan space of capitalism's peripheries (Santos, 2008) is responsible for multiple forms of deprivation, vulnerability, and segregation; but it is also related to the survival, reproduction, and stimulation of denser and hybrid ways of urban life and alternative modernities (Robinson, 2006). Therefore, the commons should not be seen only as the territory of a new generation of political activist groups: in Brazilian and other Global South metropolises, commoning has been long embedded in slums, metropolitan peripheries, and spaces assigned to the urban poor, due to the entanglement between survival strategies, informality, and social reproduction (Tonucci, 2017). Those common spaces are also part of the many "territories of resistance" found in Latin American urban peripheries, where autonomous movements from below struggle to create and maintain horizontal and emancipatory social relations (Zibechi, 2015).

As will be shown next, in new housing occupations in Brazil, the urban poor are also making collective claims to space through a series of practices of cooperation and sharing that, in many aspects, seem to underpin the production and reproduction of new urban commons in the periphery, as will be empirically explored through the case of Belo Horizonte.

"When Housing Is a Privilege, Occupation Is a Right": Housing Struggles in Brazil

Housing and the Urban Question in Brazil

On a global scale there is a close parallel between the rise of occupations, the financialization of housing, and the ongoing deconstruction of the idea of housing as a right. According to Rolnik (2015), this has been achieved through the building of an ideological and practical hegemony of individual private property over other forms of relationship with land, which is related to a model of public housing policy based on the promotion of housing markets and credit for homeownership. The imposition of this neoliberal model, in which housing is turned into a commodity and even a financial asset, is built on the deconstruction of existing social housing systems and on the criminalization and de-legitimation of other forms of property – such as land tenure and cooperative housing – many of them historically considered informal, especially in cities of the Global South. What followed has been a wave of evictions, gentrification, and dispossession, engendered by the increasing nexus between financial capital, property capital, and the state, which in turn has been answered by new forms of invasions and occupations by the urban poor in order to have minimal access to land and shelter in the city, and particularly on its peripheries.

According to Mayer (2015), organized housing occupations have multiplied in Brazilian cities over the past decade, and more intensely after the "riots of June 2013,"[3] as a way to ensure the settlement of thousands of poor families unable to cope with extreme rises in housing prices and rents in the wake of the last property boom, or even to have access to the *Minha Casa Minha Vida* (My House, My Life) federal housing program, designed to boost economic growth and save the building industry in the aftermath of the economic crash in 2008. Lourenço (2014) adds that this resumption of occupations in Brazil is also due to the weakening – and increasingly institutionalization, co-optation and lack of autonomy – of the oldest social movements, historically involved in housing struggles.[4]

The majority of recent occupiers were living in the home of relatives and friends before, often in overcrowded conditions, or under different rental arrangements, by which rent consumed from one-third to a half of all family income. A resident from Eliana Silva occupation has described housing rent as a "monster" that "devoured" her life, keeping her family in a permanent state of penury and indebtedness. Therefore, for many poor families – usually led by single mothers – the occupation is seen primarily as a way-out of the tenant condition.

Media headlines and hegemonic political discourses portray organized housing occupations as violent and illegal acts of invasion against property. However, while invasion is associated with an illegitimate, hostile act of taking by force a space used by others, disrespecting not only the law but also the unspoken rules of human coexistence, occupation means giving value to something, making useful a space that is unused (often for speculative purposes) and does not fulfil its social function (Lourenço, 2014; Nascimento & Bittencourt, 2016). This is reinforced by the motto of MLB (Movement of Struggle in Neighborhoods,

Villages, and Favelas): "When housing is a privilege, occupation is a right." In this sense, occupying a vacant lot, public or private, is a way to enforce the right to housing and the fulfilment of the social function of property principle, both of them inscribed in the Brazilian Constitution.

New Dynamics of Peripheral Urbanization: A New Cycle of Housing Occupations

Despite the contemporary nature of these disputes over land all over the world, it should be stressed that newer housing occupations have a long history. Holston (2008) notes that, in Brazil, access to land and housing for the majority of the population has always only been possible through informal means, due to the restrictions of an exclusionary social order. Thus, the expedient to settle informally – through possession, invasion, or occupations – in the absence of clear property titles, has always been the norm, not the exception, in both urban and rural Brazil.

Therefore, as urbanization gained speed and breath between the 1950s and the 1980s, most of the new urban residents had to carve their place in the city, typically through the autoconstruction of their residences (Caldeira, this volume) over plots of land acquired through illegal occupation in slums or in informal land subdivisions. Rolnik (2015) speaks of those popular territories – found all over the Global South – as marked by housing precariousness and by legal ambiguities regarding land property. The "new" housing occupations in Brazil must be seen as belonging to this broad spectrum of informal settlements.

Nonetheless, it is important to take into account that, even if they are part of a long-standing tradition of land struggles in Brazil, new housing occupations also have specific features. Differently from the slum ("favela," which is also formed through the occupation of public or private land, but in an incremental, piecemeal and "spontaneous" fashion), and from the informal subdivision (in which the settlers have paid for and thus own their land, although minimal standards of urbanization are not obeyed), occupations usually happen in a very short period of time and are conducted by some collective political force: sometimes, a mix of social movements, researchers, students, and technical collaborators.

Moreover, with the help of volunteer urbanists, some of the new occupations rely on urban planning principles, such as street layout, zoning, autoconstructed infrastructure and public facilities, and environmental concerns. This is a way to both guarantee a better urban environment and to resemble a formal neighbourhood, something that can be very important in conquering social and political legitimacy. Moreover, all houses are autoconstructed through family or community joint efforts, sometimes relying on informal paid work.

Thus, a central aspect of occupations is the weight of autoconstruction in residential production and infrastructure provision. Following the Latin American term for it, Holston (2008) and Caldeira (this volume) call autoconstruction the long-term process through which residents build their houses and neighbourhoods step-by-step, involving a great amount of improvization and continuous improvement, complex strategies and calculations, and reliance on their own labour or hired labour. Thus, in the process of autoconstruction residents become agents of urbanization, not simply consumers of spaces developed and regulated by others, be it the state, the market, or landlords. According to Nascimento (2016), it is estimated that around 70% of residential production in Brazilian cities is by autoconstruction, and since at least the 1940s it has been the most important mechanism of housing provision to the reproduction of the working class.

In Lourenço's view (2014), occupations are architectures that defy conventional definitions, according to which architecture is only what is the product of the architect as the subject and absolute author: in them, technical integrity of the project and the heteronomous role of the architect in relation to the so-called passive users of space are put in check by the practices of autoconstruction and collective deliberation. For Mayer (2015), occupations promote the construction of insurgent "common territories" in Brazilian metropolises: by challenging the hierarchical and standardized housing policies, they offer autoconstruction solutions more respectful to the problems and ways of life of the urban poor.

A Horizon of Occupations: The Case of Belo Horizonte

Metropolitan Restructuring and New Urban Activisms

The Metropolitan Region of Belo Horizonte (MRBH), formed by 34 municipalities, is the third-largest urban agglomeration in Brazil, with a population of around 5 million people. Founded as a planned city in 1897 to be the new capital of the state of Minas Gerais, its transformation from a political and commercial city into a modern metropolis dates back to its industrialization and rapid growth after the 1950s. Nowadays, it is a socially segregated and deeply unequal metropolis, where modernity and wealth live side by side with poverty and exclusion (Monte-Mór, 1994).

Like other Brazilian metropolises, the MRBH has experienced processes of deep socio-spatial restructuration in recent decades, resulting in the degradation and enclosure of many common and public spaces by the joint action of capital and the state, especially due to the expansion of the mining industry, to the impacts of large state-led developmental infrastructure projects, to the hasty sprawl of the urban fabric to suburbs and peripheries, and to the adoption of neoliberal urban policies. Both in local and state government, the political scenario experienced an explicit conservative and neoliberal turn in the 2000s (Magalhães et al., 2011).

However, many of those processes have been opposed by multiple struggles and resistances that have gained visibility and socio-political organization over the last years, such as organized housing occupations, self-managed cultural centres, occupy movements, urban agroecology and solidarity economies. In many ways, those experiences resemble urban commons, involving both entrenched and contemporary spaces of commonality, sharing, and cooperation.

Some of this new activism in Belo Horizonte extrapolated the agenda and repertoire of the most traditional movements for urban reform, strengthening the emergence of new movements also in the scope of the struggle for the right to the city, including fronts of more direct action, linked, above all, to the housing question. Nascimento & Bittencourt (2016) also relate the effervescence of new occupations to the ineffectiveness of municipal, state, and federal housing policy, to the innocuous instances of participation for housing movements, to the impacts of the great volume of urban investments without instruments to regulate land rents, and to the implementation of pro-market strategic planning at the municipal level.

Those new occupations have been organized, mainly from 2006 to the present, through the agglutination of a diverse network of social movements, Catholic Church, anarchists, independent activists, university research groups, local leaders, among other supporters. Although in the beginning occupations took place more in central areas and in vacant vertical buildings, this strategy faced many objections – from owners, judicial power, police, municipal administration, and even the homeless families, which were more used to living in houses than apartments – and so more emphasis was put on autoconstructed horizontal occupations over vacant plots of land in the periphery. There are nowadays around 60 occupations in the Metropolitan Region of Belo Horizonte, which are home to roughly 15,000 to 20,000 families (Campos, 2020), most of them still living under the threat of eviction.

Dandara and Eliana Silva: Occupy, Resist, and … Produce the Commons?

Dandara, one of the most emblematic occupations, was originally formed in 2009 with 150 families on a vacant private lot of 40 hectares on the outskirts of Pampulha, in the capital, and was chiefly organized by the *Brigadas Populares* (Popular Brigades). Today, it is the home of almost 1.200 families. Although originally conceived as a rural-urban (*rurbano*) occupation that should have been designed with larger plots to admit productive activities – an idea inspired by some occupations of this kind previously organized by the MST (Landless Worker's Movement) – due to the pressure for housing this plan had to be abandoned in favour of a more dense settlement (Lourenço, 2014). Nonetheless, Dandara managed to keep a community garden for some years, and many families were successful in growing gardens and orchards in their front or backyards. It was also the first occupation to follow a community masterplan, developed with technical advice from universities and containing land-use guidelines. And although the original repossession action was suspended, families still live without regular access to water, energy, sewers, paving, and waste collection (Nascimento & Libânio, 2016).

Since Dandara's "success story," many other occupations have taken place in Belo Horizonte. The occupation Eliana Silva, formed in 2012, currently houses approximately 350 families and is the third of a set of occupations developed in a valley in the peripheral region of Barreiro, on lands that had been donated decades ago by the state for industrial use but had been illegally held and sold for speculative purposes. In addition to the previous organization of the MLB, this occupation also relied on an important support network of students, activists, teachers, and volunteer professionals, in search of shared solutions for sanitary sewage, drainage, erosion, communal areas, and collective equipment. The community plan reserved areas for public spaces, for daycare facilities, and for environmental preservation areas. Eliana Silva is now a consolidated occupation with masonry houses and has a daycare centre, a library, and a community garden. Even if it does not have to face a repossession mandate anymore, the provision of public urban services continues to be deficient (Nascimento & Libânio, 2016).

While occupations in Belo Horizonte, in general, have a very similar history of birth, each of them has a distinct trajectory of development in what concerns tenure security, the acquisition of social rights and public infrastructure, consolidation of political processes,

Figure 8.1 Dandara occupation from above, 2013. Photo by Marcílio Gazzinelli.

Figure 8.2 Eliana Silva occupation from above, 2013. Photo by Marcílio Gazzinelli.

and so on. The same is true of the potentialities and ambivalences regarding the production of urban commons, as will be seen from now on from a sample of scenes and stories collected from the daily life of Dandara and Eliana Silva, with minor references to other occupations as well.

Meeting, Partying, and Deciding in Common: Public Spaces

As soon as one arrives at Dandara occupation, the first point of stop is the Community Center Prof. Fábio Alves. A large beaten-floor area, used for parties and local events, extends in front of the Center, where encounters and meetings take place. Apart from being the local base of the *Brigadas Populares* at the occupation, the Center also functions as an important cultural space, where musical instruments are stored and where artistic rehearsals and presentations of the community take place. The accounts of residents confirm that the Center is a very important common space, of public character, autoconstructed and maintained by some residents and by the militants of the Brigades most active in the occupation.

"Welcome to the Eliana Silva neighborhood": this is how it is written on a sign that marks the entrance to this occupation, and the visitor entering Che Guevara Avenue is soon received under the canopy of a large tree in front of the community library. This impromptu square is the main public space of the occupation, the meeting point of the residents. Bazaars are held every two weeks, where local food production is sold to raise funds for the community daycare. The square is also the venue for residents' birthday parties, traditional festivities, and year-end parties.

In occupations – and autoconstructed peripheries in general – public spaces tend to be precariously structured, incompletely organized, and bounded by tenuous borders. The absence of the state in occupations creates both the need and the opportunity for common spaces to be produced by the residents themselves. It is through the customary and everyday use of these collective spaces that they acquire legitimacy and efficiency to resist the frequent pressures of being appropriated for housing purposes and intruded by "alien" forms of power, such as drug trafficking. One striking difference between public spaces in occupations and those in the formal city is the greater intensity of political encounters in the former.

In this light, both the square in Eliana Silva and the Community Center in Dandara seem to fulfil the role of almost an ideal type of urban public space: a neuralgic point of encounter, meeting, party, and

Figure 8.3 Community Center Prof. Fábio Alves at Dandara occupation. Photo available at https://ocupacaodandara.blogspot.com/.

politics (Lefebvre, 1996). However, differently from the formal city, in occupations these places are not only public in terms of ownership and management, but mostly due to their common public use and appropriation, open to the community. The meetings held there consist of attempts to organize collective decisions in some democratic form – such as horizontal assemblies – that deviate from the more traditional deliberative models. In assemblies gathering residents, local leaders, and social movements members, important decisions are made, strategies are formulated, particular and common problems are discussed, tasks and positions are distributed, conflicts are settled, and so on. Their frequency and intensity vary in each occupation, according to its degree of consolidation and political organization.

Usually, as an occupation consolidates, assemblies tend to lose strength and to attract fewer residents. Occupations record many narratives of lost collective spaces, some of them taken up and rebuilt, others never reconstituted. Militants recognize that after the initial moments of great tension and fear of eviction, which are catalysts for collective engagement and struggle, it is common to see a demobilization of the

Figure 8.4 Children's birthday party at Eliana Silva's square. Photo available at https://ocupacaoelianasilva.blogspot.com/.

public sphere of occupation, as residents tend to face their problems. In some cases, the social movement remains highly present in the occupation, as is the case of Eliana Silva (reinforced by being the residence of one of the leaders); in others, the movements themselves lose space, reducing their mobilizing potential, as happened in Dandara.

Caring and Cooking in Common: Women and Social Reproduction

Even so-called democratic spaces are always marked, to varying degrees, by constraints, absences, and silences. Men as leading figures usually dominate the public spheres of participation, such as assemblies and councils, tending to silence subaltern voices. It is against this silence that Federici (2010) defends a politics of the commons based on an explicitly feminist perspective, centred on the collective work of social reproduction.

One of the first targets of collective effort in occupations, especially of women, is the construction of spaces for subsistence and care, such as community kitchens and daycare facilities. The role of women in occupations, many of which are already well organized in groups

Figure 8.5 Creche Tia Carminha at Eliana Silva occupation, 2013. Photo by PRAXIS-EA/UFMG.

Figure 8.6 Communal kitchen at Paulo Freire occupation, 2016. Photo by João Tonucci.

and movements, is more explicit in these collective spaces, organized around social reproduction issues, than in "instituted" political spaces, such as assemblies.

The *Creche Tia Carminha* (daycare) is a source of great pride for the residents and residents of Eliana Silva and has become one of its most important collective spaces. It serves children up to six years of age, from within the community itself, and was built – and recently expanded – by mobilization and digital collective financing, with the support of researchers and students from UFMG. The daycare is maintained by the work of volunteer women and by the contribution and financial support from inside and outside the community. It is a common space where forms of commonalization of domestic work are experienced by sharing the care of the other.

In the Paulo Freire occupation, next to Eliana Silva, the common water tank and collective kitchen are still important common spaces, mainly for the sharing of material means of reproduction – water, food, pots, utensils, gas, etc. – where commoning creates mutual ties and collective interest. Collective kitchens are key in the early months of an occupation, when residents still live in makeshift shacks without a private kitchen. Once the occupation is consolidated, communal kitchens tend to weaken, or even to be abandoned. The communal kitchen of Eliana Silva no longer exists, but its importance is recalled, and some residents rehearse its reconstruction next to a new projected community centre.

The more the occupation consolidates, the more common spaces and common-purpose practices become diluted. This poses a challenge related to the insertion of the inhabitants in the labour market, in addition to the activities of reproduction, since there are few and fragile experiences within occupations for the creation of popular and solidarity economies, of cooperatives, of alternatives for the generation of income and employment that give better living conditions to the residents who work under situations of extreme exploitation in distant and precarious underemployment. Notwithstanding the challenges and obstacles regarding the work sphere, examples of common spaces for reproduction point to more worthy and less alienated alternatives for many families and women in the occupations. When asked about, women usually express great affection toward those spaces, for they become places where ties are woven, where stories of hardship and endurance are told, where care is shared among equals.

Gardening in Common: Agriculture and Urban Nature

In the literature on urban commons, communal areas of cultivation have gained prominence by combining non-capitalist practices of production, distribution, and agricultural consumption with more collective land tenure arrangements, as well as helping to break the nature/urban dichotomy (Almeida, 2016). But in addition to this "resumption" of urban agriculture on urban bases – often incorporated into the more advanced circuits of capital – it is important to emphasize the presence and recent "rediscovery" of invisible urban farming practices and spaces in the peripheries of the southern metropolis.

Despite collective efforts and investments, the experiences of community gardens in the occupations of Belo Horizonte are quite fragile and discontinuous, exposing important limits to the production of the commons. Created at the beginning of the occupation, the community garden of Dandara, for example, no longer exists. It was located in one of the originally preserved green areas that were eventually taken for commercialization of lots, mainly by drug trafficking associated with the informal real estate market. In the Paulo Freire occupation, the community garden precedes the occupation, and although it has as its objective to make use of space and produce food (for own consumption, donations for daycare, sale, and support for demonstrations), it still has not yielded the expected outcomes, according to local accounts. In Eliana Silva, the story is not much different. The reasons for the failure of the first attempts to cultivate its community garden are both the lack of necessary water conditions and the animosity of neighbours, associated to the lack of fences to protect the area.

These experiences help to illuminate a central question in the debate over the urban commons: the tension between the opening and closing a common resource or space. In terms of more traditional economic classifications, community gardens could be classified as common goods of limited access, distinct from, for example, common goods of free access, such as public spaces. Even for Ostrom (1990) and Harvey (2012), not all commons are free and unrestricted access, since some of them require entry limits and rules of use for their protection.

The crops grown in occupations are not for profit purposes, but they serve basic social needs, destined for the use of the residents themselves, regardless of their purchasing capacity, or for future increases of income for the occupation, if there are production surpluses. The barriers and rules necessary for the maintenance of a communal garden in an occupation do not serve in this specific case to exclude someone from its fruits, or to alienate potential collaborators, but rather to guarantee some security to the space and to those who work there.

In Eliana Silva, some residents even report that the gardens and yards of their homes are "communal," since neighbours and acquaintances can use them when they need to, including to plant and harvest. These

Figure 8.7 Community garden at Paulo Freire occupation, 2016. Photo by João Tonucci.

customary practices are rooted in the notion of usufruct, that is, one's right to use and enjoy the property of others. Likewise in Dandara, and despite the loss of the collective garden, many families still grow gardens in their backyards, and there are reports of donation networks, local exchange, and sale of production surplus to families' needs for self-consumption, forming informal relations of sharing, and reciprocity that, in addition to minimizing situations of vulnerability, create community bonds.

In addition to gardens and orchards, it is possible to find green areas and gardens in occupations not necessarily related to food needs. Some houses, both in Dandara occupation and Eliana Silva, have well-tended gardens and flower beds that, beyond the boundaries of the private lot, advance over the "public" space of nonexistent sidewalks, contributing socially to the environmental and urban quality of the occupation. These examples show us how fragments of spaces neither public nor private, whose production and maintenance is intrinsically rooted in the possibilities and obstacles of the daily life of each family responsible

Figure 8.8 Vegetable and herbal gardens at Eliana Silva occupation, 2013. Photo by PRAXIS-EA/UFMG.

Figure 8.9 Flower bed at Dandara occupation's main avenue, 2016. Photo by João Tonucci.

for them, flourish especially in conditions of informality, precariousness, and lack of planning.

"Land Can't Be Sold Here": Land as Commons?

The above-mentioned scenes and narratives illustrate different facets of the production of urban commons in occupations, highlighting practices and common spaces that put into question the subordination of all spheres of life (including subjectivities) to the logic and ethos of competition and individualism (Dardot & Laval, 2015). But what is the reach of those experiences when confronted with the ideology of possessive individualism (Macpherson, 1978), with the hegemony of absolute and exclusive right to land ownership? To what extent is the struggle for housing and for the right to the city a struggle for property? Reversing the question: To what extent can we speak of land as a commons in housing occupations?

The leaders of MLB responsible for Eliana Silva are clear about the objectives of the movement to recognize the occupation as a regular neighbourhood of the city and to obtain regularization through the distribution of land titles to the residents, as happened with Corumbiara occupation before. One of the leaders was explicit in stating that it is not intended that the occupation becomes an "autonomous commune," but that it is fully recognized by the state and that the residents have their rights guaranteed. On the other hand, *Brigadas Populares* consider the issue of land ownership in occupations one of the main challenges that need to be more directly confronted and discussed by the movements, mainly due to the intensification of conflicts and the outbreak of episodes of violence related to land disputes. Hence the need to experiment with new forms of land tenure and property arrangements in occupations.

One of the first proposals that emerged in this respect in the occupations was the idea of collective lots instead of the traditional individual lot. According to Lourenço (2014), collective lots would be beneficial in terms of economies of scale and infrastructural costs, as well as in terms of creating more collective spaces for interaction in occupations. However, the implementation attempts in Dandara, Eliana Silva, and Guarani Kaiowá occupations were not successful. The author enumerates some of the main challenges and constraints to the proposal, such as the urgent need to settle the families that reached the occupations; prejudices against more collective forms of housing and property arrangements (especially because of the origin of the majority of families who lived previously, under rent or favour, in overcrowded spaces); and the

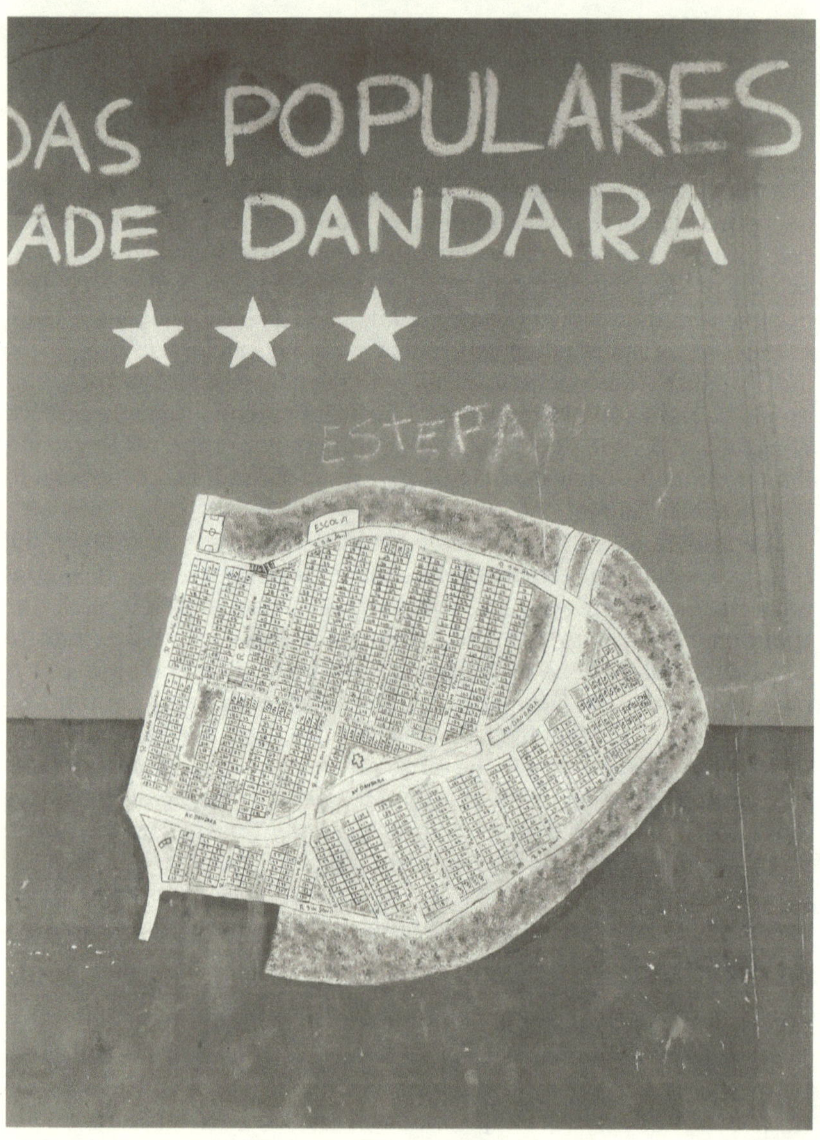

Figure 8.10 Subdivision of individual lots at Dandara occupation, 2016. Photo by João Tonucci.

Figure 8.11 "Land can't be sold here" – a sign at Irmã Dorothy occupation, 2010. Photo by PRAXIS-EA/UFMG.

absence of sufficient time to discuss this with families. None of this should be surprising since homeownership is an enduring structural block of Brazilian society.

However, the preference of family nuclei for more individualized spatial forms (such as private plots) is related, in addition to the homeownership ideology, to wider cultural values, such as the search for privacy, domesticity, and security. In addition, one of the advantages of self-construction is its individualized and open construction, which allows each family to adapt the construction and future extensions of the house to the rhythms, possibilities, and needs of the residents. In the Novo Horizonte occupation, for example, *Brigadas Populares* proposed collective yards within residential "blocks," which could be composed of four or six lots/residences: in this model, the common space could be the open space in-between the residences, guaranteeing privacy and individualization to each family. Thus, even if the individual lot is usually adopted as the main spatial arrangement of occupations, it does not follow that it needs to fit into the model of individual private property, by which the land becomes a commodity. In some occupations, experiences that point in the direction of keeping the land as a common resource have been, or continue to be, rehearsed.

By the time of the formation of the occupation Irmã Dorothy, neighbour to Eliana Silva, one could find tracks in its entry stating: "We inform you that the only way to live in the occupation is through the queue" and "Land can't be sold here. Under penalty of expulsion who sell or who buy. "At Eliana Silva, there is still a queuing system (with a preference for women and disabled people in extreme conditions of housing precariousness) and control (restrictions on the purchase and sale of the lot). If someone gives up living in the occupation, he or she can be compensated by the new inhabitant in the value of the improvement of his house, without being able to compute the value of the land. There have even been reports of some residents being evicted for not respecting those rules (and more often for violence against women).

The central role assumed by movements in the management of land and other potentially common resources in occupations bears similarities to the role of different non-state agents operating in informal real estate circuits, such as "local authorities" responsible for mediating conflicts and for ensuring coercive mechanisms, as studied by Abramo (2007). In occupations, such mechanisms are under the responsibility of the organizing movements and their leaders, especially in the occupations' initial moments. Another form of coercive power, in this case forcefully imposed, is the presence of drug dealers, which often comes into conflict with the power of movements, when it is unable to take over occupations.

In this sense, the cases of Dandara and Eliana Silva are extremely paradigmatic: while in the first case the movement (*Brigadas Populares*) lost strength and control of the land and housing market to informal agents, including drug dealers; in the second case the presence of the movement (MLB) and of "local organizers" remains strong, with some control over land allocation decisions and transactions. Two factors seem to be of importance in this case: the stark difference in the size of the two occupations, and the fact that, in the case of Eliana Silva, MLB's main leadership resides in the occupation itself, reinforcing its local leadership role.

The scenes and accounts of the occupations seem to indicate that the struggles that take place in them cannot be easily reduced to direct claims to private property, although generally this agenda is inextricably imbricated in the speeches and actions of residents and activists.

In struggles around property there are other elements, motivations, and aspirations at stake: security of tenure, decent housing, access to services, human rights, and citizenship. Whether it is in the discourse of movements, in the internal policies regarding occupations' organization, or in the vestiges and traces left by the daily practice of its inhabitants, it is possible to find some alternative and more collective formulations of land and housing property.

Yet the institutions and rules governing collective life in those occupations still under the control of movements do not seem to be embedded enough to survive the "day after" of recognition and regularization of these settlements, especially after the movements have left, or lost power. The absence of concrete and successful examples of common urban land management in Brazil certainly weighs against the longevity of such experiences in occupations, which is exacerbated by the little concern and imagination of movements struggling for urban reform to find alternative models that face and overcome the dichotomy of public or private property. There are still few and very fragile instruments, institutions, and public policies focused on cooperatives and on urban and housing auto-management.

Conclusions

As seen, there are a number of important contradictions regarding urban occupations: on the one hand, they are spaces in the making that seem very powerful for the flourishing of practices based on sharing and solidarity, in part because they are peripheral spaces where the presence of the state and capital is smaller; on the other hand, most of this autonomy comes from the fact that they exist as a product of a very heteronomous and unequal capitalist society. Therefore, occupations experience these contradictions and ambivalences living between the potentialities of autonomy and collective construction of commons, and the harsh realities of an extreme condition of exclusion, deprivation, segregation, and violence.

Occupations live in a kind of in-between situation: at the same time that social movements and residents aspire to be recognized by the state (and thus not more subjected to be living under the threat of eviction), they do acknowledge that this entering into the formal city can mean the loss of many of the collective projects and of more collective arrangements of property that were possible while all were engaged in resistance and socio-spatial experimentations (Bastos el al, 2017). Nonetheless, evidence suggest that occupations speak to a different relationship towards the commons not only through the fostering of many common practices – such as collective urban gardens, autoconstruction of community facilities, and participatory planning of the occupation with the help of movements and university – but also through the enactment of more collective claims to property that disturb the homeownership model.

Thus, we believe that urban commons in the twenty-first-century Global South metropolis (Roy, 2009) may be related to efforts in recognizing and strengthening the many alternatives and experimentations in the production of space, usually led by the urban poor, and thereby relegated to the practical and theoretical periphery (Simone, 2010). So, despite usually seeming as underdeveloped, poor, and lacking mega-problematic cities, Global South metropolises are starting to be recognized as having a lot of potential for breeding alternative future modernities (Robinson, 2006) to current trends of capitalist urbanization, and this potential should also be related to how people assemble ways of life in the peripheries also through practices of sharing, cooperation, and solidarity (Tonucci, 2017).

We want to suggest that changes in broader scales – such as the formal recognition of collective forms of (urban) landed property by law, shyly admitted nowadays in Brazil – could transform the local struggles of each particular occupation, giving them some clear alternatives beyond the capitalist public-private divide that dominates our imaginary. To introduce the commons into the urban political vocabulary and into the legal system would be an important step towards more just cities, towards a more common urban world. Resuming Marx, the French philosopher Henri Lefebvre once stated that "one day, which will indeed come, the private ownership of land, of nature and its resources, will seem as absurd, as odious, as ridiculous as the possession of one human by another" (2009, pp. 194–195). Urban occupations in Brazilian metropolises may be pointing towards this concrete utopia, as they experience the everyday challenges of producing and sustaining urban commons under peripheral urbanization.

NOTES

1 Part of the research results presented here is derived from one of the author's PhD thesis (Tonucci, 2017). The authors are grateful to professor Heloisa Costa for her guidance and to CNPq and CAPES for their financial assistance in Brazil. We also would like to thank Roger Keil, the book organizers, and the reviewers.

2 The scenes, reflections, and debates registered here are the fruit of our involvement with housing occupations in Belo Horizonte since 2013. Our varied engagements with this reality have been possible through *strictu sensu* fieldwork, technical support and advocacy to social movements and public agencies, and collaboration with colleagues who also research and work with occupations at the Federal University of Minas Gerais (UFMG) and the Catholic University of Minas Gerais (PUC-Minas).

3 The 2013 protests in Brazil, or 2013 Confederations Cup riots, also known as the Brazilian Spring or June Journeys, were public demonstrations in several Brazilian cities, initiated mainly by the *Movimento Passe Livre* (Free Fare Movement), a national autonomous movement that advocates for free public transportation. The demonstrations were initially organized to protest against increases in bus, train, and metro ticket prices in some Brazilian cities, but grew to include other issues such as the quality of public services, high corruption in the government, and police brutality used against some demonstrators. By mid-June, the movement had grown to become Brazil's largest since the 1992 protests against former President Fernando Collor de Mello. It was also captured by growing right-wing groups and middle-class participants, who seized the opportunity to protest against the alleged corruption of the left-wing PT government. As with the 2013 Gezi Park protests in Turkey, social media has played an important role in the organization of public outcries and in keeping protesters in touch with one another. Source: https://en.wikipedia.org/wiki/2013_protests_in_Brazil.

4 Some older housing movements – historically associated with the urban reform movement, and which had also organized occupations in the '70s, '80s, and '90s – became less confrontational during the Worker's Party government in Brazil (2003–2016), as many of its original members became part of the government or were increasingly involved in housing policy implementation and negotiation.

REFERENCES

Abramo, P. (2007). A cidade COM-FUSA: A mão inoxidável do mercado e a produção da estrutura urbana nas grandes metrópoles latino-americanas. *Revista Brasileira de Estudos Urbanos e Regionais, 9*(2), 25–54. https://doi.org/10.22296/2317-1529.2007v9n2p25

Almeida, D. (2016). *Isto e aquilo: Agriculturas e produção do espaço na Região Metropolitana de Belo Horizonte* [PhD dissertation]. Universidade Federal de Minas Gerais, Instituto de Geociências.

Bastos, C., Magalhães, F., Miranda, G, Silva, H., Tonucci, J., Cruz, M., & Velloso, R. (2017). Entre o espaço abstrato e o espaço diferencial ocupações urbanas em Belo Horizonte. *Revista Brasileira De Estudos Urbanos E Regionais, 19*(2), 251–266. https://doi.org/10.22296/2317-1529.2017 v19n2p251

Blomley, N. (2004). *Unsettling the city: Urban land and the politics of property.* New York, London: Routledge.

Bollier, D. (2014). *Think like a commoner: A short introduction to the life of the commons.* Gabriola Island (CA): New Society Publishers.

Campos, C. (2020). *Squatting for more than housing: Alternative spaces and struggles for the right to the city in three urban areas in Brazil, Spain, and the Basque country* [PhD thesis]. Universidade Federal de Minas Gerais.

Dardot, P., & Laval, C. (2015). *Común: Ensayo sobre la revolución en el siglo XXI.* Primera edición. Barcelona: Editorial Gedisa.

De Angelis, M. (2007). *The beginning of history: Value struggles and global capital.* London; Ann Arbor, MI: Pluto.

Dellenbaugh, M., KIP, M., Bieniok, M., Muller, A., & Schwegman, M. (Eds.) (2015). *Urban commons: Moving beyond state and market.* Basel: Birkhäuser Verlag GmbH.

Federici, S. (2010). Feminism and the politics of the commons. In C. Hughes, S. Peace, & K. Van Meter (Eds.), *Uses of a whirlwind: Movement, movements, and contemporary radical currents in the United States.* Team Colors Collective, Oakland: AK Press.

Foster, S., & Iaione, C. (2016). The city as a commons. *Yale Law & Policy Review, 34*(281) 282–349. https://dx.doi.org/10.2139/ssrn.2653084

Hardt, M., & Negri, T. (2009). *Commonwealth.* Cambridge, Massachusetts: The Belknap Press of Harvard University Press.

Harvey, D. (2012). The creation of the urban commons. In D. Harvey, *Rebel cities: From the right to the city to the urban revolution.* London, New York: Verso.

Holston, J. (2008). *Insurgent citizenship: Disjunctions of democracy and modernity in Brazil.* Princeton: Princeton University Press.

Lefebvre, H. (1996). The right to the city. In E. Kofman & E. Lebas (Ed. & Trans.), *Writings on cities / Henri Lefebvre* (pp. 61–181). Oxford, UK, Malden, USA: Blackwell Publishers.

Lefebvre, H. (2009). Space: Social product and use value. In N. Brenner & S. Elden (Eds.), *State, space, world: Selected essays.* Minneapolis, London: University of Minnesota Press.

Linebaugh, P. (2014). *Stop, thief! The commons, enclosures and resistance.* Oakland, CA: PM Press.

Lourenço, T. (2014). *Cidade ocupada* [Master's dissertation]. Universidade Federal de Minas Gerais, Escola de Arquitetura.

Macpherson, C. (1978). *Property: Mainstream and critical positions*. Toronto, Buffalo, London: University of Toronto Press.
Magalhães, F., Tonucci, J., & Silva, H. (2011). Valorização imobiliária e produção do espaço na RMBH: Novas frentes. In J. Mendonça & H. Costa (Org.), *Estado e capital imobiliário: Convergências atuais na produção do espaço urbano brasileiro*. Belo Horizonte: C/Arte.
Marx, K. (2013). *O capital: Crítica da economia política: Livro I: o processo de produção do capital*. São Paulo: Boitempo.
Mayer, J. (2015). *O comum no horizonte da metrópole biopolítica* [Master's dissertation]. Universidade Federal de Minas Gerais, Escola de Arquitetura.
Monte-Mór, R. (1994). Belo Horizonte: A cidade planejada e a metrópole em construção. In R. Monte-Mór (Ed.), *Belo Horizonte: Espaços e tempos em construção*. Belo Horizonte: Cedeplar/PBH.
Nascimento, D. (2016). Outra lógica da prática. In D. Nascimento (Org.), *Saberes [auto] construídos*. Belo Horizonte: Associação Imagem Comunitária.
Nascimento, D., & Bittencourt, R. (2016). Invadir ou ocupar? In D.M. Nascimento (Org.), *Saberes [auto] construídos*. Belo Horizonte: Associação Imagem Comunitária.
Nascimento, D., & Libânio, C. (Org.) (2016). *Ocupações urbanas na Região Metropolitana de Belo Horizonte*. Belo Horizonte: Favela é Isso Aí.
Ostrom, E. (1990). *Governing the commons: The evolution of institutions for collective action*. New York: Cambridge University Press.
Polanyi, K. (2012). *A grande transformação: As origens da nossa época*. Rio de Janeiro: Elsevier.
Robinson, J. (2006). *Ordinary cities: Between modernity and development*. London; New York: Routledge.
Rolnik, R. (2015). *Guerra dos lugares: A colonização da terra e da moradia na era das finanças*. (1st. ed.) São Paulo: Boitempo.
Roy, A. (2009). The 21st-century metropolis: New geographies of theory. *Regional Studies, 43*(6), 819–830. https://doi.org/10.1080/003434007 01809665
Santos, M. (2008). *O espaço dividido: Os dois circuitos da economia urbana dos países subdesenvolvidos*. (2nd. ed.) São Paulo: EDUSP.
Simone, A.M. (2010). *City life from Jakarta to Dakar: Movements at the crossroads*. New York: Routledge.
Simone, A.M. (2014). *Jakarta, drawing the city near*. University of Minnesota Press.
Tonucci, J. (2017). *Comum urbano: A cidade além do publico e do privado*. [PhD dissertation]. Universidade Federal de Minas Gerais, Instituto de Geociências.

Tonucci, J., & Cruz, M. (2019). O comum urbano em debate: dos comuns na cidade à cidade como comum? *Revista Brasileira de Estudos Urbanos e Regionais, 21*(3), 487–504. https://doi.org/10.22296/2317-1529.2019v21n3p487
Wall, D. (2014). *The commons in history: Culture, conflict, and ecology.* Cambridge, MA; London: MIT Press.
Zibechi, R. (2015). *Territórios em resistência: Cartografia política das periferias latino-americanas.* Rio de Janeiro: Consequência Editora.

9 Hybrid Livelihoods: Resistant Adaption in Peri-urban Bolivia

HANNAH-HUNT MOELLER

Introduction

This chapter argues that the nature of Bolivian peri-urban spaces is not one of rural-urban continuum, a continuous sequence merging "ruralism" and "urbanism," but rather one of distinct spatial tactics enhanced by heritage-based practices. The politics of the space are characterized not only by rapid population growth, but also by complex heritage-based practices that privilege mercantilism, communal land ownership, and community-based organizational activism. As such, I argue that the informality of infrastructure and governmental oversight enable dynamism and resiliency. The peri-urban space is a third option to the rural/urban dichotomy, presenting opportunity for peripheral resistant adaption in the face of neoliberal economic forces. These arguments are supported by a historical review of the Quechuan Indigenous people who inhabit Bolivia's central valley region, in addition to a case study of the present peri-urban communities in Cochabamba, Bolivia. I trace the history of Indigenous livelihood linked to territory from the Incan past to the neoliberal present. The analysis argues that the specific geography of the peri-urban enables its dwellers to be both adaptive and resistant to the dominating economic forces. Ultimately, I argue that the current space in which these people dwell, the peri-urban, poses a unique geography in which heritage-based practices of communal land use result in an adaptive resistance to singular neoliberal forces. Hybrid livelihoods in the peri-urban milieu resist clear distinction and thus engender social and political action operating at a local level.

Peri-urban Space

Our world is increasingly urbanizing; indicators estimate that by 2050, 70% of the world's population will live in greater urban regions (UN,

2011). Traditional urban cores cannot fully contain such density, and are being supplanted with active and dominant peripheries. One such border condition is peri-urbanism, a pervasive peripheral typology in the Global South (Caldeira, this volume). The hyphenated hybrid term refers to a zone both "about" and "around" the urban milieu. So, what *is* this space?

Geographically, the zone is an interstitial arena straddling urban periphery and the rural landscape. This is a space of colouring-outside-the-lines and negotiating the presumptive "urban" and "rural" delineations. Peri-urbanism challenges normative centre/urban and periphery/rural relationships, while maintained linkages complicate the dichotomy. By nature, it resists strict definition or singular delineation; it is a geographical both/and condition. Programmatically, peri-urbanization is a strategic active progression, what scholars refer to as "a process of blurring" between urban and rural (Allen, 2003). Activity in the heterogeneous zone is a patchwork of agrarian production, industrial stockpiling, residential, and small-scale commercial exchange. This border zone "encompass(es) a fragmented mixing of urban and rural worlds, transforming previously rural livelihoods" (Lerner & Eakin, 2011, p. 312), and, as I will argue, significantly influences regional dynamics.

To be clear, while I discuss the physicality of peri-urban space, my terminology of "space" borrows its depth of significance from Henri Lefebvre. "(Social) space is a social product," declares Lefebvre (2009). He dislocates space from solely the Cartesian notion of Euclidian geometry, and theorizes space to be a saturated product that is produced, reproduced, and influenced by the society's intentions. "Every society – and hence every mode of production with its subservients ... produces a space, its own space" (ibid.). This space has physical manifestations through the spatial practices of human and larger social performance through the role of the biological family, the production of labour, and the social relations in the larger societal context (ibid.). While my analysis is about the geography and location of peri-urbanism, I seek to uncover the ways in which the space is produced through human means as well as the way in which the space influences and enables further agency of the human occupants.

The premise of people-produced space is applied elsewhere in this book to the context of the urban periphery by Caldeira. She uses the term "peripheral urbanization" to refer to the relational process or a "mode of making cities" that is pervasive in the Global South. Her conceptualization builds its argument upon four major facets: (a) experiencing temporality and agency, (b) applying transversal logics, (c) generating new modes of politics, and (d) creating unequal and heterogeneous cities. I will engage with these facets of the process of peripheral urbanization; however, my analysis focuses on the spatial location and characteristics that are present within this space, both physically and socially. Caldeira argues "peripheral urbanization does not simply refer to the spatial location in the city, but rather a way of producing space" (Caldeira, this volume). I do not disagree with her claim to spatial process but am interested to evaluate the result and physicality of spatial production in the specific context of Bolivian urban fringes.

Through my analysis, I seek to situate the process of peripheral urbanization in dialogue with Zibechi's analysis of the site of peripheral territories of resistance. He complements Caldeira's assessment of transversal logics and new modes of politics but claims that peripheral territories are specific locations of resistance, spaces in which weak government authority allows space for strategically heterogeneous ways of daily living. Further, the actors in these spaces hold subversive quotidian patterns, enabling an emancipation of community traditions (Zibechi, 2011). As Caldeira and Zibechi would agree, peripheries pose the potential to facilitate resistance by localized initiatives of citizen participation challenging hegemonic political and economic forces.

This chapter examines the specific context of contemporary Bolivia and the peri-urban fringes of Cochabamba, the site of the *Guerra del Agua* (water war). In April 2000, widespread protests erupted over attempts to privatize drinking water (and increase tap fees by upwards of 200%) by foreign-owned firm Aguas del Tunari, a transnational consortium controlled by the American company Bechtel (Perreault, 2006). Tens of thousands of peasant and migrant protestors, galvanized by the ethos of *El agua es Vida* (water is life) (ibid.), clashed with Bolivian military, resulting in the death of one protestor and ultimately forcing the government to rescind the concession. The episode gained international recognition for the pointed example of neoliberal environmental governance, resource management, and decision-making happening at a scale removed from the primary users. The re-scaling of environmental resource management directly conflicts with Andean Bolivia's *usos y costumbres* (traditional customs and daily practices) of common-property arrangements, shared land use, and water as a "collective good" (Olivera, 2004). The water wars embodied the neoliberal model of increasingly complicated institutional arrangements of state, foreign development organizations, and private capital restructuring investments in natural resources.

The following chapter is structured to situate the peri-urban spatial analysis within the context of Andean Bolivia. I begin by sketching out the historical relationship of the Andean Indigenous peoples to their land, from the collective and reciprocal ecologies of the Incan era to the forceful disruption of colonial hierarchies violating traditional *usos y costumbres* of land and resource relations. I then outline a claim of resistant adaption by Indigenous peasants during the colonial era located within local trade economies and situated in the pueblerino markets. This location provides an alternative narrative to full Indigenous oppression, highlighting the geography of resistance preceding contemporary times. I proceed to create a scaffolding that situates the political and social movements related to land redistribution during the twentieth century, culminating with the rise of neoliberal policies and the presidency of Evo Morales. This provides a framework to the production of the peri-urban space as a specific result of political and economic contexts, and the potentialities at work within this milieu. At last I zoom into the specifics of the peri-urban Zona Sud of Cochabamba to unveil the features of the zone through a grassroots perspective.

Case Study: Andean Bolivia

This case study of peri-urbanism in the Bolivian context presents a poignant illustration of ethnicity in the practices and potential of livelihoods in the peri-urban. Self-identified Indigenous peoples comprise the majority (62%) of the population of the Plurinational State of Bolivia. The predominant people groups of this Andean region are the Aymaran highlanders and the Quechuan valley people. Self-distinguished by location and language, these Indigenous groups are further designated by the United Nations, which defines Indigenous as "having a historical continuity with pre-invasion and pre-colonial societies that develop on their territories." (UN, 2008). In effect, ethnic identification is directly linked to territory and the social relationship of a people-group to their land.

Quechua Identity Linked to Land: Incan to Colonial Era

Critical to understanding the present geographical situation is recognition of the historic specificity of this landscape. The peri-urbanism resultant today is not given but rather produced by complex historical trajectories. Through history, Andean heritage-based practices and values were rendered visible through the relationship of people to land, informed by and informing their geography. During the Incan Empire expansion (circa 1470 CE), land use patterns promoted systematic political and social reinforcement of the empire at large. Extended kin groups lived together on communal land organized as an *ayllu*. The ayllu were self-sustaining communities organized vertically as to span multiple microclimates: altiplano pastures for Camelidae and valley floors for crop production (Larson, 1998). Fundamental to these lineage-based ecologies was local self-government, self-sustaining communal land use, and relations of reciprocity. Each couple received a measured land plot called the *tupu*, considered sufficient nourishment to feed a married couple for a year's time (Rostworowski de Diez Canseco, 1999). The production from each *tupu* was allocated into thirds: one third of the produce was retained for the family unit; another third allocated to common cultivation to share the harvest among community members; the final third for support of soldiers, sick, aged, and orphaned children (Beals, 1973). Organized and systematic, collective land management demonstrated the communal reciprocity integrated into the social organization of Incan people and rendered visible across the Andean landscape.

Entering the Colonial era (circa 1550 CE), Spaniard conquest stripped all familial and communal land rights, forcing mass vassal labour on large hacienda estates. Such spatial consolidation and reallocation did violence to the distinct Andean Indigenous groups by collapsing heterogeneous ethnic identities to the singular classification "Indian" and flattening complex pre-Colonial hierarchies by violating existing social structures. The resultant dichotomy of conqueror/conquered subordinated diverse Indigenous cultural specificity to a futile null. Few existing societal hierarchies remained that were acknowledged by intruders, and the geographical consolidation reflected such flattening of pre-existing societal structure (Larson, 1998). Extraction of people from their communal and political land-use patterns violently damaged a social system built upon a profound connection to the land.

However, the unfolding result did not entirely destroy traditional Andean land cultivation practices. Andean historian Stern identifies the process of "resistant adaption" by which mercantile practices of European-colonial origin were appropriated and redeployed by Indigenous groups to incorporate ayllu practices (Stern, 1987). Despite vehement removal from traditional land use, a resourceful resiliency resulted. In highland regions, the colonized Indigenous continued to cultivate "micro-verticality" to diversify crop production, germane to the ayllu system. However, in valley regions where crop diversity was less available, a conglomerate of Indigenous groups pragmatically initiated a commodity trading system, akin to a horizontal *ayllu*. Although no longer lineage-based exchanges among kin and despite "sharp temporal, regional, and ethnic differences," Indigenous groups participated in a wide variety of trading and market activities within members of their own colonized class (Larson, 1998). The geographical affordances of fertile lowlands facilitated this exchange among the valley Indigenous, trading only among their fellow colonized peoples. Valley towns served as linkages and convergence points, and such settings enabled a social fraternization and cultural exchange, in solidarity against the colonial hegemony. This trade activity resulted in blurring the lines between previously discrete Indigenous community groups by maintaining the unifying Andean traditional practice of the ayllu. Resistant adaption posits the valley-based Indigenous communities as resilient agents, manipulating their conditions to maintain their customary practices.

Pueblerino: Site of Resistance

In the late nineteenth century, the resourceful trading arrangements coalesced into mercantile hubs called pueblerino barter markets, created by and for the Indigenous colonized peoples. These spaces provided zones of mercantile exchange characterized by indeterminacy: "material and discursive spaces of fluid – one might even say, strategic – in-betweenness and ambiguity, where power, meaning, and identity were rendered more indeterminate, relational, and contestable." (ibid., pp. 334–360). Such spaces were situated as hybrid zones, enabled by trade routes that benefited hacienda landowners and yet distinctly remained sites of "cultural cross over, impromptu actions, and social bonding" away from the landlord oversight. These zones of economic activity provided space for cooperation, exchange, and resiliency of pre-colonial social ecologies. The spaces of the pueblerino markets functioned strategically. The fluid, hybrid movement patterns enabled and espoused resistance to the singular homogeny. In support, I point to the origins of land reform, instigated by the resistant peasant groups in the Cochabamba Valley.

Land Reform and Resistance: 1952, Neoliberal Reform, Evo Morales

During the Colonial era, Spaniard conquerors confiscated the most desirable land, maintaining power via centralized economic control of land. During the mid-twentieth century, the national political climate further deteriorated during the 1932 Chaco War between Bolivia and Paraguay over the ostensibly oil rich Chaco Boreal region. Political instability coupled with territorial disputes contributed to the prerevolutionary climate. Civilian-based dissent organized politically into national parties and gained congressional power through the 1940s. Notably, the leftist National Revolutionary Movement (MNR) mobilized agrarian reform, nationalized natural resource industries, and established a populist state (Rivas, 2000).

The festering mobilization of anticolonial sentiments ultimately culminated in the 1952 Bolivian National Revolution, led by the MNR. This marked a significant turning point in the conceptual relationship of Indigenous people groups to land: land reform decreed in 1953 declared universal suffrage without literacy requirements, freed land labourers from servile working arrangements, and granted land to the Indigenous peoples. Officially, the transition marked a shift in the epistemological nomenclature of the "Indian." No longer a collective mass of other, the MNR mandated a symbolic transferal to the term *campesino* (Larson, 1998). The term embodied a relinkage of a people groups' connection to land use, both as farmer and labourer. However symbolic, the ethos of the radical economic reforms did not afford *campesinos* the ability to practice *usos y costumbres* as anticipated.

Adaption in the Neoliberal Era

The ushering in of neoliberal economic reforms critically altered the environmental governance of Bolivian landscapes. Beginning in 1985, Bolivia, like other Latin American countries, was subjected to the provisions of structural adjustment programs (SAPs). In Bolivia, this manifested as "the selling of state-owned enterprises, the reduction of state spending in the public sector, the active courting of foreign investment, and the aggressive exploitation of untapped primary resources." (Chasteen, 2004, p. 302). Drastic economic reforms reduced state spending and increased measures of austerity. Economically devastating, formal labour markets of manufacturing and mining crumbled and informal labour sector jobs skyrocketed to 70% by 1990 (Sanabaria, 2000).

In rural regions, a depletion of the peasant economy through the consolidation of large estates and agrarian business expansion drove small landowners from their land holdings toward urban centres (CEDIB, 2014). Additionally devastating was the closure of factories and mines, resulting in the loss of wage labour opportunities; general unemployment increased the pauperization of middle classes, causing entire rural-based communities to migrate to urban regions in search of consistent wage labour. Small-scale farmers who continued to work their land by traditional methods sustained themselves through seasonal migration to other rural regions or supplemented income with household members living in urban areas for wage labour jobs. Traditional small-plot farming was no longer economically sustainable as a primary income source, and subsistence farming was limited. The result was a significant and rapid urbanization, averaging a 4% growth rate. The mid 1980s marked the tipping point for the geographic population shift, as census data reported a majority urban population. By the early 2000s, three-fourths of the population was located in urban areas, concentrated into eight major urban regions of more than 100,000 inhabitants; and the growth trends have continued (Antequera Durán, 2007).

In urban regions, the resultant surge continued to well up on the fringes of urban areas, in the zone demarcated as the peri-urban. While the new urbanism was the result, it was not a default nor a given. Furthermore, active response to the politic and economic shifts manifested at the national and local levels. Two significant national legislative moves set the tone for regional policy frameworks to enable Indigenous peoples further political and economic agency. In 1994, Bolivian Congress enacted the *Ley de Participación Popular* (Law of Popular Participation) to decentralize political power and management of resources to the municipal level, and provide Indigenous political representation through the instrumentalization of Indigenous-based Grassroots Territorial Organizations (*Organizaciones Territorial del Base*, OTB).[1] In 1996, National Agrarian Reform Law began the process of surveying land to redistribute land from wealthy landholders back to the Indigenous populations.

Nationally, Bolivia elected its first self-proclaimed Indigenous president, Mr. Evo Morales, in 2006, who foregrounded land redistribution and Indigenous rights as his administration's primary agenda for his party, Movement for Socialism (MAS). His party's socialist project was realized with a new constitution in 2009, seeking so-called plurinational unity and elevating the voice of the Indigenous land workers to the national level. The constitution espoused the United Nations Declaration into formalized legislation as a means to grant greater self-determination to its Indigenous peoples. The following year, the Autonomy and Decentralization Framework Law introduced procedures for Indigenous communities to establish autonomy, as a form of self-governance and recognition of self-determination (Kohl & Bresnahan, 2010). While nationally President Evo Morales ostensibly provided a platform for the slow work toward the redistribution of land titles, it is the community-based grassroots action rapidly unfolding in the peri-urban outskirts that is the locus of resistant adaption to maintain connection to their *usos y costumbres*. Insomuch as peripheral zones resist the dichotomy of urban/rural, these spaces allow for Indigenous community activism to unfold. In the case study of Cochabamba's peri-urban milieu one can appreciate how this reality is already occurring.

Peri-urban Cochabamba: Zona Sur

In Cochabamba, the southern extents of the city limits are blurred. The vast lands that stretch south in the valley are affectionately christened with the catchall term Zona Sur. The Zona Sur extends 3 million hectares south of the proper city limits, comprising 43% of the total municipal population and composed of districts 5, 6, 7, 8, 9, and 14 (Antequera Durán, 2008). Families and entire communities migrate to this zone in search of cheap land and access to urban amenities. Alexander Zenteno of the Alcaldia Cochabamba explains that basic services arrive in the peri-urban in reverse order than that of the city. First, newcomers must claim a lot by constructing a small, contained room as a living space. This is constructed at the back of the lot to allow for further building toward the fronting property, and to provide sufficient space for animal husbandry. Subsequently, families apply for a land title to own their property. Next, transportation access is created to link the lot to larger road networks leading into the urban centre. Once a minimum of eight privately owned blocks are constructed, the national Law of Popular Participation mandates communities to form an OTB and elect a leader for a two-year term in order to begin engaging the municipality for basic resources provided by Cochabamba's municipality: healthcare, roads, electricity, water lines, access to education (Zentero, 2013).

Once officially recognized as an OTB, the designation of community land can be annexed into the city as urban territory, and OTBs are complicit to the municipality's terms of engagement. Municipalities are all but overwhelmed by the scope of post-planning oversight needed in the Zona Sur, and progress is slow. Zenteno admits that municipal water lines would likely never reach the southern extents of the Zona Sur. In process, generators provide limited electricity to adjacent structures and *aguatero* (water trucks) provide paid water services at upwards of 700% the cost of 200L in the urban districts (Moeller, 2011). Tension is thick between municipalities and the OTBs. Non-governmental organizations supplement the resources shortage by fortifying houses, providing water lines, schools, and health centres throughout the Zona Sur. Peri-urban residents seek access to urban economies, yet seek to maintain agricultural processes, ties to rural kin, and reciprocal relationship with their land to maintain heritage-based connections. Constant, strategic, subtle – these are the daily practices that enervate the class struggle for Indigenous self-governance that can only arise from grassroots momentum, and what Scott describes as "everyday forms of peasant resistance" (Scott, 2008). These are the informal trade networks, piecemeal transportation routes, seasonal migration to facilitate harvest and offer ritual prayers, maintenance of Indigenous festivals, and animal husbandry that occurs just out of view of street storefronts. It is the children who identify themselves not as Cochabambino nor Campesino, but a third option: "one of the Zona Sur" (Kraft, 2013).

District 9, Zona Sur (Place)

District 9 is the southern most zone of the Zona Sur and lacks a distinct southern boundary. It is the largest of the peri-urban districts, encompassing more than half of the Zona Sur. Population figures approximate 46,000 inhabitants on 1,400 square kilometres of land, yet the numbers grow daily. A group of non-governmental organizations – Pro Casha, Agua Tuya, Habitat para la Humanidad, and Pro Habitat – work in the zone to supplement resources to residents. District 9 is the most recent extension of the other zones and the least dense, primarily encompassing agricultural production. It is composed of 104 distinct compartments: OTBs, neighbourhood boards, and agricultural unions (Antequera Durán, 2008). In short, District 9 captures the essence of the hybrid and heterogeneous territorial expansion of the Zona Sur.

Community Maria Auxiliadora is itself a locus of the conflicts at play in District 9. A group of six women purchased the 16.8 hectares of land from an agricultural labour union to form the community, maintaining the status of an agricultural union. Rosmary Irustra led the group, initiating a communal land trust organized by all women as a place for women and children to live. Irustra structured the community to provide a safe environment for families seeking to escape domestic violence and a machismo culture pervading traditional Indigenous domestic arrangements. Fortified by the national laws promoting Indigenous autonomy and self-organization, Irustra established her community, foregrounding communal ownership, reciprocity, and autonomous organization.

The community continues to organize and functions on its own terms, in alignment with traditional ayllu practices: communally owning land and maintaining relational and economic ties to rural kin. The first Friday of each month the pungent smell of incense rises as the traditional *k'oa* ritual is practised as an act of reciprocity to the Pachamama, Mother Earth. Maintaining traditional connection to land is primary to Maria Auxiliadora; however, the community has clashed with urban political requirements. Municipal authorities jailed Irustra for failing to issue individual land titles. The current president of Maria Auxiliadora, Sra. Maria Eugenia Veliz, explains: "We like [to maintain] the agricultural union designation because you don't have to divide the land." Veliz defends the communal land ownership as a heritage-based practice, resisting the designation of OTB that necessitates private land ownership. She cites the Autonomy and Decentralization Framework Law that allows for Indigenous groups to self-organize and self-govern as they see fit. "[Municipalities] should read the law better to see that this is how we want to organize!" she exclaims (Veliz, 2013).

For Veliz, the OTB requirement of private land ownership plays into the hand of the capitalist model, and she is urgent to resist. She claims the system limits Indigenous peoples' full embrace of their collective autonomy. On the other hand, the municipality recognizes the OTB system as an essential organizing tool to provide resources to informally planned settlements albeit on urban terms. Nationally, the Law of Popular Participation established the OTB as a method to decentralize governance; however, such governance has become complicit with neoliberal models. Maria Auxiliadora is a microcosm for the peri-urban territories that are festering resistant adaption, maintaining tension with local authorities, and providing a platform for actors to realize their agency as Indigenous peoples.

Conclusion

The focus of this research is not only the physical geography of the peri-urban zone, but also the hybrid spatial tactics that unfold in this locale. The peri-urban settlements in Cochabamba, Bolivia present conditions that are informal and enabling. Without normative urban infrastructure of sanctioned utilities and organized transit, or a strictly rural land-use pattern, inhabitants rely on self and community inertia to establish and sustain livelihoods. The interface is a threshold in which peri-urban dwellers maintain social ties to rural communities and also access the city. The fluid nature of this space allows for varied levels of porosity – economic, political, social, and ethnic – that animate this space of exchange. This research narrows to focus upon the ethnic heritage dimensions taking place in the peri-urban zone. Critically, this zone must be understood not only as the geographic manifestation of a particular economic climate, but further informed and fortified by the ethnic heritage-based practices of those who live in this interstitial space.

This research claims that the space of the peri-urban can be a site of resistance to the hegemonic forces of neoliberal globalization. Wrought with contradictions and complexities of land use patterns, economic motives, and livelihood practiced, this zone is situated peripherally to the penetrating powers of the global markets. Slippery and in-between, peri-urbanism is understood and defined in apophatic terms, by what it is not. It is not rural; it is not urban. Capitalism is inherently linked to the dichotomy of rural versus urban as locations of extraction and economic production. Therefore, it does not fall into the clear dichotomy against the global capitalist, but rather poses a third option. It is other. It resists clear distinction and thus serves as refuge for social and political action operating at a ground level.

This model of communal land ownership indirectly challenges the capitalist model of private land ownership and the rural/urban dichotomy. In this peripheral in-between space, community members maintain active ownership of the land. Not maximized for capital profit, this land is considered an integral component to maintaining livelihood, linked to heritage-based practices. Maria Auxiliadora's relationship to land presents a highly distinct value system in contrast to the capitalist model of land as commodity. Urban centres function within the capitalist global framework as do rural regions leveraged for agricultural production. The peri-urban spaces are thus a hold-out for resurgence of truly collective sites of Indigenous self-governance and resistant adaption to the neoliberal context.

In the particular context of Bolivia, the historical specificity amplifies the necessity of a zone such as this. As Indigenous people have been repeatedly torn from their land heritage, this new zone provides potential to regain a collective and communal practice of community making. Heritage-based linkages to the Incan past and the *ayllu* practice once again have a space to develop. And yet, the challenge and the practice is active. The fluid and dynamic nature of this space posits a third alternative to direct opposition to hegemonic forces. Direct opposition cannot compete with the global power dynamic; instead, a peripheral resistant adaption may supply the subversive tactics needed to counter the neoliberal forces. It is here that heritage can take shape and form again, through communal land use practices that engender a collective livelihood, responsive and reactive to global capitalist forces at large.

NOTE

1 Participacion Popular: Que Son Las Organizaciones Territoriales Del Base. Cooperación Alemana En Bolivia, 97ADAD, www.bivica.org/.

REFERENCES

Allen, A. (2003). Environmental planning and management of the peri-urban interface: Perspectives on an emerging field. *Environment and Urbanization, 15*(135), 135–136. https://doi.org/10.1177%2F09562 4780301500103

Antequera Durán, N. (2007). *Territorios urbanos*. Cochabamba: CEDIB.

Antequera Durán, N. (2008). *Carpeta de datos de la Zona Sur de Cochabamba*. Cochabamba: CEDIB.

Beals, C. (1973). *The incredible Incas: Yesterday and today*. New York: Abeland-Schuman.

CEDIB. (2014). Presentacion Escarley Torrico.

Chasteen, J. (2004). Problems in modern Latin American history: Sources and interpretations: Completely revised and updated. In J.C. Chasteen & J.A. Wood (Eds.), *Latin American silhouettes*. Wilmington, Del: SR Books.

Kohl, B., & Bresnahan, R. (2010). Introduction: Bolivia under Morales: National agenda, regional challenges, and the struggle for hegemony. *Latin American Perspectives, 37*(4), 5–20. https://doi.org/10.1177%2 F0094582X10370172

Kraft, A. (2013). Interview by Hannah-Hunt Moeller, 2 April 2013.

Larson, B. (1998). *Cochabamba, 1550–1900: Colonialism and agrarian transformation in Bolivia*. Durham and London: Duke University Press.

Lefebvre, H. (2009). *The production of space* [Trans. Donald Nicholson Smith]. Oxford: Blackwell.

Lerner, A., & Eakin, H. (2011). An obsolete dichotomy? Rethinking the rural-urban interface in terms of food security and production in the global south. *The Geographic Journal, 177*(4), 311–320. https://doi.org /10.1111/j.1475-4959.2010.00394.x

Moeller, H.-H. (2011). *Encuesta: Sistemas de agua reclidado*. Foundation for Sustainable Development.

Olivera, O. (2004). Leader of the coordinator of defense of water and life. Cochabamba Interview by Thomas Perreault, 12 February 2004.

Perreault, T. (2006). From guerra del agua to the guerra del gas: Resource governance, neoliberalism and popular protest in Bolivia. *Antipode, 38*(1), 150–172. https://doi.org/10.1111/j.0066-4812.2006.00569.x

Rivas, A.S. (2000). *Los hombres de la revolucion: Memorias de un lider campesino*. La Paz.

Rostworowski de Diez Canseco, M. (1999). *History of the Incan realm*. Cambridge: Cambridge University Press.

Sanabaria, H. (2000). Resistance and the arts of domination: Miners and the Bolivian state. *Latin American Perspectives, 27*(1), 56–81. https://doi.org /10.1177%2F0094582X0002700105

Scott, J.C. (2008). *Weapons of the weak: Everyday forms of peasant resistance*. New Haven: Yale University Press.
Stern, S. (1987). New approaches to the study of peasant rebellion and consciosness: Implications of the Andean experience. In S.J. Stern (Ed.), *Resistance, rebellion, and consiousness in the Andean peasant world, 18th to 20th centuries*. Madison.
UN. (2008). *United Nations declaration on the rights of Indigenous peoples*. http://www.un.org/esa/socdev/unpfii/documents/DRIPS_en.pdf.
UN. (2011). *World urbanization prospects, 2011 Revision*. https://www.un.org/en/development/desa/population/publications/pdf/urbanization/WUP2011_Report.pdf.
Veliz, M.E. (2013). Interview by Hannah-Hunt Moeller, 5 April 2013.
Zentero, A. (2013). Alcaldia Cochabamba, interview by Hannah-Hunt Moeller, 3 April 2013.
Zibechi, R. (2011). *Territorios en resistencia: Cartografia politica de las periferias urbanas Latinoamericanas*. CGT.

10 Blurring the Urban-Rural Divide: Urban Peripheries as Sites of Food Sovereignty Construction in Caracas

CHRISTINA M. SCHIAVONI AND ANA FELICIEN

Introduction: Food Sovereignty and Urbanization

This chapter explores how communities in Venezuela are blurring the urban-rural divide as they reconfigure territory, relationships, and identities in the attempted construction of food sovereignty.[1] This exploration is embedded in broader discussion and debate around food sovereignty and urbanization. What is the relevance of food sovereignty, a concept known for its peasant origins, to urban struggles and movements? And what are the prospects for overcoming longstanding tensions characterizing urban-rural relations, such as the need for cheap food provisioning to feed the urban proletariat versus farmers' needs for fair pricing for what they produce?[2] Such questions are increasingly pertinent with over half of the world population living in cities at the same time that the problems of the global food system are becoming ever more pressing, particularly in the face of climate change and a global resource rush (De Schutter, 2014; Murphy & Schiavoni, 2017). Among scholars and advocates of food sovereignty is the widely held conviction that transforming today's deeply flawed food systems will require the convergence of movements across a number of traditional divides, including geographical and sectoral ones (Holt-Giménez & Shattuck, 2011; Brent et al., 2015; Tramel, forthcoming). Not only is the task of food sovereignty construction too great for rural populations to shoulder alone, but urban and rural populations find themselves inextricably linked both in their relationship to the predominant global food system, from global commodity flows to circuits of migration, and in their forms of resistance to it, such as the development of alternative models of food provisioning (McMichael, 2009; Figueroa, 2015). Approached in this light, food sovereignty is no more of a rural question than an urban one, requiring the engagement of the populace as a whole.

A barrier to broader engagement in food systems transformation, however, lies in the reality that many urban inhabitants find themselves distanced, both spatially and relationally, from processes of food production and from those who produce it (Robbins, 2013, 2015). This condition is reflected in the widely applied classifiers of "producers" and "consumers," largely divided along rural-urban lines, in which the former is reduced to serving the latter, while the latter is reduced to being "merely shoppers rather than political, active agents in the food system" (Robbins, 2013, p. 24). Dismantling the binaries of urban/rural and producer/consumer thus becomes key to meaningful and sustained food systems transformation. This is already happening, however gradually and unevenly, in numerous spaces across the globe, where such long-standing divisions are becoming increasingly blurred. This is especially the case in urban peripheries that, out of necessity, are becoming sites of food sovereignty construction as marginalized urban inhabitants take matters into their own hands through piloting new (and old) forms of food provisioning tailored to their particular histories and identities. In doing so, they are tapping into elements of their not-so-distant agrarian pasts and renewing points of connection with the countryside and its inhabitants (Figueroa, 2015; Schiavoni, 2015, 2017).

Several features of urban peripheries lend themselves well to organic forms of food sovereignty construction that are blurring the urban-rural divide. As Caldeira (this volume) describes, "Everything [there] happens in a kind of alternate market that specializes in the needs of the urban poor and bypasses some dominant official logics." This in turn opens up spaces for creative and resourceful forms of "autoconstruction"; for "new discourses of rights"; and for "the invention of new democratic practices" (Caldeira, this volume). A vivid demonstration of these points in relation to food sovereignty construction is the work of Figueroa, premised on how "[t]he sheer scale of unemployed, superfluous labor in urban peripheries the world over has created vast spaces where self-reliance has become a vital necessity" (Figueroa, 2015, p. 505). Within these spaces, she argues, food practices of *everyday life*, often born out of survival techniques under capitalism, contain the seeds for transcending capitalism's grip on the food system and other realms of life, and the seeds for building contextually meaningful forms of food sovereignty. Figueroa (2015, p. 505) conceives of "everyday life" as "an ongoing, living process [that] is continually "leaking out the sides," so to speak, of capitalist structures; its "residue" confounding the attempts of abstraction and alienation to contain it." Focusing on Chicago's predominantly African-American South Side neighbourhood, she examines how survival techniques passed down from the

Jim Crow-era rural Southern US,[3] such a food-purchasing clubs, are forming the basis for food sovereignty construction today. In doing so, she makes the case for a "people-centred" approach in which food is "perceived as an *ensemble of relations*, a kind of nexus in and through which social processes at varied spatial and temporal scales converge and interact" (emphasis added; Figueroa, 2015, p. 502).

Figueroa's exploration into food sovereignty efforts in an urban community, looking in particular at the innovations arising out of spaces that have not been fully penetrated by capitalist logic, speaks to Caldeira's work on peripheral urbanization in demonstrating how creative forms of autoconstruction among marginalized urban inhabitants are, as Caldeira (2017, p. 9) frames it, "generat[ing] new modes of politics through practices that produce new kinds of citizens, claims, circuits, and contestations." Particularly relevant here is Figueroa's assertion that

> In the spaces where people resist, or are discarded by, the march of capitalist development, the diverse social networks, practices, and resources they have always marshaled for daily subsistence become salient building blocks for new social configurations of collective survival that ... can potentially emerge as practically viable, culturally meaningful, and self-determined pathways to food sovereignty as a means of transcending life under capitalism. (Figueroa, 2015, p. 506)

With upwards of 90% of its population living in cities, Venezuela provides a fascinating backdrop for exploration into the blurring of the urban-rural divide via processes of food sovereignty construction within urban peripheries and beyond. On the one hand, given the sheer extent of its urban population, Venezuela exemplifies many of the challenges of food provisioning vis-à-vis "mega-urbanization," from inequalities in food access to a prevalence of diet-related disease. Such challenges are compounded by a macroeconomic context of import dependency sustained through a petroleum-centred economy, with all of the instability that entails, along with high rates of inflation and a thriving parallel market, each with deeply political as well as economic dimensions (Curcio Curcio, 2017). At the same time, Venezuela's urban spaces, particularly its urban peripheries such as the many hillside barrios of Caracas, are part of a living experiment in food sovereignty construction over the past nearly two decades of the Bolivarian Revolution, with food sovereignty enshrined in law as well as a priority of many social movements and citizen groups. Both the barriers to and opportunities for food sovereignty construction in Venezuela are thus immense.

But what does food sovereignty even mean – and what do attempts toward it look like – for such a highly urbanized population? To explore this question, we will look at a variety of grassroots efforts in Venezuela that are blurring the boundaries of "urban" and "rural," drawing from the frameworks of Caldeira (this volume) and Figueroa (2015), as well as Schiavoni's (2017, p. 3) historical-relational-interactive approach to food sovereignty construction as "a historically embedded, continually evolving set of processes that are interactively shaped by state and societal forces, reflecting competing paradigms and approaches."

Constructing the Divide

Employing the above lenses to Venezuela, particularly a historical lens, we can examine the ways in which social processes around food have been fundamental to the construction of the urban-rural divide there, just as they are fundamental to its deconstruction today. The role of food as a means of social and geographic differentiation dates back to the period of colonization by the Spanish, from the late fifteenth to early nineteenth centuries. This involved not only the development of an extractive agro-export complex of products such as cacao, coffee, and sugarcane, supported in part by slave labour, but also differentiations between the foods of Indigenous, *campesino*, and Afro-descendent communities living mainly in rural areas and the foods of the colonial elite, concentrated more in urban areas (Lovera, 1998). One example is the introduction of wheat into Venezuela in the colonial period, which remained largely limited to the urban elite through the 1940s, while the rural majority of the population continued to consume maize, cassava, and other root crops as their main carbohydrates (Carbonell & Rothman, 1977). These indigenous crops were produced mainly through family and community plots known as *conucos*.

Throughout colonization and into the independence period, the majority of the population remained in the countryside, largely subsisting from its own production. The situation took a major turn, however, in the early 1900s, when Venezuela began to exploit its vast petroleum reserves, becoming the world's largest petroleum exporter by 1935 (Wilpert, 2006). As attention turned to oil, both the land-owning elites and the government lost interest in agriculture and stopped investing in land. Few jobs remained on large landholdings for rural workers, while small-scale farmers were met with little to no support. The flight of capital from the countryside was thus accompanied by a mass exodus of *campesinos* (peasant farmers and rural workers) into the cities (ibid.), particularly Caracas, which grew from 80,000 inhabitants at the start of the twentieth century to 3 million by the end (Rauseo, 2014). With little work to be found, many *campesinos* were pushed to the edge of existence, living in extreme poverty. For those remaining in the countryside – just over 10% of the population by 1999 – the situation was equally tenuous. Reflecting Venezuela's colonial legacy, 75% of the land was concentrated among 5% of the largest landowners while 75% of the smallest landowners shared only 6% of the land (Wilpert, 2006). These small landowners also faced a lack of basic public services and received little to no technical or material support to engage in agricultural production.

The abandonment of its agriculture sector led Venezuela to become among the most urbanized countries in Latin America and the first country in the region to be a net importer of food (Wilpert, 2006). Along with the emergence of the oil economy emerged a relationship between the steady decline of domestic food production on the one hand, and the increasing reliance on cheap imported food items on the other, in order to feed an increasingly urbanized population. This was facilitated by an ongoing flow of revenue from oil exports and the availability of cheap food made possible through subsidized overproduction in the US, beginning in the 1950s (Carbonell & Rothman, 1977). Such an arrangement corresponds to what Friedmann and McMichael have characterized as the "second food regime" during the 1950s–70s, which "rerouted flows of (surplus) food from the United States to its informal empire of post-colonial states on strategic perimeters of the Cold War" (McMichael, 2009, p. 141). A particularly significant development at this moment was the (re)introduction of wheat into Venezuela, this time not as a food for the elite, but for the population at large, facilitated through trade arrangements between the US and Venezuela.

The results of these measures were widely felt. With the exception of a few pockets in the Andean region, wheat is not grown in Venezuela, as it is not adapted to the country's agroecosystems. By the 1950s, imported wheat from the US entered into a dynamic competition with corn produced through the national agro-industry as the country's top-consumed staple foods. Culinary traditions and dietary patterns were radically altered while dependency on imported US agricultural goods grew. By the end of that decade, imported wheat officially displaced corn as the country's top staple, although corn consumption would again rise with the introduction of industrially produced "pre-cooked" corn flour in the early 1960s (Carbonell & Rothman, 1977). Pre-cooked corn flour remains the country's most widely-consumed carbohydrate to date, particularly in the form of round corn cakes called *arepas*, while wheat has also remained a prominent component of the Venezuelan diet, as one of the top five most consumed foods (Ablan & Abreu, 1999; Vida Agro, 2017). More importantly, Venezuela's food system would continue to be characterized by a prevalence of industrially processed foods supporting cheap food provisioning for new urban habitants. A 2004 study found that the majority of Venezuelans' calories are derived from pre-cooked corn flour, vegetable oils, and refined sugar delivered through the processed foods industry (Abreu & Ablan, 2004). The prevalence of processed foods in the Venezuelan diet is believed to be a key factor in the high incidences of diet-related disease in the country, with over a third of the population overweight or obese (Briceño-Iragorry et al., 2012).[4]

These developments marked a major loss of food self-sufficiency for Venezuela, which was importing upwards of 80% of its food supply by the 1990s (FAO, 2002). Another impact of Venezuela's turn to imports of industrially processed foods was the consolidation of a powerful food importation and distribution complex, controlled by several national and transnational firms that have dominated Venezuela's food system in an oligopolistic fashion (Rios de Hernández & Prato, 1990; Curcio Curcio, 2017; Felicien et al., 2018), particularly Cargill, Nestle, and Polar (Gavazut, 2014). Of these firms, the country's largest private food conglomerate, Polar, controls more than half of the country's supply of pre-cooked corn flour, along with a number of other items comprising Venezuela's basic food basket (Felicien et al., 2018). As Polar is owned by a prominent member of the political opposition to the government, many see it as no coincidence that these very food items have been missing from supermarket shelves as of late, particularly following the death of former President Hugo Chavez Frias in 2013. While such shortages have intensified in recent years, they are nothing new. Venezuelan economist Pasqualina Curcio has documented a correlation between shortages of basic food items such as corn flour and politically heated moments such as election periods over the course of the Bolivarian Revolution (Curcio Curcio, 2017). Although the causes of ongoing shortages of basic food items remain contested, they point to some severe cracks in Venezuela's food system.

The story of Venezuela's food system does not end here, however. Amidst waves of urbanization and depeasantization over the twentieth century, some *campesinos* did stay in the countryside and continued to farm, against the odds, while others brought pieces of the countryside, and their rural identities, with them into the cities. An example of the former are *campesinos* in the Andean region, who not only stayed on their terraces, but continued to grow native potato and other tuber varieties cultivated by their ancestors since before colonization. Such producers are playing an important role in the reviving of local food systems and in broader efforts toward food sovereignty today (Romero et al., 2016). Those who flooded into cities, the majority into vertical makeshift shantytowns covering the hillsides of Caracas, were met with few job opportunities and a lack of basic services. Practices brought from the countryside, from rearing poultry and other small livestock, to growing fruit and coffee trees, to maintaining traditional culinary techniques, were among their tools for survival. Additionally, some of these urban migrants maintained links to the countryside though family members remaining there, making periodic visits and bringing back supplies such as food, seeds, medicinal plants, and building materials into the city. Remnants of their rural origins thus remain strong among many of the inhabitants of Venezuela's urban peripheries, arguably fitting Jacobs's (2017) characterization of an "urban proletariat with peasant characteristics."[5] The variations of dishes prepared from one household to another in Caracas' barrios, for instance, will often reveal families' distinct geographical origins, based on diverse local culinary traditions brought from the countryside.

Long flying below the radar, today such "everyday" food practices, as framed by Figueroa (2015), are being looked toward as key building blocks of food system transformation, particularly among working-class communities, both urban and rural, who are the hardest hit by the current shortages and other challenges facing Venezuela's food system (Felicien et al., 2018). That is, as once-ubiquitous processed foods have become harder to access through manufactured food shortages, there is a resurgence of interest among the working-class communities affected by the shortages in home-grown foods and traditional cooking techniques, such as making *arepas* from freshly ground corn and other starches such as cassava, plantains, and sweet potatoes. We will explore this resurgence in the pages to follow, looking at attempts to scale such efforts upward and outward. We will also examine how, in the process, the longstanding binary of urban and rural is being challenged as the lines between the two are becoming increasingly, and intentionally, blurred. While these efforts have been, and remain, of a highly political nature, they are becoming of increasing material significance as well, as they take on a new level of urgency and importance in the face of the major challenges presently confronting Venezuela's food system.

Deconstructing the Divide

At the start of Venezuela's Bolivarian Revolution in 1999, with rural communities in dire straits and over half of its largely urban population facing hunger and poverty, food production and provisioning were identified as strategic priorities of both the government and citizens. This is reflected in article 305 of Venezuela's constitution, passed by popular referendum the same year, guaranteeing the population a stable and secure food supply based on a prioritization of domestic production (Ministerio de Comunicación e Información, 1999). Toward these ends, a variety of state-sponsored initiatives have been carried out, in tandem with citizen efforts. On the production end, the state has made substantial reinvestments in agriculture, including an agrarian reform process to redistribute large landholdings and various forms of support for small- and mid-scale farmers and fishers (Wilpert, 2006, Lavelle, 2014, Enríquez, 2013). On the distribution end, strategies have included increased availability of basic food items at subsidized and regulated prices and provisioning of free meals via school and workplace programs as well as community-based feeding sites (Schiavoni, 2015).

While such efforts have made historic gains in food security, earning Venezuela multiple recognitions by the by the UN Food and Agriculture Organization (FAO, 2013, 2015), many of the state's food and agrarian policies and programs have functioned along the urban-rural binary. That is, faced with hunger rates among its largely urban population at the start of the Bolivarian Revolution, the government turned to the longstanding food importation complex to import more food to feed people at the same time that it rolled out its agrarian reform programs toward the longer-term vision of greater domestic food production. But these were approached as largely separated projects rather than systemic shifts. Thus, even with domestic production reinvigorated and the population better fed, a weak link has been in connecting the two, that is, in building alternative channels connecting production and consumption at the national scale. This weakness is reflected in the continued entrenchment of the country's powerful, longstanding food import and distribution complex, which has remained largely unaltered to date, and the periodic food shortages associated with it (Curcio Curcio, 2017).

Where state policies have faltered, however, social movements have stepped in, working to forge direct links between producers and consumers that bypass many of the bottlenecks in Venezuela's food system. As one food sovereignty activist explained, "Now that a lot of food is being produced in the countryside, we're trying to structure a proposed alliance between popular movements in the city and in the countryside, in order to cut the 'destructive distance' that lies between us" (Schiavoni, 2015, p. 474). Another added, "We know that food security is achieved through resources. But *food sovereignty* has to be a process coming from the bottom up – from the peasant, from the communities" (ibid.). Such grassroots efforts have received a boost in recent years in the form of a push, coming from both above and below, for the construction of new citizen-led social institutions intended to advance direct citizen participation in both politics and governance. While these are diverse, spanning sectors and geography, the most ubiquitous and widely recognized are communal councils and comunas. Numbering more than 45,000 across the country, communal councils are democratically elected community bodies that practice self-governance, while interfacing with the state. Increasingly, communal councils are linking up across territories to form comunas, as vehicles for strengthening popular power, many of which have self-managed socio-productive projects in a move toward greater autonomy (Foster, 2015). Communal councils and comunas are officially recognized by the state via a series of popular power laws enacted over the past decade, although the existence of such citizen bodies precedes the law (Ciccariello-Maher, 2016).

Today, there are more than 1,500 comunas organized as delimited territorial units across Venezuela. As comunas link together horizontally through economic and social interactions, this is creating a new communal network that is expanding across territory, in which both urban and rural comunas are part. One such mechanism for the expansion of communal territory is the creation of "communal corridors" that join together multiple comunas in social, economic and political experiments in autoconstruction, including in food production and exchange. Such configurations are challenging the longstanding urban-rural divide in Venezuela. Angel Prado, member of the rural comuna El Maizal and part of the national communal movement, has stressed that while the majority of the population is concentrated in cities, the majority of the national surface area remains rural, and much of this rural surface area is under communal construction.[6] This gives rural spaces a new level of importance in the largely urban-dominated political scene of the Bolivarian Revolution, and weaves together urban and rural movements in the reconfiguration of territory via comunas, under the shared identity of *comuneros* and *comuneras*. Rey Torres (2014, p. 184) describes "comunas" as having a fluidity that "extends across the whole social body, producing new spatial modalities that break with the historical countryside/city dichotomy, proposing a new form, a new relationship." This in turn opens up new potential for comunas as vehicles for the construction of food sovereignty across the urban-rural divide. Prado's vision is both to achieve greater food self-sufficiency in his own comuna, as well as support the food needs of urban comunas with less production capacity. The vision, however, is not of a one-way flow of goods from the countryside into the cities. Prado stresses, for instance, that in cities such as Caracas there are both the people power and the infrastructure for food processing and distribution enterprises, and ample possibility for partnership between rural and urban groups in this area, among others.

A glimpse into such possibility can already be seen in the urban comuna El Panal 21 of Caracas' 23 de Enero barrio, a well-known hotbed of political organizing and autoconstruction. El Panal has a food packaging microenterprise as one of its core socio-productive activities and has partnered with rural movements for the processing of a number of food items for local distribution, as well as partnered with them to grow food outside of the city. Robert Longa of El Panal explains: "We're building national points of connection between the urban and the rural that allow us to break capitalist chains of distribution and production" (Schiavoni, 2015, p. 475). Longa explains that the comuna has several projects underway in the countryside, which include training and educational components that enable comuna members to connect (or in some cases, reconnect) to processes of production. These efforts are complemented by an urban agriculture effort within El Panal that includes raised garden beds, a nursery, and worm composting, among other components. Longa explains that it is a process of ongoing learning that combines life in the city with life in the countryside.

Scaling Alternatives

At present, El Panal is among a growing number of comunas involved in bringing such grassroots efforts for food system transformation to a larger scale through an initiative known as Plan Pueblo a Pueblo (Plan People to People). True to its name, Pueblo a Pueblo is a people-to-people effort to forge direct connections between Venezuela's small-scale farmers and urban communities. Organizer Ricardo Miranda, a 23 de Enero native who has long worked at the intersection of urban and rural struggles, explains that the countryside and rural issues have tended to be invisibilized in the public consciousness, even when they have direct bearing for urban dwellers: "Take the question of seeds. If urban people want particular foods that farmers can't access seeds for, then this is just as much an urban concern as a rural one."[7]

The basic premise of Pueblo a Pueblo is relatively simple, at least on the surface. Rather than rely on intermediaries who often undercut producers while overcharging consumers, urban communities partner directly with rural ones to bring fresh food into the city at affordable prices. By mid-2016, in just over a year since its inception, this initiative was already reaching upwards of 40,000 families in the urban peripheries of Caracas and several other cities with affordable farm-fresh goods. Today, it is reaching upwards of 60,000 urban families, and growing. Among its keys to success is working through comunas. That is, Pueblo a Pueblo works with communities that are already self-organized. It is up to the community organizers involved in the comunas to do an inventory of community food needs by household, to physically get the food from the farmers (as most comunas have their own vehicles such as pick-up trucks, a barrier for many small-scale farmers), and to organize distributions. Meanwhile, the farmers focus on growing the crops based on communities' needs, and the Pueblo a Pueblo teams helps to build essential linkages, both linkages among urban and rural communities and strategic linkages with the state.

Among the many noteworthy aspects of Pueblo a Pueblo is its origins. Its core organizing team is comprised of both urban and rural social movement leaders who came to hold key positions in state institutions earlier on in the Bolivarian Revolution, including overseeing the distribution of land, credit, and other forms of support to small-scale farmers and landless workers through the country's agrarian reform process. Upon witnessing the limits of working from within the state, however, they have opted to return to grassroots organizing as what they see as the most effective vehicle for social transformation, especially via the comunas. At the same time, they have maintained important links with the state that have helped support the efforts on the ground.

Another notable aspect of Pueblo a Pueblo is their deliberate distinction from market-based models. First, they are intentional about determining cost structure based on the realities of the farmers and consumers involved, not on pricing arbitrarily determined by the market. Ricardo Miranda explains:

> Food should not be treated as merchandise. We don't go by supply and demand. Why should potatoes cost more than carrots just because there's more demand for them right now? Who determines that potatoes cost more? The market dictates it. But we don't go by "the invisible hand of the market." We go by cost structure determined by the groups involved. We're not vendors, nor are we Santa Claus – we're into politics.[8]

Reflecting this philosophy, the prices paid by urban comuna members at Pueblo a Pueblo distributions are substantially lower than going market rates (on average, one-third of the cost of typical market prices), while the small-scale farmers involved are guaranteed fair prices that they themselves have helped to set, rather than being at the mercy of what the intermediaries offer them. The farmers are also relieved of the cost and burden of transportation to get their goods into the city. This is significant, as Pueblo a Pueblo works with some of the country's most remote and resource-limited farming communities, such as those who cultivate via animal traction on steep Andean mountainsides.

Second, Pueblo a Pueblo is intentional in their focus on the working-class communities, or what are known as the "popular sectors," of Venezuela's urban peripheries. These communities are seen as natural partners, in that they are already self-organized and engaged in various forms of experimentation in alternative-building via radical forms of democratic participation, or what Caldeira (2017, p. 9) has characterized as "the experimentation with new forms of local administration, and the invention of new approaches to social policy, planning, law, and citizen participation." They are also among those most deeply impacted by the food shortages and other problems of Venezuela's food system and, out of necessity, are already engaged in grassroots responses to these. This approach of working through comunas situated in urban peripheries has significantly boosted Pueblo a Pueblo's reach. Every family within a given geographic area covered by a given comuna is included in the initial census and is then automatically eligible to participate, and prices are structured to be accessible to the community at large.

The accessibility of Pueblo a Pueblo is a significant departure from the market-based models pursued by many mainstream food movements, which have increasingly come under fire for feeding into as opposed to dismantling inequities in the food system. That is, within many of these models, good food is available to those who can afford it rather than approached as a universal human right (Brent et al., 2015; Trauger, 2014; Billings & Cabbil, 2011; Holt-Gimenez and Shattuck, 201). While such trends are particularly evident in the Global North, similar patterns can increasingly be seen throughout the Global South as well, where organic and other alternative foods are on the rise, but largely targeted toward wealthier "conscious consumers"/"foodies" and largely out of reach of the poor (e.g., Woo, 2015). Finally, as stressed in Miranda's quote above, Pueblo a Pueblo is explicitly political. The vision is food system transformation, as part of broader societal transformation, across the urban-rural divide. The food distributions are not seen as an end goal, but rather as a means of joining urban and rural inhabitants in a common effort while addressing the immediate needs for affordable food for consumers and liveable incomes for farmers. Toward this end, in the Pueblo a Pueblo exchanges, a term being increasingly used is that of *prosumidor(a)*, a combination of the words for "producer" and "consumer" in Spanish, and an intentional blurring of the two.

Involvement in Pueblo a Pueblo does not necessarily imply identifying as a *prosumidor(a)*, however, nor are all 60,000+ participants involved out of a political commitment to change the food system. Some are simply concerned about healthy and affordable food for their families, and the organizers understand and embrace this. The accessibility of Pueblo a Pueblo is what sets it apart from many other initiatives of a similar nature, in Venezuela and elsewhere, presenting both unique opportunities and challenges. On the one hand, in working through the vehicle of the comunas, Pueblo a Pueblo has rapidly reached a scale largely unparalleled by similar efforts, while bypassing issues of elitism and exclusivity that have plagued local food initiatives elsewhere. Community members of all walks of life are involved, including in some of the most historically excluded inhabitants of Caracas' urban peripheries. And that's the very idea. On the other hand, operating at such a scale presents challenges in advancing the political dimensions of Pueblo a Pueblo, which are the main motivation among those who conceived of it. The organizers explain that this is why education and outreach are so central to their work. They see this as an ongoing process of construction, and "as a new way of relating to each other, and to food – that food is not a commodity."[9]

A key component of their political work, according to the Pueblo a Pueblo team, is supporting communities in changing predominant food cultures and associated eating patterns that have permeated urban areas, such as a dependency on processed foods and preferences for uniform and "perfect-looking" produce. One means of this is through recipe-sharing, including of dishes that are still commonly consumed in the countryside, but that, it is perceived, "Caraqueños don't eat." These recipes and food preparation techniques brought in from the countryside are combined with those of urban inhabitants who have held onto culinary practices passed down over generations, in what has become an ongoing process of reclaiming traditional foodways. Also key in shifting habits and emphasizing the political dimensions of Pueblo a Pueblo has been the organization of *encuentros* [encounters] in the countryside, fostering direct interaction between urban comuna members and their rural counterparts, those who are growing their food. Once they return to the city, many of these urban dwellers "begin to look at their food – and their relationship to it – differently."[10]

Blurring Boundaries

Another initiative that has gone perhaps even further in blurring the urban-rural divide is the Feria Conuquera, a large monthly alternative market in Caracas featuring agroecologically produced fresh foods and artisanal versions of many of the products missing from supermarket shelves. The Feria was initiated by the same network of food sovereignty activists who had mobilized around Venezuela's progressive new Seed Law passed at the end of 2015, which banned transgenic crops and protected locally held seed varieties (Camacaro et al. 2016; Felicien et al. 2020). They realized that they needed to be building models in the here and now as concrete steps toward the long-term goal of food system transformation reflected in the Law. Another motivating factor was the shortage of basic food items that had begun to intensify in 2014, when the Feria began. The idea was to provide alternatives to these products while supporting local producers and actively working to build an alternative food system, in a form of prefigurative politics.

The Feria has since evolved into both a space and a collective. As a space, it functions as a vibrant and highly visible monthly market in a public park in downtown Caracas with sustainably produced and artisanally made goods at affordable prices. It also features free hands-on workshops on topics such as seed saving and food preservation, as well as serving as a space of political debate. Much more than a market, the Feria has become a monthly hub of education and organizing around food sovereignty. Additionally, the Feria functions as a collective in which urban, peri-urban, and rural producers, herbalists and artisans of different backgrounds and levels of experience support each other in a variety of ways, including bartering of goods, seed exchanges, skill sharing, and work exchanges. Those with larger operations, for instance, will plan work days called *cayapas* in which other collective members come and share their labour, while gaining hands-on experience and learning new skills.

Part of what makes the Feria so interesting is that it brings together urban, peri-urban, and rural inhabitants in a highly organic and fluid manner. Some are life-long food producers while others are looking to get their hands in the soil. What unites them all is the common identity of *conuqueros* and *conuqueras*. Translated literally, a *conuquero/a* is a person who works a *conuco*, a form of traditional polyculture with Indigenous origins, as mentioned above. But this term is deeply political as well. Following decades of agricultural modernization, depeasantization, and urbanization, the *conuco* has come to represent a form of resistance and a reclaiming of traditional foodways and cosmologies. Taking on the identity of *conuqero/a*, as those involved in the Feria refer to themselves, is thus at once a political stance, a reminder of shared origins, and an invitation to others to tap into their own inner *conuqero/a* as well. Indeed, many are. As one of the Feria collective members explained:

> There's always been a lot of urban agriculture and artisanal food production in Caracas, but for years it's flown below the radar. There are something like 300 urban producers in the community of La Vega (Caracas) alone. We're not doing anything new. We're just bringing visibility to what has long existed. We're highlighting the underground agriculture of the city. And in doing so, others are getting onboard. It's like a connection to the countryside that everyone already had inside them.[11]

In the midst of the crisis currently facing Venezuela's food system, this connection to the countryside, long nascent and under the radar, is not only coming to the fore at the household and community levels in Venezuela's urban peripheries, but is increasingly emerging as a potent force for broader food system transformation. The Feria can be understood in this light, reflecting Figueroa's insights into how "how the infinite array of relationships, resources, histories, struggles, and aspirations that express themselves in the everyday experience of food are the raw material from which many possible paths to a just and sustainable food system can be built" (Figueroa, 2015, p. 510).

Conclusion: Food as Dividing, Food as Uniting

Through the prism of food as a nexus of social relations, it is evident how the urban-rural divide in Venezuela was constructed over time and the ways in which it is being deconstructed today. This involves a reconceptualization of both territory and relationships, and a construction of common identities, such as *comuneros/as*, *prosumidores/as*, and *conuqueros/as*. While the Venezuelan case could be easily dismissed by scholars of food sovereignty for its seemingly inhospitable conditions for food sovereignty construction, it is these very conditions that, out of necessity, are proving to be fertile ground for seeds of food sovereignty today. This is particularly the case in the country's urban peripheries, where connections to the countryside, deeply embedded in communities' collective consciousnesses, are manifesting in everyday practices around food and being scaled upward and outward through grassroots efforts aimed at blurring the urban-rural divide. As the Feria Conuquera collective member quoted above reflected, "Food is part of what's been separating us and now is what is bringing us together."

The Venezuelan case points to the need for further inquiry into food sovereignty vis-à-vis urbanization, particularly into urban peripheries as sites of food sovereignty construction.

NOTES

1 A concept first popularized by transnational agrarian movements in the 1990s in response to the impacts of globalization on food and agriculture systems, particularly with the rise of the World Trade Organization (WTO), food sovereignty is most commonly defined (in short) as "the right of peoples to healthy and culturally appropriate food produced through ecologically sound and sustainable methods, and their right to define their own food and agriculture systems" (Nyéléni 2007).
2 For further elaboration of such debates, see Bernstein (2014).
3 The Jim Crow era was marked by a series of local and state laws enforcing racial segregation in the southern United States from the 1870s to the 1950s.
4 These trends have likely shifted in the face of ongoing shortages of key food items from 2014 onward, although reliable data on the dietary impacts of the shortages is not available at the time of writing.
5 Although with many lacking formal employment, perhaps they might be characterized as a "surplus population," as described by Li (2010), or as what Figueroa (2015) describes as "not-quite-proletarians," with peasant characteristics.
6 Personal communication, July 2015.
7 Personal communication, April 2016.
8 Personal communication, April 2016.
9 Personal comunication, April 2016.
10 Personal communication, April 2016.
11 Personal communication, August 2017.

REFERENCES

Ablan, E., & Abreu, E. (1999). The cereal flour enrichment program in Venezuela: Some results during a decreasing food purchasing power stage. *Food Policy, 24*(4), 443–45. https://doi.org/10.1016/S0306-9192(99)00044-5

Abreu, E., & Ablan, E. (2004). ¿Qué ha cambiado en Venezuela desde 1970 en cuanto a la disponibilidad de alimentos para el consumo humano? *Agroalimentaria, 9*(19), 13–33. http://ve.scielo.org/scielo.php?script=sci_arttext&pid=S1316-03542004000200001&lng=es&tlng=pt

Bernstein, H. (2014). Food sovereignty via the "peasant way": A sceptical view. *Journal of Peasant Studies, 41*(6), 1031–1063. https://doi.org/10.1080/03066150.2013.852082

Billings, D., & Cabbil, L. (2011). Food justice: What's race got to do with it? *Race/Ethnicity: Multidisciplinary Global Contexts, 5*(1), 103–112. https://doi.org/10.2979/racethmulglocon.5.1.103

Brent, Z.W., Schiavoni, C.M., & Fradejas, A. (2015). Contextualising food sovereignty: The politics of convergence among movements in the USA. *Third World Quarterly, 36*(3), 618–635. https://doi.org/10.1080/01436597.2015.1023570

Briceño-Iragorry, L., Gabriela Valero, B., & Briceño, L.A. (2012). Obesidad ¿Es una realidad en Venezuela? Epidemiología Pandemia del siglo XXI. *Colección Razetti, 120*(2), 59–90. Retrieved from http://saber.ucv.ve/ojs/index.php/rev_gmc/article/view/17903

Caldeira, T. (2017). Peripheral urbanization: Autoconstruction, transversal logics, and politics in cities of the Global South. *Environment and Planning D: Society and Space, 35*(1), 3–20. https://doi.org/10.1177%2F0263775816658479

Camacaro, W., Mills, F., & Schiavoni, C. (2016). *Venezuela passes law banning GMOs, by popular demand*. Counterpunch. http://www.counterpunch.org/2016/01/01/venezuela-passes-law-banning-gmos-by-popular-demand-2/ (Accessed on 4 September 2017).

Carbonell, W.J., & Rothman, H. (1977). An implicit food policy: Wheat consumption changes in Venezuela. *Food Policy, 2*(4), 305–317. https://doi.org/10.1016/0306-9192(77)90040-9

Ciccariello-Maher, G. (2016). *Building the commune: Radical democracy in Venezuela*. London: Verso.

Curcio Curcio, P. (2017). *The visible hand of the market: Economic warfare in Venezuela*. Caracas: Ediciones MinCi.

De Schutter, O. (2014). *The transformative potential of the right to food: Report of the special rapporteur on the right to food to the UN human rights council*. A/HRC/25/57.

Enríquez, L. (2013). The paradoxes of Latin America's "Pink Tide": Venezuela and the project of agrarian reform. *Journal of Peasant Studies, 40*(4), 611–638. https://doi.org/10.1080/03066150.2012.746959

FAO. (2002). *Feature: FAO in Venezuela: Ploughing oil into the land – Venezuela fosters rural development to bolster food security*. http://www.fao.org/english/newsroom/news/2002/9788-en.html (Accessed 16 September 2017).

FAO. (2013). *38 countries meet anti-hunger target for 2015*. http://www.fao.org/news/story/en/item/177728/ (Accessed 17 September 2017).

FAO. (2015). *Venezuela and FAO create SANA, a new cooperation programme to eliminate hunger*. http://www.fao.org/americas/noticias/ver/en/c/283757/ (Accessed 16 September 2017).

Felicien, A., Schiavoni, C.M., Ochoa, E., Saturno, S., Omaña, E., Requena, A., & Camacaro, W. (2020). Exploring the "grey areas" of state-society interaction in food sovereignty construction: The battle for Venezuela's seed law. *The*

Journal of Peasant Studies, 47(4), 648–673. https://doi.org/10.1080/03066150.2018.1525363

Felicien, A., Schiavoni, C.M., & Romero, L. (2018, March 17–18). Food politics in a time of crisis: Corporate power vs. popular power in the shifting relations of state, society and capital in Venezuela's food system [Conference paper]. *Emancipatory Rural Politics Initiative Conference Paper No. 9.* The Hague: Institute of Social Studies.

Figueroa, M. (2015). Food sovereignty in everyday life: Toward a people-centred approach to food systems. *Globalizations, 12*(4), 498–512. https://doi.org/10.1080/14747731.2015.1005966

Foster, J. (2015). Chávez and the communal state: On the transition to socialism in Venezuela. *Monthly Review, 66*(11), 1–17. http://doi.org/10.14452/MR-066-11-2015-04_1

Gavazut, L. (2014). *Dólares de maletín, empresas extranjeras y modelo económico socialista: Un análisis inédito que le sorprenderá.* Análisis Estadístico de la Base de Datos Publicada por CADIVI 2004–2012. https://www.aporrea.org/media/2016/10/la_cosecha._leg.pdf.

Holt-Giménez, E., & Shattuck, A. (2011). Food crises, food regimes and food movements: Rumblings of reform or tides of transformation? *Journal of Peasant Studies, 38*(1), 109–144. https://doi.org/10.1080/03066150.2010.538578

Jacobs, R. (2017). An urban proletariat with peasant characteristics: Land occupations and livestock raising in South Africa. *Journal of Peasant Studies, 45*(5–6) http://doi.org/10.1080/03066150.2017.1312354.

Lavelle, D. (2014). A twenty-first century socialist agriculture? Land reform, food sovereignty and peasant-state dynamics in Venezuela. *International Journal of Sociology of Agriculture & Food, 21*(1), 133–154. https://doi.org/10.48416/ijsaf.v21i1.159

Li, T. (2010). To make live or let die? Rural dispossession and the protection of surplus populations. *Antipode, 41*(1), 66–93. https://doi.org/10.1111/j.1467-8330.2009.00717.x

Lovera, J.R. (1998). *Historia de la alimentación en Venezuela: Con textos para su estudio* (segunda edición). Caracas: Editorial Torino.

McMichael, P. (2009). A food regime genealogy. *Journal of Peasant Studies, 36*(1), 139–169. https://doi.org/10.1080/03066150902820354

Ministerio de Comunicación e Información. (1999). *Constitution of the Bolivarian Republic of Venezuela.* Retrieved 1 October 2016 from http://venezuela-us.org/live/wp-content/uploads/2009/08/constitucioningles.pdf

Murphy, S., & Schiavoni, C.M. (2017). Ten years after the world food crisis: Taking up the challenge of the right to food. *Right to Food and Nutrition Watch,* 16–27.

Nyéléni. (2007). *Declaration of Nyéléni.* http://www.nyeleni.org/spip.php?article290 (Accessed 16 September 2017).

Rauseo, N. (2014). La producción de la ciudad en la Venezuela del siglo XX. In N. Rauseo & P. Sanz (Eds.), *Pensar la ciudad: Realidades, procesos y utopías* (pp. 57–84). Caracas: Fundación Editorial el Perro y la Rana.

Rey Torres, E. (2014). Caracas: Territorialidades en disputa, aproximaciones etnográficas a lo urbano. In N. Rauseo & P. Sanz (Eds.), *Pensar la ciudad: Realidades, procesos y utopías* (pp. 163–188). Caracas: Fundación Editorial el Perro y la Rana.

Rios de Hernandez, J., & Prato, N. (1990). *Las transformaciones de la agricultura Venezolana: de la agroexportación a la agroindustria.* Caracas: Fondo Editorial Tropykos.

Romero, L., Torres, B., Silva, B., & y Toro, J. (2016). Semillas de tubérculos andinos en mérida: Rescate y revalorización en la Venezuela bolivariana. In M. Pérez, A. Felicien, & S. Saturno (Eds.), *Semillas del pueblo, luchas y resistencias para el resguardo y reproducción de la vida* (pp. 142–153). Codition El perro y la Rana y Estrella Roja. Retrieved from http://www.elperroylarana.gob.ve/semillas-del-pueblo/

Robbins, M.J. (2013). *Locating food sovereignty: Geographical and sectoral distance in the global food system.* ISS Working Paper No. 557. The Hague: Institute of Social Studies.

Robbins, M.J. (2015). Exploring the "localisation" dimension of food sovereignty. *Third World Quarterly, 36*(3), 449–468. https://doi.org/10.1080/01436597.2015.1024966

Schiavoni, C.M. (2015). Competing sovereignties, contested processes: Insights from the Venezuelan Food sovereignty experiment. *Globalizations, 12*(4), 466–480. https://doi.org/10.1080/14747731.2015.1005967

Schiavoni, C.M. (2017). The contested terrain of food sovereignty construction: Toward a historical, relational and interactive approach. *Journal of Peasant Studies, 44*(1), 1–32. https://doi.org/10.1080/03066150.2016.1234455

Tramel, S. (forthcoming). *Convergence as political strategy: Social justice movements, natural resources, and climate change.*

Trauger, A. (2014). Toward a political geography of food sovereignty: Transforming territory, exchange and power in the liberal sovereign state. *Journal of Peasant Studies, 41*(6), 1131–1152. https://doi.org/10.1080/03066150.2014.937339

Vida Agro. (2017). Trigo Ruso que llegará al país alcanzará para solo 15 días de consumo. Available at: http://www.vidaagro.com.ve/trigo-ruso-que-llegara-al-pais-alcanzara-para-solo-15-dias-de-consumo/ (Accessed 20 September 2017).

Wilpert, G. (2006). Land for people not for profit in Venezuela. In P. Rosset, R. Patel, & M. Courville (Eds.), *Promised Land* (pp. 249–264). New York: Food First Books.

Woo, W. (2015). There is no alternative ... is there? Organic food provisioning in Jamaica [MA Research Paper]. The Hague: International Institute of Social Studies (ISS).

PART 4

Extended Urbanization between New Rurality and Operational Landscapes

11 Planetary Urbanization, Agro-Exports, and Informality: Making Sense of the Expanding Peripheries and Emerging Cities in Coastal Ecuador

GUSTAVO DURÁN, JONATHAN MENOSCAL,
AND MANUEL BAYÓN

Introduction

During the last years, descriptive analysis has pointed out new urban phenomena in Ecuador, with the emergence of medium-sized cities as well as the importance of the coastal region in the urban process, with rapid and explosive dynamics in comparison to the rest of the country (Barrera, Olmedo, Muñoz, & Guevara, 2015). City networks are identified, thus forming territorial relationships on different levels and scales (Brenner, 2013, p. 61). In the space between the urban networks of the Andes mountains and the coast, a new network has emerged (see figure 11.1), consisting of nine new cities with more than 10,000 inhabitants, with high economic momentum in the region that dominates the network of pre-coastal Ecuadorian cities, which consists of cities such as Santo Domingo (northeast of the network) and Durán (southeast of the network), along the interprovincial road network that connects them (Menoscal, 2016, p. 47).

Thus, this study aims to answer the following questions: Which territorial processes influence the production of urban space in the network of pre-coastal Ecuadorian cities? And what kind of relationship exists between the production processes of urban space and the suburbanization dynamics at the urban-rural peripheries of these cities? The study of the Ecuadorian coast and its urban flows makes it possible to conclude that the cycles of increased expansion of agricultural exports have been echoed in the same cycles of explosive urban development in the intermediate cities of the Ecuadorian coast. The importance of the process of land concentration on the Ecuadorian coast at the time of the export boom is one of the causes of a migration from the most rural places to the nearest population centres. Therefore, urban growth can be explained by two related territorial effects: the departure of farmers with own land

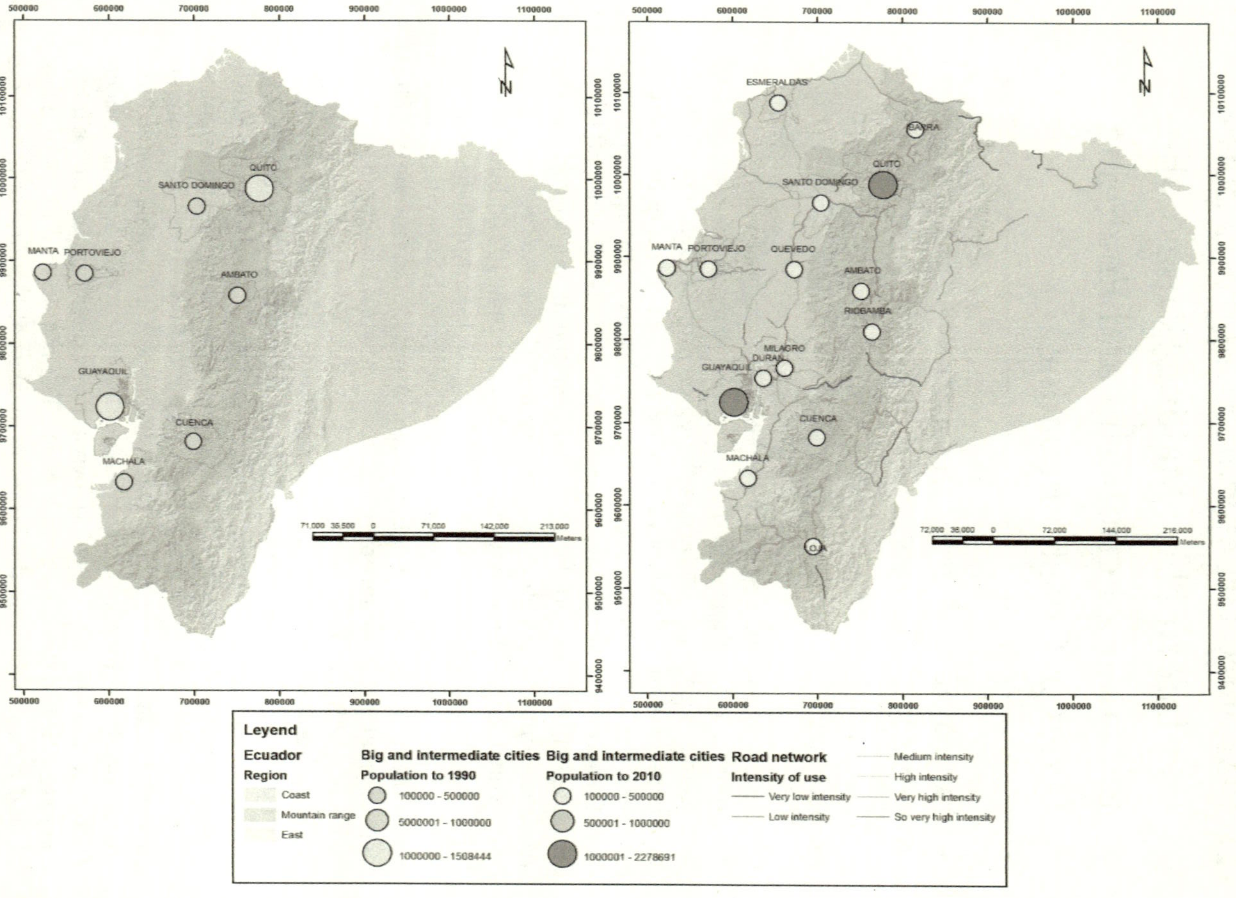

Figure 11.1 Territorial development and networks of Ecuadorian cities. Map created by the authors.

in rural areas because of the arrival of agro-industrial monocultures in areas close to urbanization, and the attraction of migration from other parts of the country sparked by the search for wage labour and a bigger opportunity for small and medium-sized merchants. In addition, the agro-industrial sector generates jobs and new businesses in other areas of the country that are located near the areas of agro-export crops, in spaces that can be consolidated or in incipient centres.

To be clear, our focus on territory refers to the territorial processes that result from the expansion of export agroindustry, through which population and capital are anchored in urban spaces with the relationship between implosion and explosion in this extended urbanization. The extension of urbanization implies that small nuclei towns are condensed in emerging cities, and progressively in intermediate and in consolidated cities. This process is created by the inflow of capital that is in the most remote and rural spaces.

The concept of territory is not merely spatial: when spatializing the theoretical understanding of social processes, this is fundamentally related to power. For Foucault, the territory is, "without doubt a geographical notion, but it is in the first place, a legal-political notion: what is controlled by a certain type of power" (Foucault, 1979, p. 86). Thus, a territorial process is understood as the transformation over time of the implementation of powers in dispute over territory, in which the social subjects that are constituted as territorial powers will be the centre of analysis (Porto-Gonçalves, 2002, p. 127). Furthermore, the term "territorial process struggle" can also be studied because it explains how social groups fight in order to impose, consciously or not, their own way of spatial planning (Sevilla, 2014). In this sense, the territorial processes will be related to the implementation of agro-export capital as the main producer of space, which will have in other subjects dispute relations with the territory (Capel, 1983), as a new way of extended urbanization.

The Dynamics and New Forms of Urbanization: A Theoretical Discussion

The global urbanization process has reached the point of no return (Borja & Sánchez, 2013, p. 20). In 2018, 55% of the global population lives in urban areas. It is estimated that by 2050, the global population will have increased by 68% (UN Department of Economic and Social Affairs, 2019). Because of this urban expansion, it is estimated that urban areas will triple between 2000 and 2030, from 400,000 km^2 to 1.2 million km^2, which represents a challenge and a substantial opportunity from an urban management perspective (UN-Habitat III, 2015, p. 2).

Cities go through many changes and transformations over time, depending on the political, social, economic, and demographic characteristics, among others, that affect them at a specific moment (De Mattos, 2006). Among these characteristics, global economic restructuring, political, social, technological, and cultural transformations, the weakening of nation states in the framework of globalization, and the freedom of capital have led to important spatial transformations, with a tendency to form new territories and territorialities, in rural as well as urban areas (Ciccolella, 2012, p. 10).

The rise of medium-sized cities presents new dynamics and forms of land occupation, reactivating local economies and the functioning of regional society in general. This type of city is defined as follows for demographic reasons: between 100,000 and 1 million inhabitants (Comisión Económica para América Latina y el Caribe, 1998, p. 75) and with spatial extension. However, in reality, characteristics such as the capacity to form relationships and to create urban networks, the openness to the consolidation of other political levels, such as the regional, national and international levels, are common (Bellet & Llop, 2004). In reality, medium-sized cities respond more to the necessity of urban-rural synergies and territorial cohesion (Precedo & Miguez, 2007).

According to Bellet and Llop, medium-sized cities are starting to be considered as mediators between the flow of goods, information, innovation, administration, etc., and between urban-rural relationships and local, regional national, and global relationships (Bellet & Llop, 2004). Such mediating functions are expressed in economic, political-administrative, socio-cultural, and environmental dimensions, among others, and partially depend on their spatial location (Balay, Rabinovich, & De La Porte, 2004, pp. 10–11). Thus, explosive cities create new networks, paths, and human hierarchies to the extent that they expand and relate with other territories (Davis, 2006, p. 15).

This tendency does not only lead to an increase in the size of cities, but it also changes their morphology (Salcedo & Dear, 2012, p. 2); it generates changes in the locational behaviour of families and businesses, thus forming growth processes towards peripheral areas, and the generation of new centres according to the city's size, type, and specialty (De Mattos, 2006, p. 59). As such, it is important to introduce the concept of planetary urbanization as a new form of understanding urbanization processes, eliminating firm distinctions between the rural and the urban. Classic definitions of cities are now obsolete, which is why it is necessary "to decipher the emerging patters of planetary urbanization" (Brenner, 2013). Furthermore, the object of analysis is no longer the space defined as urban, becoming instead an analysis of socio-spatial processes and their implementation in urbanization axes.

The globalization of spatial processes brings us to consider the inseparable relationship between the areas of concentration and the extension of capital flows. "Globalization has led to an increase in the movement of people, goods, services, information, news, products and money. This is where the presence of urban characteristics in rural areas comes from, as well as rural features in urban centers," (Davis, 2006, p. 23) mainly in medium-sized cities, where urbanism tends to go outwards, thus creating external and peripheral zones which previously consisted of urban zones, agricultural land, forests, or suburban areas (Brenner, 2013, p. 41).

Planetary urbanization is not an automatic process, but it becomes reality through specific governmental strategies, which produce the material space required for the global accumulation of capital (Brenner, 2013, p. 42). For this reason, transportation infrastructure plays a fundamental role in planetary urbanization (Brenner, 2014). According to Harvey (2007), it is the form in which "unequal geographic development" is generated, through the production of spatial configurations, such as the active moment within the temporary global dynamics of accumulation and social reproduction (Harvey, 1990, pp. 376–377).

The importance of these processes is that they are supported by capital export investments, where consumption and the implementation of planetary urbanization take place thousands of kilometres away from the place where the investment in infrastructure is made. Planetary urbanization has its origins in the removal of barriers to the mobility of goods and money, for which the progressive liberalization of the global market has been fundamental, especially during the last three decades, leading to the equalization of the world (Duran, Bayón, & Menoscal, 2016).

Thus, the capitalist system in which planetary urbanization takes place generates a "series of issues, syndromes, processes, transformations and emerging struggles that generalize the inequality of urbanization on all scales" (Brenner, 2013, p. 49). With the growth and expansion of cities as a part of planetary urbanization, processes of occupation and the differentiated appropriation of land and peripheral displacement take place, forming poor and informal enclaves, and diffused and segregated places (Sousa, 2012, p. 55).

The production of cities based on the logic of needs and inequalities, related to urban poverty, causes a series of individual and collectivist actions that produce informality in the sense of the occupation and use of urban land (Abramo, 2012, 36). "The geography of capitalism is more varied than ever: modern urbanization processes partially reflect the significance of uneven spatial development and territorial inequality on all scales" (Brenner, 2013, p. 56). Therefore, the logic of needs increases in current urban processes (Brenner & Schmid, 2014).

Informality, a process that is common in the planetary urbanization context, refers to the need for the use of urban land, and the informal market that commercializes goods outside of the regulatory government framework, in relation to property rights and the ownership of land and housing (Abramo, 2012, pp. 40–41) as a "product of the scarcity of economic, social, cultural, institutional and political resources, which affect the popular sectors and are primarily associated with conditions of instability, low salaries, job insecurity," among others (Ziccardi, 2003, p. 11).

It is estimated that in Latin America, almost 70% of human settlements are constituted in an informal way, compared to the remaining minimal percentage of legal and regulated settlements (Castro et al., 2015, p. 110, quoting Mertins & Kraas, 2008).

The informality that exists in these cities is perceived as a product of substantial inequality in the form of urban poverty, evident in the social gaps between groups that must accept minimal levels of life and groups that live in opulence. This difference characterizes cities as divided or fragmented, causing socio-residential urban segregation processes (Ziccardi, 2003, p. 12).

Socio-residential urban segregation refers to the process by which a group of inhabitants with similar characteristics places itself in areas of homogenous composition (Kaztman, 2001, p. 78) – separated, marginalized, and isolated from groups with different characteristics. Socio-residential segregation finds importance in the construction of identities and social differences that can exist in a city, aiming to distance certain groups from others or from prior social conditions (Sabatini et al., 2014, p. 40).

This segregation is related to a lower level of interaction between the social groups that live in the city, favouring social disintegration and negatively impacting the perception that the urban poor have of their conditions (Sabatini, 2006, p. 9). While spatial differences between the zones of the higher class and the lower class have decreased, "a process of social mixing stands out on a large scale, while on a micro-level, the segregation pattern is enhanced" (Janoschka, 2002, p. 22).

As such, based on the city model or a network of neoliberal cities, cities grow uncontrollably towards their peripheries, along the large road axes, influenced by market activities. This increases inequality and social gaps, forming poor, informal, and segregated spaces in cities, which distinguish and consume the city in a certain way, and of a different standard than the other inhabitants of the city.

Summarizing, cities are living in a phase of planetary urbanization, where the intensification of urbanization and the extension of infrastructure and productive networks are reaching increasingly more remote places. The growth of capitalist urbanization is related to the extractive activities, such as oil, mining, and agro-industrial activities (Wilson & Bayon, 2017). The immediate consequences of these are deep territorial transformations, such as the explosive growth of intermediate cities that contribute to a chaotic development that produce the strong emergence of the "logic of necessity" (Abramo, 2012) represented in informality and, in some cases, in forms of segregation.

Methodology

This chapter is divided into two parts. The first part corresponds to the historic analysis of the occupation process of the Ecuadorian coast since the 1950s up to now, particularly around the network of pre-coastal cities. The second part includes an analysis of the dimensions of urban expansion in peripheries.

The New Geography was the neopositivist current of geography that emerged in the 1950s from the generation of territorial models under a purely quantitative logic of spatial relations. This form of geography implicitly proposed the depoliticization of social relations and had a strong imprint in the industrial world, and in business, commerce, and administration. The absence of spatial power relations, the lack of a historical view of the processes or the territorial homogenization in its analysis elicited strong criticism regarding the geographical understanding that its tools provided, which led to its decline as a branch in the rise of geography in the 1970's (Sheppard, 2001).

For the first part of this study, we deviate from methodologies based on forms stemming from the New Geography, to try to understand the occupation process and the expansion of cities from the perspective of the planetary urbanization concept. The methodology is based on the historic analysis from secondary sources, which is complemented by multiscale geo-statistics of population and housing censuses provided by INEC, displayed in thematic maps using geographic information systems and socio-spatial analysis methods.

For the second part of the study, the dimensions of urban expansion and informality are considered. Two cases are included in this study, for which the same dimensions and analysis methodologies were used, based on socio-spatial methods. It was carried out with the purpose of trying to explain the differences and similarities, tendencies and changes of identified patterns of the constructive elements that exist in the units of analysis or in specific cases (Pliscoff & Monje, 2003, p. 6).

The study of the dimension of urban expansion in peripheries was carried out while taking the variable of urbanization into consideration, meaning that the rate of change of the use of land during the period of 2000 to 2010 was measured, in order to determine the changes that have taken place at the urban-rural peripheries over time, and up to which point the periphery has expanded. It is a multi-temporal analysis with spatial characteristics (Ruiz, Savé, & Herrera 2013, p. 117).

Once it was determined up to which point and in which way the cities expanded, two new neighbourhoods in each city were taken as specific points, as a micro-study. They fulfil the criteria of being established during the last ten years, being in the process of densification, and being of informal type.

Within the dimension of informality, the variables of poverty and socio-residential segregation were considered, which are common components of this type of land occupation.

To identify poor expansion areas, the indicator of unsatisfied basic needs was used, making it possible to determine whether homes or communities satisfy several basic needs (Feres & Mancero, 2001, p. 65). These needs include the availability and quality of housing, health services, education, and basic services.

To make an estimate of the socio-residential segregation in the study areas, the Duncan segregation index was applied, enabling the measurement of the distribution of an interest group within a specific area and whether this distribution is equal and equitable among the different social groups that form the city (Buzai et al., 2003, p. 9).

An Historic Analysis of the Occupation Process of the Pre-coastal Ecuadorian Network

Banana Boom 1950 to 1962: The Beginning

During the 1950s, Ecuador continued to have an urban network similar to the one established by the Spanish colony: two strong urban polarities between Quito in the mountains and Guayaquil on the coast, and the Andean axis with its connection to Guayaquil as the primary drive behind urbanization (Deler, Gómez, & Portais, 1983, p. 197). The Banana Boom (1948–65) changed the coastal territory in a profound way. The need for goods to be transported over land, together with

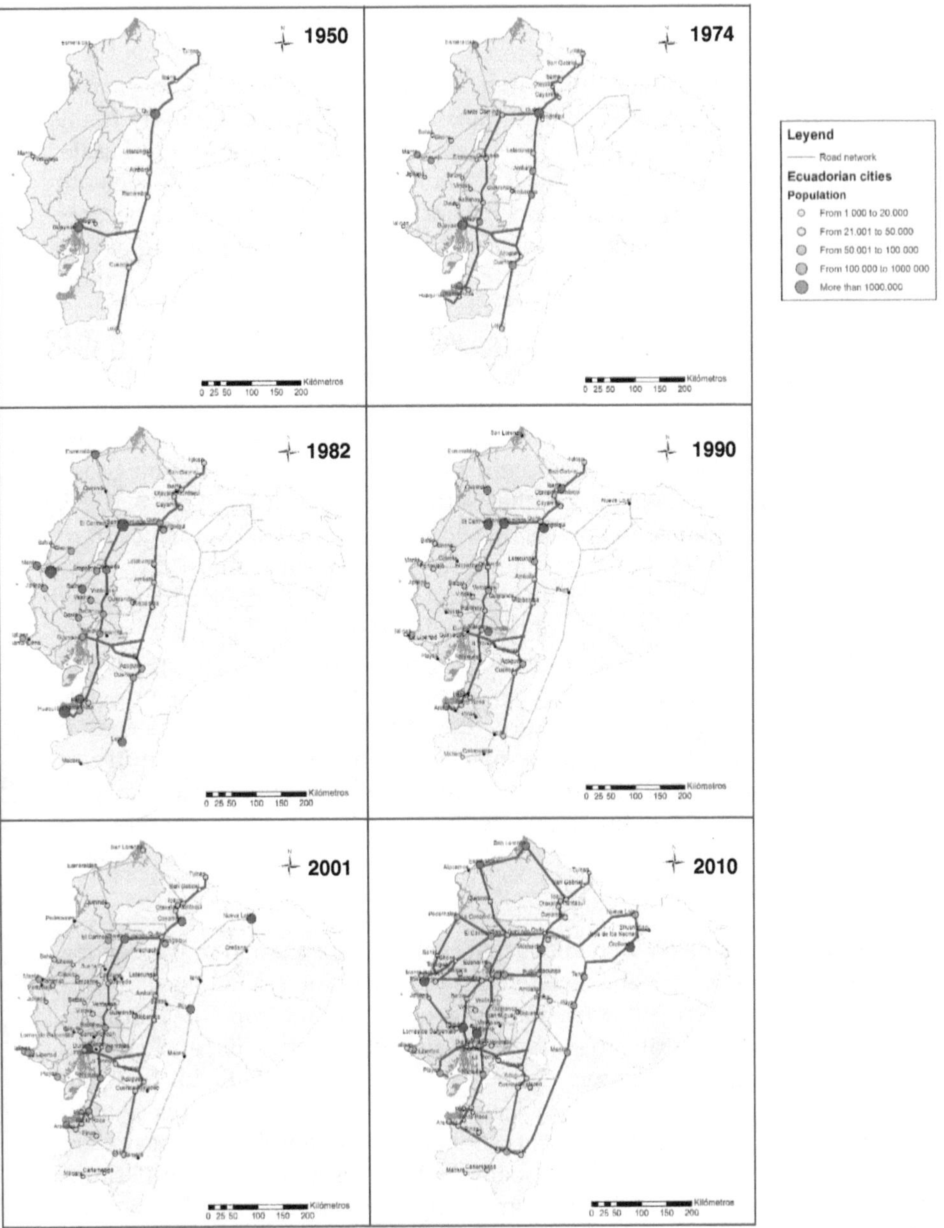

Figure 11.2 A comparative evolution of the formation of new cities in Ecuador. Maps created by the authors.

the developmental period of military juntas, fostered considerable road development in the entire region.

Endorsed by Guayaquil, a high number of improvement works were carried out in the 1950s, which transformed coastal connectivity to the south, towards the emerging banana-centre of Machala, and to Pasaje, a few kilometres away. Towards the north, connections were made with the new city of Quevedo, and Babahoyo, which managed to become a city with its old colonial centre, and the noticeably growing city of Milagro. Guayaquil consolidated its position as a substantial export harbour, establishing itself as the main city of Ecuador.

The importance of these farmland properties resulted in a small banana bourgeoisie on the one hand, as well as high employment of manual labour based on capitalist relationships, with relatively higher salaries that generated an internal market; and on the other hand, the activities of the supply of agricultural input and of consumption goods for the migrant population that settled in these cities (Larrea, 1985, p. 83). The Banana Boom was developed by capitalist companies from the United States, such as United Fruits, but was dominated by a new player in the coastal agricultural sector: the medium-sized capitalist farmer, forming a small rural bourgeoisie (Larrea, 1986, p. 115).

A typical characteristic of these new urbanizations in the pre-coastal network is the importance of being located near roads, as the presence of trade and agricultural supply became a constant. The pre-coastal region had the form of a whale during this phase, where Guayaquil was the large harbour that absorbed all the agro-export production, mainly of bananas, in the entire region, with Quevedo as the fluttering tail.

The Banana Crisis1965 to 1972: The Formation of Oligopolies

Although, as of 1965, there was an economic stagnation in the banana sector because a plague resulted in a drastic reduction in tons of exportation, the urban and demographic growth did not stop. The construction of new roads due to migratory flows towards new expansion areas contributed to this growth (Allou et al., 1987, p. 170). "Roads and railways are not just technical objects ... they encode the dreams of individuals and societies and are the vehicles whereby those fantasies are transmitted and made emotionally real," (Wilson & Bayon, 2017 quoting Larkin, 2013, p. 333). Fundamentally, new centres formed in the new interconnections, such as El Empalme and Balzar on the southern coast, and Santo Domingo de los Tsáchilas as a main communication node between Quito and the different coastal regions (Allou, 1985, p. 31).

The agricultural reforms that took place in 1964 and 1973 brought along a process of relative abandonment of the banana, and a drop in the inequality index of the division of land, reducing growth in the network's cities. At the same time, Quito secured its position as the administrative capital of the petroleum boom (Larrea, 1987, p. 114).

The Modernization of the Banana Sector and the Slowdown in Agro-Export Urbanization: 1974 to 1982

This period concerns the reconversion of the banana sector on the precoastal axis, a situation that noticeably reconfigured the form of urbanization. The process generated two simultaneous processes: as more people own land, there is less proletarianization; as fewer people own land, there is more proletarianization. The banana sector, which is currently owned locally or by large, foreign-owned businesses, along with medium-large Ecuadorian banana producers, responded with a reconcentration of property and with the modernization of exportation. However, it set in motion a process that would take several decades. The first process that occurred was was the disappearance of the small banana bourgeoisie in the ownership structure that generated a source of consumption for the region: as the bourgeoisie shrinks, more farmers with their own land emerge and urban centres lose dynamism in places; the capture of agrarian land rent is the fundamental characteristic of the economic structure. Furthermore, urban unemployment rates increased and salaries dropped, with an increase in the concentration of surplus appropriation (Larrea, 1987, p. 59).

The structure of land ownership that had existed before the banana boom once again returned, with an increasingly strong separation growing between the countryside and the city, with trade, export, financial, and real estate activities re-concentrating in Guayaquil, thereby strengthening a dynamic of unequal geographic development. The process is complex. Several authors argue that, on the one hand, the failure of the agrarian reform prevented farmers from getting out of credit cycles; on the other hand, indebtedness to banks that belong to the main landowners favoured the enrichment of a privileged few. Furthermore, two bad harvests along with the fact that the land was no longer theirs contributed to the process of land loss and migration. Guayaquil surpassed one million inhabitants during this period. The exportation structure became an oligopoly, where 10 exporters controlled 90% of the market, creating conditions in which producers offered them actual control of the entire chain (ibid., p. 59).

This sum of factors caused a notable loss in urban dynamics along the internal coastal axis, with features and characteristics of the previous period reasserting themselves. Only Santo Domingo continued with a strong momentum, maintaining its expansion based on services and trade, and aided by its connection to the new centres of El Carmen in the Manabí Province and Quinindé in Esmeraldas. The supremacy in the growth of Quevedo is based on a commercial class that dominates landowners in Los Ríos and a new type of poverty that emerges from the difficulty of farming sectors, increasingly indebted (Pérez and Mogroviejo, 1983, pp. 120–121). New centres linked to banana production emerged along the Guayaquil-Quevedo road, such as Ventanas, or Naranjito, right on the axis of communication with the mountain region.

A General Slowdown of Urbanization in the Period 1982 to 1990

The end of the petroleum boom and the start of the indebtedness that led to the debt crisis marks this period. However, the liberalization of agriculture was slow during this period, and as a result there were no big changes. On the one hand, the modernization of the banana sector that started in 1977 continued its course. From the perspective of the advantage of having the lowest taxes on exports among countries focused on the banana, and the lowest salaries, the sector supported its recovery (Larrea, 1987, pp. 82–83).

The company REYBANPAC emerged as a new banana producer in Los Ríos province, defining the social structure of the area (Larrea, 1987, p. 104). Founded by descendants of Chinese migrants who arrived in Quevedo, and through an innovative strategy of selling bananas of a lower quality to countries of East Asia, they converted themselves into the second largest exporting company, after the Noboa Group (Roberts, 2009, p. 191). In the Quevedo area, a limited export quota was imposed, which sidelined the exclusion zones, leading to an increased concentration of banana production (Pérez & Mogroviejo, 1983, pp. 115–116). In Milagro, Ingenio Sugar Company Valdez, which had gone through a considerable crisis during the previous decades, was bought and modernized by the Noboa Group, as part of a diversification strategy of the food market (Roberts, 2009, p. 207).

During the census period between 1982 and 1990, the pre-coastal network remained practically the same as during the previous period, with certain new urbanization nuclei of lesser importance, such as Durán, the suburban area of Guayaquil, and Naranjal in the South of the Guayas Province. Santo Domingo established itself in a stronger way, rising to become the fifth largest city in the country, based on the number of its inhabitants.

Regarding growth, two groups of cities with high growth stood out: on the one hand, the transportation node cities, such as Santo Domingo and el Carmen; and on the other hand, cities such as Durán and Sangoloquí, suburban areas of Guayaquil and Quito, respectively. Quito and Guayaquil noticeably decreased in growth, which did not exceed 4%. The same happened with Quevedo, Babahoyo, and Milagro, which were the main urbanization nodes of the pre-coastal network during the banana boom.

The Neoliberalization of Agriculture during the Last Decade of the Twentieth Century

The arrival of the 1990s saw the implementation of measures for the liberalization of the agricultural sector, such as the Trole Laws, which prohibited the taking of land by farmer movements. The government repealed the agrarian reform law that would allow the legal purchase of land from the local cooperatives, formed mostly by farmers and poor people. For the banana sector, it involved its revitalization and the achievement of its demands, even counting on the support of Minister Gonzáles Portés, a great man in favour of liberal reforms. These improvements were accompanied by high banana prices and low cacao prices (Tamayo & Cepeda, 2007, p. 168). The recovery of the banana sector in the Los Ríos sector that started in the previous decade was reinforced, as this sector received the majority of investments. The opening of new markets by REYBANPAC, such as China, contributed to this expansion. From 1983 to 1997, the export volume increased from 800,147 to 4,456,275 tons (Roberts, 2009, p. 206). This period of expansion ended with the phenomenon of El Niño that destroyed most of the harvest. From 1986 to 1994, the production doubled (Campaña, 1993).

With the new framework, Quevedo became the case with the highest land property concentration (FAO, 1994, p. 85). A study carried out by the FAO during the first part of the decade, showed how different members of the family that owned REYBANBAC bought large properties of over 500 hectares in just one year. The liberalization caused medium-sized properties to disappear. Cooperatives and impoverished settlers were the main parties that offered land. After decades of accumulated debt with exporting companies, and the absence of government loans, neoliberal economic measures were the last straw for the unfeasibility of the banana cooperatives (Guerrero, 1993, p. 299).

The expansion of banana production into areas where diversified crops were grown previously was also reflected in hectares, with an increase from 138,190 hectares in 1990 to 180,331 hectares in 2001 (Martínez, 2004, p. 129). The province of Los Ríos continued to stand out with 50,419 hectares. Furthermore, in 2001, the occupation structure, altered by an entire decade of liberalization and concentration of land, added to the previous decades of the process: the small and medium-sized producers did not achieve 30% of land ownership put together, while 12.2% of the large owners owned more than 51% of the land. The concentration of exports went from 10 main companies to only three: Noboa, Ubesa, and REYBANPAC (Martínez, 2004, pp. 132–133).

The structure of the pre-coastal axis became denser, through new, small cities that emerged in the intensified areas of agro-export, where liberalization and concentration had stronger effects. In the province of Los Ríos, cities such as Buena Fé emerged, and Valencia, close to Quevedo, and the new city of La Maná, already in the coastal part of Cotopaxi province. These cities did not emerge along the road Guayaquil-Quito but, rather, they grew as new population centres linked to the new migrant population, migrating from the nearby countryside. This caused the relative population of the rural province of Los Ríos to drop, which was a source of countryside to city migration (Cueva, 2008, p. 137).

The last census period available in Ecuador is for the period of 2001–2010, starting with Ecuador's economic crisis and ending in the middle of the petroleum boom and economic growth that would continue until 2015. Thus, the data shows the process of economic recovery and the new petroleum boom. The first tendency shown by the data is that Quito and Guayaquil had a yearly growth rate of 1.65%, the lowest rate of the study period. The two metropolitan cities continued growing, as cities with almost 300,000 inhabitants and around 500,000, respectively, but with a net importance that is exceeded by medium-sized cities of between 100,000 and 239.000 inhabitants in Cuenca, the country's third largest city. In total there are 13 cities, with eight on the coast. On the pre-coastal axis, there is Santo Domingo (fourth in the country) and Durán (sixth) with over 200,000 inhabitants, Quevedo (twelfth) and Milagro (fourteenth) with over 100,000 inhabitants.

The urban axis of the pre-coastal region has a specific importance in the configuration of the regional difference. The network represents a type of urbanization that has a specific history and with particular sources regarding the intensity of its connection to agro-exports. From a cartographic perspective, one can appreciate how the results of 2010 represent the area with the greatest urbanization density, in nuclei as

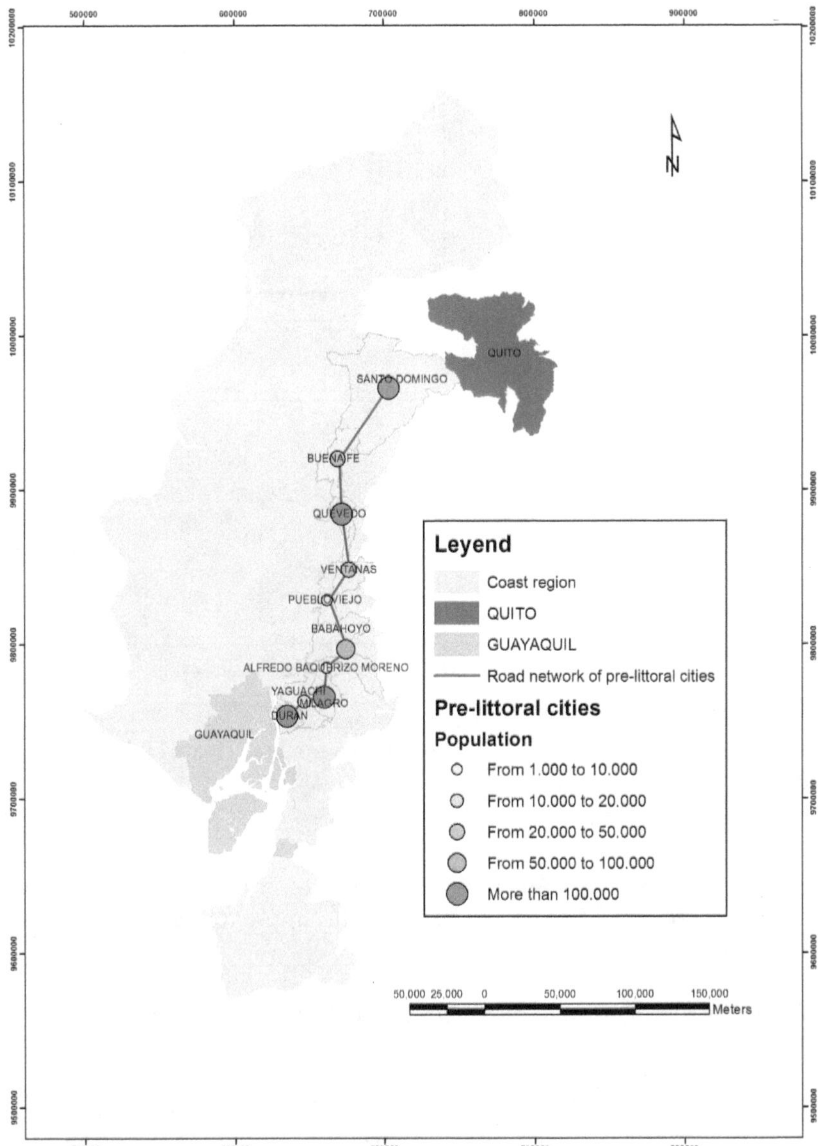

Figure 11.3 The network of pre-coastal Ecuadorian cities. Map created by the authors.

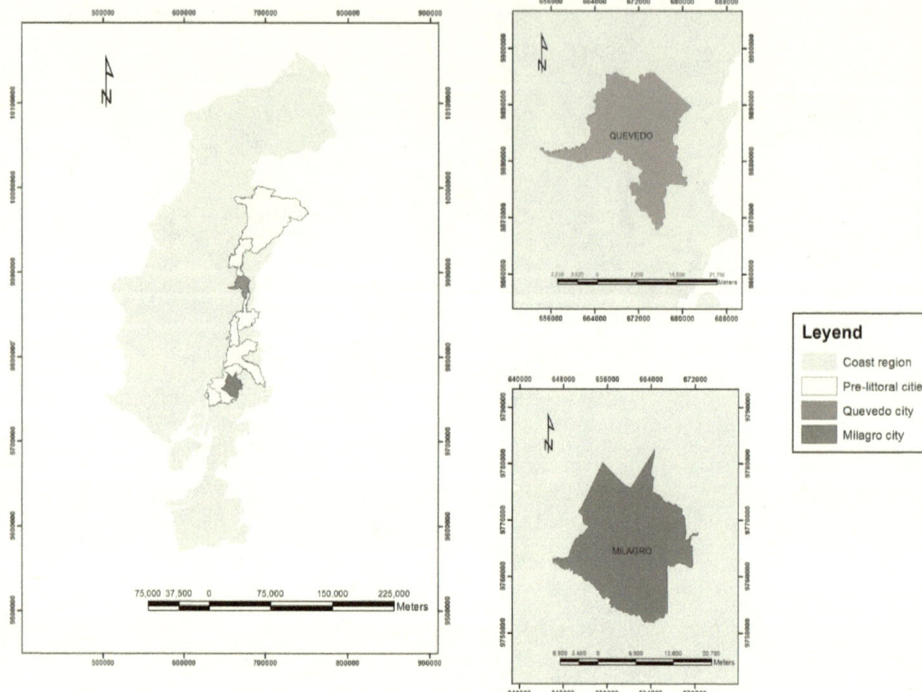

Figure 11.4 The locations of Quevedo and Milagro. Map created by the authors.

well as in territorial continuity. Furthermore, it concerns a period of strong expansion of roads, as well as the improvement of the main network.

The Suburbanization of Medium-Sized Cities in the Pre-coastal Ecuadorian Network

On a micro-territorial level, Quevedo and Milagro appear as cities with more than 100,000 inhabitants that form part of the urbanization process based on specific agro-export booms. That means that both cities are related to agro-exports with most of their population working on the land.

Both cities show similar changes in configuration and growth rate of their population and therefore in the morphology of the urban core which has expanded in a diffused way towards the peripheral areas.

These cities started to be considered as medium-sized after exceeding 100,000 inhabitants and, therefore, the expansion pattern of the urban stain is similar in both cases.

According to data from the INEC, there were 86,910 inhabitants in Quevedo in 1990, and by 2010 there were 150,827. During the study period, 2000–2010, the calculation of the population growth rate in Quevedo was 2.26%.

The growth rate of its urban stain during the study period was 12.35%. Currently, the city has a surface area of 2,988.76 hectares, meaning that more than one thousand hectares went from being rural to urban land in 10 years. Today, the city consists of nine urban parishes and two rural ones. San Camilo is the most populated urban parish of Quevedo with the fastest spatial growth, located on the eastern shore of the Quevedo River. Within the parish, in the sector of Cruz María, there are recent diffused and informal expansion processes. Therefore, it was chosen as micro-study location. By 2010, Cruz María had 2,920 inhabitants.

Meanwhile, the city of Milagro, according to the analysis of this data, had a population growth rate of 1.63% between 2000 and 2010. According to data from the Ministerio de Agricultura Ganadería y Pesca MAGAP, the growth rate of the urban stain in Milagro during the study period was 8.97%, taking on a form without regular growth patterns. Around one thousand hectares went from being rural to urban areas during the study period. The city currently comprises four urban parishes and four rural ones. Enrique Valdez is the urban parish with the highest urban and demographic growth in the city, and is where the micro-study area, Rosa María 2, is located.

By analyzing the poverty data in Quevedo as well as in Milagro, using the neasures in Unsatisfied Basic Needs (UBN), we can see that the central areas of both cities contain the lowest poverty levels, increasing in a concentric way from the centre towards the peripheral areas, which are occupied by the poor residents.

Specifically, in the city of Quevedo, based on UBN data, the areas near the urban centre, where the city initially originated, present the lowest poverty levels, which do not exceed 20% of the population.

In the Cruz María micro-study zone, the poverty levels according to the UBN are very high, surpassing over 90% of the population. This is basically due to the precariousness of constructions, the lack of access to basic services of decent quality, and the distance to educational centres and health centres.

Meanwhile, in Milagro city, in accordance with the UBN data, it can clearly be seen that there are levels with less poverty, affecting up to 40% of the population. Poverty levels increase in a concentric way from

Figure 11.5 A photographic comparison of poor housing in the micro-study sectors. Photos by the authors.

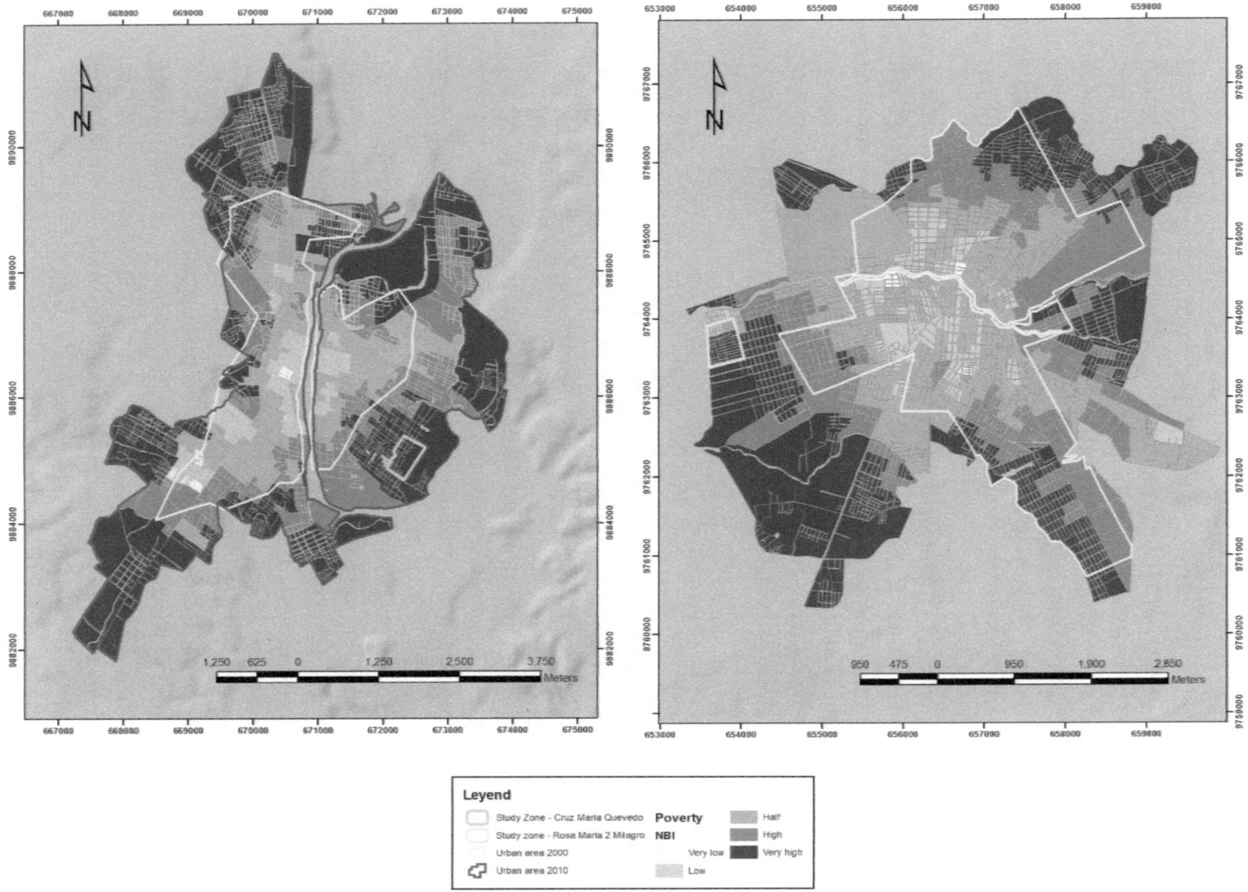

Figure 11.6 A comparative poverty map measured by unsatisfied basic needs in the studied cities. Map created by the authors.

the centre outwards; however, the increase is not in similar proportions towards each side, because the city itself is a heterogeneous mosaic on a large scale, but highly segregated on a smaller scale.

On the level of the census sector, in the micro-study area of Rosa María 2, part of the parish Enrique Valdez, UBN data show very high levels of poverty, in over 80% of the area. However, in its northern part, the oldest and densest area of the sector, the UBN data is low and very low. Due to the housing agglomeration they are equipped with higher quality services and their housing meets higher standards of construction and conservation, observing real economies of agglomeration in this specific sector.

Taking the levels of UBN of the census sectors belonging to Cruz María and Rosa María 2 as a basis, the socio-residential segregation was calculated for each of these, obtaining very similar values for both cases, showing that it concerns neighbourhoods that are very segregated and marginalized, mostly occupied by people in a situation of poverty in the new urban, diffused, and unorganized periphery.

Conclusions

Studies regarding the network of pre-coastal Ecuador cities allow us to see which territorial processes influence the production of urban space in the context of planetary urbanization, applied to the specific conditions of Latin America.

The diffused and materialized socio-spatial processes in emerging urban networks, such as in the case of pre-coastal Ecuador, are historic proof of the concentration and extension mechanisms of capital flows, central to the production of urban space in neoliberal cities.

This space is required for the global accumulation of capital, due to the territory subsumed to international networks of transnational commerce. In the area of the study, the space is incorporated through the infrastructure intended for agro-industry, generating urban nodes that depend on these flows, but above all, creating unequal geographic development.

In terms of the relationship between the production of urban space and the dynamics of suburbanization at the urban-rural peripheries of these pre-coastal Ecuadorian cities, this study enables us to confirm that the processes of differentiated occupation towards the inside of the urban nodes of the network, with peripheral displacement of the social sectors of low income, are the bases of a generalized formation that has a high amount of poor and informal enclaves.

These informally and explosively shaped urban structures produce highly segregated spaces within the small and medium-sized cities of the pre-coastal Ecuadorian network, and as such, reproduce the constituent general tendencies of urbanization in Latin America.

REFERENCES

Abramo, P. (2012). La ciudad com-fusa: Mercado y producción de la estructura urbana en las grandes metrópolis latinoamericanas. *EURE, 38*(114), 35–69. http://dx.doi.org/10.4067/S0250-71612012000200002

Allou, S. (1985). *Las ciudades intermedias en el Ecuador: La problemática municipal y los casos de Santo Domingo de los Colorados, en Ciudades Intermedias del Ecuador, boletín 14.* Quito: Centro de Investigaciones Ciudad.

Allou, S., et al. (1987). *El espacio urbano en el Ecuador.* Quito: Centro Ecuatoriano de Investigación Geográfica.

Balay, J.C., Rabinovich, A., & de la Porte, Cherryl. (2004). *Interfase urbano – rural en Ecuador: Hacia un desarrollo territorial integrado.* LaSUR 5.

Barrera, A., Olmedo, P., Muñoz, M., Paz, B., & Guevara, C. (2015). *Transformaciones demográficas y proceso de urbanización en Ecuador.* Quito: CITE-FLACSO

Bellet, C., & Llop, J. (2004). *Ciudades intermedias: Entre territorios concretos y espacios globales.* Ciudad y Territorio. Estudios Territoriales CYTET.

Borja, M., & Sánchez, D. (2013). Ciudadanía informacional: Gobernanza inclusiva en la ciudad informal. *CIDOB d'Afers Internacionals, 104.* https://www.cidob.org/es/articulos/revista_cidob_d_afers_internacionals/104/ciudadania_informacional_gobernanza_inclusiva_en_la_ciudad_informal

Brenner, N. (2013). *Tesis sobre la Urbanización Planetaria.* México D.F.: Nueva Socidad.

Brenner, N. (2014). Introduction: Urban theory without an outside. In N. Brenner (Ed.), *Implosions/explosions: Towards a study of planetary urbanization* (pp. 14–35). Berlin: Jovis.

Brenner, N., & Schmid, C. (2014). Planetary urbanization. In N. Brenner (Ed.), *Implosions/explosions: Towards a study of planetary urbanization.* Berlín: Jovis.

Buzai, G., Baxendale, C., Rodríguez, L., & Escanes, V. (2003). Distribución y segregación espacial de los extranjeros en la ciudad de Luján: Un análisis desde la Geografía Cuantitativa. *Signos Universitarios, 23*(39), 29–52.

Campaña, I. (1993). *La industria bananera en la encrucijada, en Boletín Economía 69.* Quito: Universidad Central del Ecuador.

Capel, H. (1983). Positivismo y antipositivismo en la ciencia geográfica: El ejemplo de la geomorfología. *Cuadernos críticos de geografía humana.* Barcelona.

Castro, C., Ibarra, I., Lukas, M., Ortiz, J., & Sarmiento, J. (2015). Disaster risk construction in the progressive consolidation of informal settlements: Iquique and Puerto Montt (Chile) case studies. *International Journal of Disaster Risk Reduction, 13,* 109–127. https://doi.org/10.1016/j.ijdrr.2015.05.001

Comisión Económica para América Latina y el Caribe. (1998). *Ciudades intermedias de América Latina y el Caribe: Propuestas para la gestión urbana.* Comp. R. Jordan & D. Simioni. Italia: Cooperazione Italiana. https://repositorio.cepal.org/bitstream/handle/11362/31024/1/S9800066_es.pdf

Ciccolella, P. (2012). Revisitando la metrópolis latinoamericana más allá de la globalización. *Buenos Aires: Revista Iberoameticana de Urbanismo, 8,* 9–21. http://hdl.handle.net/2099/13012

Cueva, V., et al. (2008). Desplazados por agroexportación, la concentración de la tierra por multipropiedad y fracturación: el caso de Quevedo. In *¿Reforma agraria en el Ecuador?: Viejos temas, nuevos argumentos* (pp. 133–152). Quito: Sistema de Investigación sobre la Problemática Agraria en el Ecuador (SIPAE).

Davis, M. (2006). Planeta de ciudades – miseria: Involución urbana y proletariado informal. *New Left Review, 26,* 5–34. https://newleftreview.es/issues/26/articles/mike-davis-planeta-de-ciudades-miseria.pdf

Deler, J.P., Gomez, N., & Portais, M. (1983). *El manejo del espacio en el Ecuador.* Quito: Centro Ecuatoriano de Investigación Geográfica.

De Mattos, C. (2006). Modernización capitalista y transformación metropolitana en América Latina: Cinco tendencias constitutivas. In *América Latina: Cidade, campo e turismo* (41–73). Buenos Aires: CLACSO.

Duran, G., Bayón, M., & Menoscal, J. (2016). *Eje urbano interno-costero: Hacia una comprensión del proceso urbano ecuatoriano desde la urbanización planetaria.* (En revisión).

FAO. (1994). *Campesinos y mercado de tierras en la costa ecuatoriana.* Food and Agriculture Organization of the United Nations.

Feres, J.C., & Mancero, X. (2001). *El método de las necesidades básicas insatisfechas (NBI) y sus aplicaciones en América Latina.* CEPAL.

Foucault, M. (1979). *Microfísica del poder.* Madrid. La piqueta.

Guerrero, R. (1993). Voluntad de Dios: Los campesinos y la producción del banano. *Ecuador Debate, 28,* 293–302. https://repositorio.flacsoandes.edu.ec/bitstream/10469/9082/1/REXTN-ED28-14-Guerrero.pdf

Harvey, D. (1990). *Los límites del capital y la teoría marxista.* México D.F.: Fondo de Cultura Económica.

Harvey, D. (2007). Notas hacia un desarrollo geográfico desigual. *Revista GeoBaireS, Cuadernos de Geografía.*

Janoschka, M. (2002). El nuevo modelo de la ciudad latinoamericana: Fragmentación y privatización. *EURE, 28*(85), 11–20. http://dx.doi.org/10.4067/S0250-71612002008500002

Kaztman, R. (2001). Seducidos y abandonados: El aislamiento social de los pobres urbanos. *Revista Cepal, 75,* 171–189. https://repositorio.cepal.org /handle/11362/10782

Larkin, B. (2013). The politics and poetics of infrastructure. *Annual Review of Anthropology, 42,* 327–343.

Larrea, C. (1985). El sector agroexportador y su articulación con la economía ecuatoriana durante la etapa bananera 1948–1972: Subdesarrollo y crecimiento desigual. In L. Lefeber (Ed.), *La economía política del Ecuador.* Quito: Flacso.

Larrea, C. (1986). Crecimiento urbano y dinámica de las ciudades intermedias del Ecuador (1950–1982). In D. Carrión (compilador), *Ciudades en conflicto.* Quito: El Conejo-Ciudad.

Larrea, C. (1987). *El banano en el Ecuador: Trasnacionales, modernización y subdesarrollo.* Quito: Corporación Editora Nacional.

Martínez, L. (2004). Trabajo flexible en las nuevas zonas bananeras del Ecuador. In T. Krovkin (Compiladora), *Efectos sociales de la globalización* (pp. 129–156). Quito: Abya Yala.

Menoscal, J. (2016). *El pre litoral ecuatoriano y sus dinámicas de urbanización: Informalidad y construcción social del riesgo en sus ciudades intermedias: Quevedo y Milagro.* Flacso Ecuador.

Mertins, G., & Kraas, F. (2008). Megastädte in Entwicklungsländern: Vulnerabilität, informalität, regier-und steuerbarkeit. *Geographische Rundschau, 60*(11) 4–10.

Pérez, C., & Mogrovejo, J. (1983). Quevedo: Espacio comercial y alternativa campesina. *Ecuador Debate, 3,* 115–124. Quito: Centro de Arte y Acción Popular.

Pliscoff, C., & Monje, P. (2003). Método comparado: Un aporte a la investigación en gestión pública. *VIII Congreso Internacional del CLAD sobre la Reforma del Estado y de la Administración Pública.* Panamá: CLAD.

Porto-Gonçalves, C.W. (2002). Da geografia às geo-grafias: Um mundo em busca de novas territorialidades. In *La guerra infinita Hegemonía y terror mundial.* Buenos Aires: CLACSO.

Precedo, A., & Miguez, A. (2007). La evolución del desarrollo local y la convergencia territorial. In M. García (Ed.), *Perspectivas teóricas en desarrollo local* (pp. 77–110). España: Netbiblio.

Roberts, L.J. (2009). *Empresarios ecuatorianos del banano.* Quito: Iberia.

Ruiz, V., Savé, R., & Herrera, A. (2013) Análisis multitemporal del cambio de uso del suelo, en el Paisaje Terrestre Protegido Miraflor Moropotente Nicaragua, 1993–2011. *Ecosistemas,* 22(3), 117–123. http://hdl.handle .net/10045/34856

Sabatini, F. (2006). *La segregación social del espacio en las ciudades de América Latina.* Washington: Banco Interamericano de Desarrollo Departamento de Desarrollo Sostenible División de Programas Sociales.

Sabatini, F., & Trebilcock, M.P. (2014). Desigualdades, clasismo y mercados de suelo. *Mesas Hurtadianas: Exclusión socio espacial en Chile*. Universidad Alberto Hurtado.

Salcedo, R., & Dear, M. (2012). La Escuela de Los Ángeles y las metrópolis sudamericanas. *Bifurcaciones, 11*. http://www.bifurcaciones.cl/bifurcaciones/wp-content/uploads/2012/12/bifurcaciones_011_Salcedo-y-Dear.pdf

Schteingart, M., & Salazar, C. (2003). Expansión urbana, protección ambiental y actores sociales en la Ciudad de México. *Ciudad de México: Estudios Demográficos y Urbanos, 18*(3)(54), 433–460. http://dx.doi.org/10.24201/edu.v18i3.1155

Sevilla, Á. (2014). Hegemonía, gubernamentalidad, territorio: Apuntes metodológicos para una historia social de la planificación. *Revista de Metodología de Ciencias Sociales, 27*, 49–72. http://doi.org/10.5944/empiria.27.2014.10862

Sheppard, E. (2001). Quantitative geography: Representations, practices, and possibilities. *Environment and Planning D: Society and Space, 19*, 535–554. http://doi.org/10.1068/d307

Sousa, E. (2012). *De la ciudad a la metrópolis prematura: Los tres procesos intervinientes*. Concepción: URBANO 25.

Tamayo, C., & Cepeda, D. (2007). El dilema constante del productor bananero en tiempos de brete: Asociatividad o individualismo. In M. Vaillant (Ed.), *Mosaico Agrario*. Quito, IFEA.

UN-Habitat III. (2015). *17 Cities and climate change and disaster risk management*. Habitat III Issue papers. United Nations.

UN-Department of Economic and Social Affairs. (2019). *World urbanization prospects: The 2018 revision* (ST/ESA/SER.A/420). United Nations.

Wilson, J., & Bayon, M. (2017). Fantastical materializations: Interoceanic infrastructures in the Ecuadorian Amazon. *Environment and Planning D: Society and Space*, 1–19. https://doi.org/10.1177%2F0263775817695102

Ziccardi, A. (2003). Pobreza y exclusión social en las ciudades del siglo XXI. In *Procesos de urbanización de la pobreza y nuevas formas de exclusión social: Los retos de las políticas sociales de las ciudades latinoamericanas del siglo XXI* (pp. 9–33). Bogotá: Siglo del Hombre Editores, Clacso-Crop.

12 Worlding the Atacama Desert: Peripheral Urbanization and Transnational Resource Extraction Urbanism in Antofagasta, Chile

MICHAEL LUKAS

Introduction

Almost wherever you stand in Antofagasta, a city of roughly 400,000 inhabitants that stretches over 30 kilometres along a coastal plain in Northern Chile, you will be aware of its three defining geographical features: the Pacific Ocean that delimits the city to the west, the hills of the Atacama Desert that delimits it to the east and the mobile and immobile infrastructures of the national and transnational copper mining industry. Antofagasta is a clear example of the long-standing logistical, social, and political importance of Latin American cities in the global production networks of the extractive industries. While from the late nineteenth century Antofagasta was part of the "resource extraction network" (Correa, 2016, p. 39) of the nitrate industry, today it is one of the major urban hubs in the global production network of the copper industry (Arias, Atienza, & Cadematori, 2013). The copper (and some gold) mines are located up to 180 kilometres from Antofagasta, to where the extracted minerals are brought via trucks, a railway system, and pipelines. From its two port facilities – Antofagasta Terminal Internacional (ATI) and Coloso – the concentrated minerals are shipped to their destinations around the world. The city of Antofagasta is thus a kind of bottleneck through which all productive and mineral products of resource extraction have to pass. Its urban fabric and governance are of strategic importance, hence, not only for the Chilean state, whose principal export product is copper, but also for the several national and transnational companies of the extractive industries that operate the major mines and the vast infrastructural networks that crisscross the city.

A privileged spot from where to observe the urban environmental and socio-spatial impacts of mining-related activities in this "global city from off the map" (Robinson, 2002) is Hotel Antofagasta, located at the shoreline in the centre of the city. Here, in October 2017, the first OECD Meeting of Mining Regions and Cities was held. Around 275 participants from across 14 countries discussed for two days the challenges related to "business, well-being and governance," among them CEO's of transnational mining companies, the Chilean minister of mining, high rank officials of the OCDE, scholars, consultants, and urbanists. Characteristic for the debate at the OCDE meeting was its highly technocratic view on the relations between business, well-being, and governance, and a huge gap between what was discussed as the main problems of mining cities and what one could see, smell, and feel being physically around in Antofagasta those days. From one of the hotel's terraces, for instance, where the coffee breaks and evening receptions took place, you could not only have a view on a small private beach but also constantly see mega containerships entering or leaving the Antofagasta Terminal Internacional (ATI), one of the main mining ports of the region, located only some blocks south of Hotel Antofagasta. Since at least 2013, when social mobilization against the extension of its infrastructure formed, ATI has been made responsible by public health experts and citizen groups alike for the extremely high rates of pollution (copper, lead, zinc, arsenic) in its immediate surroundings (Sanhueza, 2016). In June 2018, *The Clinic*, an influential weekly magazine of national circulation, called the area of influence of ATI, where the highest cancer rates of the country are detected, the "Chilean Chernobyl" (Bañados, 2018).

Leaving Hotel Antofagasta through the main door the view opens towards the city itself. From here the mushrooming of new land takings and autoconstructed settlements can be observed, on the eastern limits where the arid hills of the Atacama Desert and the existing hazard risks limit the advancement of the "official city." With the super-commodity cycle between 2008 and 2011 (Arboleda, this volume) Antofagasta became attractive for immigrants from other regions of Chile and Latin American looking for employment opportunities at the mines or the transport and service sectors that cater to the industry. If Antofagasta as a city has the highest per capita GDP of the country, at the same time it has the highest poverty and strongest growth rates of land takings and new informal settlements (in Chile called *campamentos*), it is a city of stark contrast and high segregation (see figures 12.1, 12.2, and 12.3). Only between 2011 and 2016 the number of campamentos in Antofagasta grew by 487%, counting today for at least 50 informal settlements and where around 5,000 families live in autoconstructed houses on the urban fringe (Cooperativa, 2018).

Figure 12.1 Container ship entering the ATI-port, as seen from the Hotel Antofagasta. Photo by Michael Lukas.

Figure 12.2 New land takings on the eastern part of the city. Photo by Michael Lukas.

Figure 12.3 The ATI-port complex. Photo by Sanhueza, 2016.

While being a superlative in many aspects, Antofagasta is only the most extreme representation of the crisis of Chile's development model, which is based almost exclusively on the extraction of natural resources and its rapid export. Many cities and territories that are operationalized for resource extraction, energy production, and the dump of waste today show very high degrees of segregation, social inequality, and industrial pollution (Vásquez et al., 2017). In the context of increasing critiques of this social, spatial, and environmental injustice both in Antofagasta and other cities that function as urban nodes of global extractive networks, in the last decade Chilean and foreign multinational corporations have begun to enter the arena of urban planning and governance (Lukas & Brueck, 2018). Under the umbrella term of urban master plans (*planes maestros*), these global players of the extractive industries are setting up public-private alliances for, officially, the long-term planning of urban development to enhance quality of life, sustainability, and civic participation.

While the model originated with the Plan for Sustainable Reconstruction (PRES) Constitución after an earthquake and tsunami in 2010, the most ambitious of these master plan initiatives is called CREO Antofagasta (from here CREO) and was set up in 2012 by BHP Billiton, the world's leading company of mineral extraction, operating a large copper mine in the hinterland of Antofagasta and important mining-related infrastructure in Antofagasta itself. The stated goal of CREO is to "meet the challenges of urban growth, with a strong sense of improving the quality of life of the community" and to design the city "in which we want to live" (CREO Antofagasta, n.d.). The plan is based furthermore on the objective of "involving regional government, the city, and organized civil society so that in a participatory dialogue, the needs and expectations of all those who belong to this city converge" (ibid.). In order to transform Antofagasta into a "model city on a national and international level" urban planners of high international reputation have been hired, and an important advising role was given to the OCDE. CREO Antofagasta is thus not only a case of multinational corporations of the extractive industries that shift the centre of their corporate social responsibility activities from the actual sites of extraction to the city and thus the field of urban governance, but also one of the privatization and transnationalization of planning.

In this chapter I analyse why and how the extractive industries get involved in urban governance and planning, and what that means for the political and spatial dynamics of peripheral urbanization. Based on conceptual work on the dialectical relationship of concentrated and extended urbanization, the changing politics of global extraction networks, and the literature on urban policy mobilities, I argue that the CREO master plan can be seen as a platform for the "strategic coupling" (Yeung, 2009) of city space and its institutional landscape to the needs of lead firms of the extractive industries. Rather than meeting the challenges of urban growth in a truly inclusive manner and thus addressing pressing problems as the access to quality housing and stop mineral pollution, the privatization and transnationalization of planning through CREO aims at material, symbolic, and political dimensions of strategic coupling: first, the retrofitting of city space as infrastructural production force in a changing organizational and technological environment of resource extraction; second, the de-politicization of the inherent violence of planetary urbanization through technocratic discourses on consensual decision-making and ecological modernization, trying to gain social legitimacy for their operations; and third, a long-term role for private companies in urban governance. While not all these goals are actually achieved, what is taking place is to move urbanism towards spectacle: where the dust of badly handled minerals is contaminating children's bodies and thousands of people are pushed into insecure dwelling conditions, the urbanists responses are cultural festivals, cycling lanes, and, most importantly, an urban waterfront redevelopment inspired by Barcelona and San Francisco. What I conclude in this context is that Antofagasta and its master plan represent a special type of peripheral urbanization and peripheral urbanism: one where urban processes, infrastructures, and governance dynamics are subordinated to the particular logics of capital accumulation by the global networks of resource extraction, sidelining the needs of a great number of urban dwellers, especially the poor. Importantly, together with other similar initiatives realized in Chile in the last years, the CREO master plan, by consultants, planners, and international organizations such as the OCDE, is packaged as best practice and mobilized to be emulated around the world through the global circuits of urban policy making. In methodological terms, the article is based on qualitative methods and extensive fieldwork in Antofagasta. Around 50 expert interviews were conducted with policy makers, planners, local officials, corporate representatives, and civil society organizations between 2016 and 2018; furthermore, events such as the OCDE meeting were attended.[1]

The chapter is structured as follows. In section two I develop the conceptual perspective by tracing the intersections of work on concentrated and extended urbanization in the context of resource extraction: how multinational companies seek to produce strategic coupling between cities and global production networks, and how the arenas of urban governance and policy in cities around the world are ever more dominated by a global intelligence corps of urban expertise that detects problems and offers solutions, often in a highly simplified way.

In section three I turn to the case of Antofagasta. First, I draw out how two multinational companies of the extractive industries, the Chilean Luksic Group and the Australian BHP Billiton, design and operationalize the urban fabric for their global value chains and production networks. While already operating important heavy mining-related infrastructure in the city, in the last years they embarked on strategies to expand their extractive activities in Chile as well as widen their global outreach. However, I also show that this is taking place in an urban environment turning ever more complex for their operations in social and political terms, since the negative economic, environmental, and socio-spatial impacts – the housing crisis and land takings, contestation of mineral pollution – of the mining business are intensifying and provoking resistance. It is in this context that CREO Antofagasta emerged – an initiative for strategic coupling through strategic urban planning.

In section four I thus turn towards describing CREOs governance scheme, its tactics and impacts, and how the initiative, despite its insistence on consensus-based decision-making and ecological modernization is clearly biased towards the material imperatives of resource extraction, leaving aside – or organizing out of the agenda – the real pressing problems of urban development in Antofagasta.

The chapter concludes with reflections on what the findings mean for our understanding of the governing dynamics of peripheral urbanization in Latin America's extractive cities.

Operational Landscapes, Strategic Coupling, and Transnational Resource Extraction Urbanism

Since colonial times, cities and urban infrastructural networks in Latin America have been designed to sustain the extraction of gold and silver, nitrate, and copper in cities such as Potosí, Belo Horizonte, and Antofagasta. However, this operationalization of city space and its associated spatial, political, and metabolic dynamics have rarely been addressed in the urban studies literature. Only within the recent upsurge of work on planetary urbanization (Brenner, 2014; Brenner & Schmid, 2015) have scholars begun to scrutinize the entanglements of resource extraction and urbanization, especially focusing on those places that had largely been overlooked by urban theory because they were considered to be non-cities (Wilson, 2013; Arboleda, 2015; Arboleda, 2020), or because they were "global cities from off the map" (Robinson, 2002; Kanai, Grant, & Jianu, 2018). The conceptual framework of planetary urbanization, with is emphasis on the dialectical relationship between concentrated and extended urbanization, not only allows for the analysis of the explosion of spaces and the operationalization of remote hinterlands for extraction, but also to consider how the re-scaling of urbanization towards the planetary and the concomitant extension of the extractive frontier change the spatial, social, and political dynamics of city space itself. Authors such as Bridge (2009), and Arboleda (this volume) put forward a materialist-relational perspective, which allows us to understand mines, holes, shafts, pits, ports, interoceanic corridors, railway systems, entire cities, and governance landscapes as inextricably linked and dialectically related elements and processes at the interface of resource extraction, capital circulation and urbanization. Looking at the explosion of such "infrastructural space" and their technologies in times of planetary urbanization, Easterling (2015) states that "some of the most radical changes to the globalizing world are being written not in the language of law and diplomacy, but in these spatial, infrastructural technologies – often because market promotions or prevailing political ideologies lubricate their movement through the world."

As Harvey (1982) argues, capital seeks to represent itself in the form of a physical landscape "created in its own image," a second nature aligned with the logics of accumulation and circulation at specific moments in time, only to have to destroy it in later rounds of capitalist modernization, a process coined as "creative destruction." With regard to the relation between the circuits of resource extraction and urbanization, this has two aspects. On the one hand, urban agglomerations and infrastructural networks are sinks for excess liquidity (Harvey, 1982), whereby the skyscrapers of the global cities can be understood as "inverted mines," physical landscape created in the image of capital and produced through the "cycling of mineral wealth through the city, and its fixation in space" (Bridge, 2009). On the other hand, as Arboleda (in this volume) states, the urban fabric "with its multiple networks of communication and exchange becomes a part of capital" and "in this context, the spatial arrangement of a city, a region, a nation or even a continent, increases productive forces in a similar way as do the machinery and equipment of a factory." Importantly, in order to cycle mineral wealth through the city on an ever expanding scale and to optimize the spatial arrangement of urban space for resource extraction and capital circulation, strategic interventions are needed in such different fields as taxation, finance and urban-regional planning and governance. Until now, analyses of the relationship between new rounds of capital accumulation, urbanization, and creative destruction have largely focused on the role of the neoliberal state and its "surge of infrastructural investments, enclosures and large-scale territorial planning strategies" (Brenner, 2014, p. 20). The role of multinational corporations of the extractive industries in processes of institutional and territorial creative destruction, particularly in the resource peripheries, has largely been neglected in this strand of literature.

Insights on how multinational corporations mediate with cities and regions can be found in the literature on global production networks (GPN) and specifically on the concept of strategic coupling (Bridge, 2008; Yeung, 2009; MacKinnon, 2013; Coe & Yeung, 2015; Breul & Revilla Diez, 2018). Similar to the procedural and relational understanding of space in planetary urbanization, within the GPN literature, cities and regions are understood as "porous territorial formations" whose "national boundaries are transcended by a wide array of network connections" (Coe et al., 2004, cited in MacKinnon, 2013, p. 307). While older work in globalization-related economic geography has dealt with the impact of FDI on recipient countries on a national scale, the GPN approach is about conceptualizing the entanglements of cities and regions with the increasingly network-based organization of global production processes.

A concept within the GPN literature for understanding the interdependent relationship between cities and regions and global production networks in its governance implications is that of "strategic coupling" (Yeung, 2009; MacKinnon, 2013; Breul & Revilla Diez, 2018). Yeung (2009) defines strategic coupling as "the dynamic processes through which actors in cities and/or regions coordinate, mediate, and arbitrage strategic interests between local actors and their counterparts in the global economy. These transurban and transregional processes involve both material flows in transactional terms and nonmaterial flows (e.g., information, intelligence, and practices)" (Yeung, 2009). For Yeung (2009) strategic coupling "is strategic because the process does not happen without active intervention and intentional action on the part of the participants," and the concept aims to explain "how key actors in specific cities and regions become articulated into the imperatives of lead firms in GPNs; it is about dynamic relational processes that mediate their collective action and common interests" (Yeung, 2009).

As conceived by Yeung (2009), the design and negotiation of the relationship between GPN and cities and regions through processes of strategic coupling contains both material and symbolic-discursive dimensions. With regard to the material dimensions, two main issues can be identified: first, the exploitation and disciplining of labour force (Herod, 2009), and second, the spatial-infrastructural organization and optimization of urban space. For instance, the mining industry is under constant pressure to minimize the cost of transporting extracted minerals and increase the speed of their circulation (Arboleda 2016, this volume). With the commodity-super cycle, mining companies have undergone a shift from an operational focus on the immediate sites of extraction (i.e., mines, shafts, and mining settlements) to a concentration on global production and value chain. For Chile, this means reducing the time of capital turnover by accelerating the transport of copper from the high Andean mines to the final consumers through organizational, logistical, and political measures on different spatial scales (Arboleda, 2020). Regarding the symbolic-discursive dimensions of strategic coupling, of increased importance are aspects of social legitimacy. As mistrust towards multinational corporations has grown worldwide, companies have learned to take care of the "soft" and intangible factors of the production process, all along the production chain. In that context, since the 1980s the discourses and practices of corporate social responsibility (CSR) have emerged, focusing on local stakeholder groups and the management of local conflict, and on improving corporate reputation in the consumer markets (for instance through international certification mechanisms such as the Forestry Stewardship Council). The global mining industry over the last decades has adopted "a distinctive environmental narrative" and "has become an advocate for the environment and has embraced concepts of sustainability and stewardship" (Bridge & McManus, 2000, p. 34). In the last years there has been a rapid evolution of concepts that suggest a move from instrumental rationalities – managing local conflict and secure good stakeholder relationships – to an imaginary and rationality of "disinterest" (Tironi & Zenteno, 2013). Under the banner of concepts such as Shared Value (Porter & Kramer, 2011) and *Political* Corporate Social Responsibility (PCSR) corporations increasingly position themselves as holding a corporate citizenship (Ehrnström-Fuentes, 2016; Scherer et al., 2016). The management literature, strategic consultancies, and corporate headquarters in that context recommend that companies should become more deeply involved in urban and regional decision-making processes, in order to close "governance gaps," participate in the provision of public goods and enter new forms of public-private partnerships (Scherer et al., 2016, p. 273). Interestingly, here a strong ideological familiarity emerges between the corporate discourses and those of "global sustainable urbanism" (Rapoport, 2015).

In the last years in urban studies there has been a huge output on the transnational dynamics of global sustainable urbanism and urban policy making, rooted in the same epistemological turn towards processual and relational thinking which informs the concepts of planetary urbanization and global production networks (McCann & Ward, 2011; Künkel, 2015; Baker & McGuirk, 2017). Of particular interest here are the globally interconnected processes of assembling and circulating policies, plans and programs and the actor networks and epistemic communities that sustain it. Today a tightly networked global informational infrastructure exists, through which ideas, policy fragments, and best practice models are produced and diffused transnationally. Rapoport (2015, p. 111) here refers to the "global intelligence corps" of urban development expertise – a loose and expanding network of politicians, planners, consultants, and activists who not only mobilize ideas, practices, and models, but are often also globally mobile themselves. Moving from one conference to the next consultancy job, and from that keynote address to the next business meeting, these "traveling technocrats" (Larner & Laurie, 2010), "international master planners" (Rapoport, 2015), and "persuasive practitioners" (Montero, 2017) are specialized in reconciling the languages and interests of city administrations and business leaders, and also in diagnosing urban problems and offering solutions. They act as "transfer agents" (McCann & Ward, 2011) that in their form of communication rely heavily on techniques of "storytelling" (Montero, 2017) and the construction of "parables" (Wilson, 2016). In that context, a characteristic strategy of these storytellers is to discursively simplify the urban problems to be addressed and the solutions recommended, often relying on a set of international best practices. For a long time Barcelona was the key reference when it came to strategic planning and citizen participation; in the last ten years or so the Medellin-model of social urbanism is a major reference for any discussion about urban planning and governance innovation in Latin America. A key role for the (re-)production and circulation of models and ideas do play the meetings and conferences – together with the resulting documents and policy recommendations – of international organizations such as OECD, World Bank, or UN-Habitat (Jessop, 2002, Kaika, 2017). The fact that UN-Habitat's World Urban Forum in 2014 was held in Medellín, the Habitat III-conference 2016 in Quito, and Chilean architect Alejandro Aravena in the same year received the Pritzker Prize indicates not only that Latin American cities and their planners are an integral part of the global circuits of urban knowledge production and policy diffusion, but also that their strategies of branding themselves as hubs and actors of innovation have been highly successful. As Justin McGuirk (2014, p. 25) shows, Latin American cities and its persuasive practitioners have been particularly successful in positioning themselves as oriented towards issues of socio-spatial inclusion, social urbanism, and participatory planning, offering a whole new "urban repertoire." From another perspective, however, they are part of the "global urban intellectual and professional technocracy [that] has spurred a frantic search for a 'smart' socio-ecological urbanity" (Swyngedouw & Kaika, 2014). "Under the banner of radical techno-managerial restructuring, the focus is squarely on how to sustain capitalist urbanity so that nothing really has to change!" (ibid.). In the remainder of this chapter, I show how the global urban intellectual and professional technocracy of international master planners is being involved in processes of strategic coupling of the extractive industries. Although not all corporate goals are achieved in CREO Antofagasta, through strategic privatization and transnationalization urban governance and planning are increasingly delinked from the pressing urban problems in Antofagasta.

Antofagasta: Between the Operationalization of City Space for Resource Extraction, Autoconstruction, and Social Mobilization

Two facts about Antofagasta's relation to the global circuits of mineral extraction stand out: that the actual mines – the holes, shifts, and pits where minerals are extracted – are located in huge distance from the city in the Andean Cordillera from where the minerals have to be transported to the ports in Antofagasta; and that both the principal mines and the urban support infrastructure are controlled by few multinational corporations, the most important being Antofagasta Minerals and Minera Escondida. While the former belongs to the Chilean Luksic Group, the latter is a subsidy of BHP Billiton, the world's largest mining company, with headquarter in Melbourne, Australia. For both companies Antofagasta is a strategic terrain where they organize several crucial operations in a vertically integrated manner.

The Luksic Group, one of the most powerful conglomerates in Chile, in the last decade has embarked on a process of diversification and global expansion (Leiva, 2017). It controls companies in a wide array of sectors, from mining, finance, and banking to telecommunications and transportation. Its origin lies in Antofagasta, where the group operates an integrated production and logistic chain that encompasses activities such as extraction, transportation, loading, and shipping, and that spans from the heights of the Andean mountains, through the Atacama desert, the urban fabric of Antofagasta and the Pacific Ocean, to those places where the minerals are unloaded, mainly in North American and Chinese port cities. The starting point of its logistics chain is over 3,000 metres above sea level and 180 kilometres from the city of Antofagasta, where, through Antofagasta Minerals, the largest private mining company of Chilean origin, Luksic operates the copper and gold mines Zaldivar and Zentinela.[2] After extraction, the minerals are transported to the Antofagasta port facilities via pipelines, highways and the railway Ferrocarril Antofagasta Bolivia (FCAB). FCAB, founded in 1888 in the nitrate epoch and since 1980 owned by the Luksic Group (through Antofagasta plc), operates a rail network of more then 900 kilometres and is entirely dedicated to the transportation of mining inputs and products, such as sulphuric acid and copper cathodes. One of the principal destinations of the minerals is the ATI-port. As a railway, it caters exclusively to the mining industry and has been controlled by Luksic since 2011, the same year a contested extension of the port began to be implemented (see below). At ATI, the minerals are first stored and then loaded on to large container ships, principally those of the Compañia Sudamericana de Vapores (CSAV). While Luksic took control of CSAV in 2011, in 2014 another important step towards global integration was made with the purchase of the German shipping company Hapag-Lloyd. Today Luksic has a fleet of 200 ships and the world's fourth biggest company of container transportation. Between 2003 and 2015 Luksic also controlled Aguas Antofagasta, the principal water company in the city.

While the Luksic Group is an impressive example of how Latin American conglomerates are not only part of global production networks but seek to be their dominant players, thereby operationalizing vast hinterlands and city space through complex infrastructural networks, the role of BHP Billiton in Antofagasta represents the power of foreign owned multinational companies and their grip on urban regional space. Due to a legislation extremely friendly to Foreign Direct Investment (FDI), since the early 1990s the mining industry and with it the region of Antofagasta have seen process of far reaching transnationalization. Between 1990 and 2008 the region received between 15% and 30% of all FDI in Chile and the share of copper production controlled by multinational companies as BHP Billiton, Anglo American, and Barrick Gold increased from 24% of the national output in 1990, to 65% in 2010 (Atienza & Aroca, 2013). BHP's significance becomes clear with the fact that with Minera Escondida[3] it operates the world's largest private copper mine and between 2010 and 2016 inverted more than USD 10 billion in Chile.[4] In 2016 Escondida produced 1,243,524 tons of fine cooper which corresponds to 22% of national and 6% of worldwide copper production (BHP, 2016).[5] Escondida is located 3,100 metres above sea level, and 170 kilometres southeast of Antofagasta, close to the mining town of Sierra Gorda from where the concentrated mineral via two tubes is directed to the Coloso Port complex at the southern edge of Antofagasta, also property of BHP. At Coloso BHP since 2006 operates a desalination plant which prepares sea water for industrial use and pumps it to the mines via a 166-kilometre-long aqueduct. In 2013 BHP kicked of the construction of a $3.43 billion second mega desalination plant in order to secure water supply at Escondida and to support future expansion.

According to Arboleda (2015) Antofagasta can be understood as a "metabolic vehicle for planetary urbanization" (Arboleda, 2015), an urban-regional landscape operationalized for natural resource extraction which fuels urban agglomeration processes around the world, but does little for local economic development. The almost exclusive orientation of urban development in Antofagasta around the needs and logics of global circuits of mineral extraction has sharpened its historic condition as mining enclave with a lack of economic diversification, a housing crisis leading to the upsurge of land takings, and social mobilization around mineral pollution. As a first case in point, Antofagasta is economically the least diversified region of Chile, and in the last decades it has actually deepened its dependence on mining, which accounts for 60% of the regional product (Atienza & Aroca, 2013, p. 414). Only a poor local entrepreneurial environment exists, and the region is highly dependent on foreign capital, especially since since public investment is low.[6] Today Antofagasta is considered to be one of the best places to work (for the high salaries in the mining industry and related services) and the worst to live. The mining companies find it difficult to retain a skilled workforce in Antofagasta; almost none of the higher executives of the industry lives in Antofagasta, further aggravating the flight of mining wealth (Atienza & Aroca, 2013, p. 415). Calama, a major neighbouring mining town with similar problems related to of public investment and of green spaces and cultural activity, experienced an outburst of social discontent in 2011, led by the local mayor, accusing the central state and its copper company – CODELCO – of radical abandonment.

Despite being considered a bad city in which to live by people who can afford otherwise, Antofagasta has been attracting important numbers of immigrants, and it is seeing high rates of population growth. Through a combination of the circulation of mineral wealth through the city and associated real estate speculation Antofagasta has become one of the most expensive cities in Chile, especially with regard to the cost of housing (ibid. 416). Due to a parallel absence of social housing construction, access to the formal land and housing markets for the poor has become ever more difficult, and has transformed Antofagasta into the city with the strongest growth rate of land takings and campamentos in the whole country. Immigrants from other Latin American countries, especially from Colombia and Peru, have been attracted by the commodity boom since 2008, finding residential solutions often only in overcrowded and abusive rental apartments in the city centre or in the campamentos towards the eastern hills of the city (figure 12.4). Conceiving land takings to be a problem of public security, the regional government set up a public-private initiative in 2013, named the *Plan de Superación de Campamentos* (Plan for the overcoming of informal settlements), with financial participation from Minera Escondida (BHP Billiton). While the goal was to develop guidelines and provide permanent housing solutions for the *pobladores* (informal settlement dwellers), in the end very few concrete results have been achieved. As was the case before the program started, still today some land takings are tolerated while others have been cleared by police, which leads to a high level of insecurity for the *pobladores*. The answer is similar to that of previous phases of peripheral urbanization in Latin America: people do not only autoconstruct houses but begin to organize politically and embark on transversal logics and relations with other actors – with not only the state, but also the Catholic church, NGOs, or universities (Caldeira, 2017). The NGO Fractal and academic activists of ORDHUM, for instance, currently are conducting an initiative involving local participatory planning in the Los Arenales settlement funded by the NGO Slum Dwellers International, in order to achieve permanence and legalization of land tenure. In

Figure 12.4 Autoconstruction in the hills of Antofagasta. Photo by Michael Lukas.

2017 and 2018 several of these groups organized marches calling for the right to housing and the right to the city, putting forward arguments and agendas that are radically different from those promoted by the extractive industries.

While the housing crisis might be seen as a real socio-spatial problem, but with only an indirect impact on the operationalization of city space by the mining companies, a very direct impact comes from mineral pollution. While historically there have been high levels of arsenic in potable water leading to public health problems, such as very high cancer rates in Antofagasta, new epidemological studies have shown that the biggest problem is the bad handling of minerals during transportation, storage, and loading. Both the FCAB-railway that crosses the entire city with copper products as well as the ATI-port, surrounded by schools, childcare facilities, and hospitals, are identified in studies as sources of contamination. When in 2013 the Luksic Group began the construction of a new storage copper facility at its ATI port the movement *Este polvo te mata* (This Dust Kills You) formed, organizing marches and making demands, such as, first and foremost, to stop the construction of a new copper storage facility and, later, to remove the entire ATI-port and the FCAB-railway from town. Due to Luksic's political influence, however, in 2015 in the face of huge criticism the storage facility (in Antofagasta known as "el galpón") was completed, and the protest movement lost momentum.

CREO Antofagasta, Strategic Coupling, and the Transnationalization of Resource Extraction Urbanism

In this context of the deepening of the operationalization of city space for resource extraction and circulation in Antofagasta through new infrastructure on the on the one hand, and ever more evident social and environmental problems on the other, in 2012 BHP Billiton (through its Minera Escondida Foundation) and the Luksic Group (through FCAB) set up the CREO Antofagasta master plan initiative. Formally organized as "a public-private multi-stakeholder body led by the region's Intendente and the city's mayor," the private sector would finance the elaboration of the plan (OCDE, 2013, p. 251). Three levels of steering were installed. At the strategic level the mayor and intendente led a public-private council of around 40 members from the public, private, and civil society sectors that "promote, facilitate and guarantee the accomplishment of the objectives of the plan." On a tactical level the executive committee, formed by ten members (two local universities, six business organizations including Minera Escondida and FCBA, four representatives of the citizenship council and the executive secretary), organized the operation of CREO and plans and implemented initiatives.

In practice the most important function is the executive level. Here a director leads the executive secretariat, a team of architects, urbanists, sociologists, and journalists, among others. It organizes the strategic planning process, from the visioning exercises, project design, community relations, communication to the implementation of a wide range of projects. While CREO started in 2012 with only three professionals (two architect-urbanists and one anthropologist in charge of setting up the citizen participation process) in 2017 a group of around 25 young, academically skilled and highly motivated people worked in an open space, a start-up-like office right in the city centre. From its beginnings and until 2017 CREO was led by an internationally renowned Chilean urbanist with best connections to the global network of sustainable urbanism. Before starting at CREO he worked as a director at ARUP, in London, a leading global firm of urban planners and engineers (Rapoport, 2015) and, among many other projects, he was involved in the Ecocity Dongtan in Shanghai, one of the major global flagship projects of ecological modernization. The manpower and know-how concentrated in CREO thus has been remarkable. Few other planning agencies in Chile are comparable in size and professional scope; the municipal planning authority of Antofagasta, for instance, most probably does not have more than five professionals.

As a strategic planning initiative, the work of CREO was sequentialized. In the first phase, from 2012 to 2013, trust was to be built among the different participating actors, and both participatory and expert-led diagnosis of the urban challenges were elaborated (*línea base*), as well as early projects implemented (*obras de confianza*). In the second phase, from 2013 to 2014, the goal was to build a common vision, to identify strategic projects, and to edit the actual master plan document. The third phase then would be developed from 2015 to 2035, dedicated to the full implementation of the master plan and its strategic components. In practice, however, CREO Antofagasta took a different route, first and foremost because the actual master plan, the principal goal and main instrument of the whole initiative has not been made public, and it is not clear if it ever will happen. As shown in the following, the reason for this seems to be related to what Easterling (2015) describes as a characteristic of such "spatial technologies" as strategic urban master plans: they are based on "stories" that "foreground content to disguise or distract from what the organization is actually doing."

The empirical findings indicate that CREO was never intended, nor was it able, to re-think Antofagasta in an integral way in which "all the needs and expectations of all those who belong to this city converge." Rather, it seems that CREO from the outset was conceived as a flexible platform for organizing strategic coupling processes, that is, to bring urban space and its governance landscape in line with the imperatives of lead firms in global production networks (Yeung, 2009). A first point that supports this hypothesis is found in what one CREO planner said in response to the question about how decision making is organized between the public and private actors: "The donors council is the actual directorate; they approve projects and its financing." BHP is "the principal of CREO," not the mayor or the intendente (interview with CREO planner, June 2017). Asked why BHP set up CREO, a high-ranking manager explained that the initiative was an expression of the company's corporate culture, developed in its headquarters in Australia. This culture, with a view to community relationships, would not be merely instrumental, as CSR-initaitives in the past have been designed, but rather seek to "make a real contribution to improving the quality of life where the company is active" (interview with Vice President of Corporate Affairs of BHP Billiton, March 2018). In symbolic terms CREO thus aims to position the mining company as a responsible member of the community, as one citizen "among many others that belong to this city" (CREO Antofagasta, n.d.). A leading CREO planner, however, is clearer: "Social responsibility is important, but even more so, that the city functions for productive activities, among which BHP is one of the most important" (interview with a high-ranking CREO planner, May 2016). He went on to explain that the city is the support system on which all productive activity is based and "that the quality of the city is critical to their [BHP's] business." Another planner said that the objective of CREO was "to attract global business, to catalyze the global market in the city, so that the city responds to the global business that takes place in it" (interview with CREO planner, June 2016). The mining companies' view that the city does not respond (anymore) to the needs of their activities becomes clear in a document of the Chilean Sustainable Business Network (Devenin, 2017), where CREO Antofagasta (and the neighbouring initiative Calama PLUS) is coined as a best practice, to be emulated in other contexts. From the perspective of capital, of course, the main problems to be addressed are not the housing crisis or industrial pollution, but "high employee turnover and loss of significant human resources investment by the companies" (Devenin, 2017, p. 5). Because Antofagasta is such an unattractive place to live for those that can afford otherwise, "people lived in Antofagasta only temporarily, planning to return to their original cities" (ibid.). In this perspective the problem to be solved is that "despite incentives, the city did not help the industry to attract and retain talent; companies' competitiveness declined" (ibid.). And it was "because the mines were expected to have a long useful life" that "companies saw issues in Antofagasta as strategic" (ibid.) and recognized "that they must create shared value and long-term community relationships in order to gain and maintain social legitimacy" (ibid., pp. 5–6). At least the higher ranking CREO planners are well aware of this interest of its principle, one resumed the initiative's main objective to be improving Antofagasta's "international competitiveness vis-à-vis other mining cities" in order to attract "highly qualified human capital" (interview with CREO planner, August 2017). Here two goals of strategic coupling become evident: to physically, socially, and culturally adapt city spaces to the needs of lead firms in global production networks; and to gain and maintain social legitimacy through discourses about corporate citizenship.

From its beginning CREO placed high emphasis on the participatory and consensual character of the initiative. For the elaboration of the so-called citizen baseline, which would feed the vision and the will of the citizenry into the strategic guidelines of the master plan, throughout 2013 tactics of citizen activation and participation were enacted, such as *malones urbanos* (typical Chilean neighbourhood meetings), open forums, and territorial meetings. These activities culminated in a citizens' forum where a citizens' "manifest for quality of life" was presented, which is very much in line with the problems and guidelines at the same time elaborated by the OCDE in a territorial review (OCDE, 2013; see below). A representative of one of the local universities that participated in those events states that "the issues are previously established. These events work like focus groups where the objectives are clear and the goals defined; these workshops are organized to validate information that they already have" (interview with a university academic, May 2016). Further sustaining this view is that in CREO Antofagasta there is no room for radical voices or dissent. Members of Este Polvo Te Mata or the NGOs working for the Right to the City in the campamentos are highly critical of CREO and do not participate in its activities. On the other side, CREO usually does not take any public position in controversial issues such as the construction of the copper storage facility at ATI, mineral pollution, or the housing crisis. The principal projects executed in the first two phases of CREO, therefore, have been oriented unsurprisingly towards organizing uncontroversial activities such as marathons, bodyboard and skateboard competitions, social innovation networking events, the construction of 40 kilometres of cycling lanes, and the participative cleaning of micro waste dumps.

New about CREO is the far-reaching re-scaling of planning away from the local and regional towards the transnational scale of global sustainable urbanism. While from the outset one of the stated goals was to transform Antofagasta "in a model city on a national and international level," in this storyline it was not mentioned in how far this would delink the whole initiative from local communities and its pressing problems. The first step in this de-linkage was the designation of the international master planner Alejandro Gutierrez as executive director. Presumably through his networks the permanent "expert council" has been set up to sharpen the international profile of the initiative through its members, including Richard Burdett, from the London School of Economics, and Enrique Peñalosa, former mayor of Bogotá and outstanding persuasive practitioner (Montero, 2017). On the one hand these and other members of the global intelligence corps of ecological modernization gave talks in Antofagasta; on the other, on several occasions delegations of local politicians, planners, and community leaders have been send to Medellín to learn from its miracle of urban reinvention, an example of the role that international "policy tourism" (Gonzalez, 2010) plays in contemporary global sustainable urbanism. The most important move towards the transnationalization of planning, however, was the role assigned to the OECD. While in the diagnosis of urban problems – on which the definition of measures, projects, and the master plan would be based – was designed to consider the existing local development plan (*Plan de Desarrollo Económico Local*) and community participation, by far the most influential input was the 300-page "territorial study on Antofagasta" conducted by OECD-professionals. As a CREO planner stated, the goal was to "raise the quality of the urban problem diagnosis," "to bring Antofagasta up to international circuits," and to establish a bottom line based on expert knowledge, an "international standard that could not be ignored" (interview with a CREO planner, June 2018). As Kitchin et al. (2015) show, this kind of technical report is a very specific way "of knowing and governing" a city and, unsurprisingly, the OECD-report reflects a particularly technocratic view of urban development and governance. Although it digs deep into some business-related problems, such as the lack of economic diversity, it almost completely leaves out the problems of mineral pollution, the housing crisis, or the problems of water scarcity and conflict. In the sections on urban governance issues it neither addresses the huge power asymmetries between the multinational companies, local communities, and governments. Rather, it is an excellent example of "process benchmarking that compares the practices, structures and systems of a place" to other places whereby the very different qualities of those places are not worked through (Kitchin et al., 2015, p. 10). For example, in the OECD-report, London and Toronto are the shorthands for successful public-private partnerships that could be emulated in Antofagasta. Considering that it was CREO that hired and paid the OECD – formally via the regional government – it seems like an act of decision-based evidence making that the report states that "a long-term vision for urban form could be an important first step towards building a forward-thinking urban planning culture and an integrated agenda for Antofagasta's urban development," and furthermore, that "the private sector could – and should – be a key partner in developing this vision, in addition to other stakeholders" (OECD, 2013, p. 165). It seems evident that the role of OECD was to make the CREO initiative itself legitimate, relying on its reputation reflecting knowledge, expertise, and international standards that cannot be ignored in a city such as Antofagasta. With a view on the few major projects that are being implemented by CREO, the most important being a huge project of waterfront redevelopment, of particular importance in the OECD report is the section on building "broad-based support to capitalize on key natural assets, particularly the coastline" (ibid.). Here again strategies of benchmarking and "urban interreferencing" are applied, with references to Sydney, Toronto, Bilbao, or San Francisco as best practices of waterfront redevelopment to be emulated in Antofagasta. By far the biggest project CREO is proposing for Antofagasta, which is in its first phase of implementation, then, is the construction of a new waterfront. While this is not in itself a bad idea in a city with a poorly developed 30-kilometre shoreline, it is neither what can solve the high cancer rates nor the access to affordable housing. What it might be able to achieve is to make Antofagasta more attractive to highly skilled labour.

Conclusion

In this chapter I addressed the causes, modalities, and consequences of multinational corporations of the extractive industries entering the fields of urban governance and planning in contexts of peripheral urbanization. While the case analysed has been Antofagasta, it is important to remember that the CREO initiative is only the most ambitious of a series of master plans that, since the birth of the model in Constitución in 2010, have been developed in other Chilean cities that are hubs and nodes of the global extractive industries. Two main findings regarding the literature on planetary urbanization, global extraction networks, and urban policy mobilities shall conclude this chapter. The first finding is that multinational corporations in the extractive industries enter urban governance and planning for reasons of strategic coupling in both material and symbolic dimensions. In material terms, attempts are made to adjust the city space and the urban fabric to the needs of lead firms in global networks of resource extraction. In the case of Antofagasta and its mining companies, that means making the city more attractive to advanced human capital, with the most important project for achieving this being a new waterfront, proposed by CREO. Thus, it is not only or principally the state that organizes creative destruction for capital accumulation and planetary urbanization, as authors such as Brenner (2014) suggest; there are also powerful private players involved, and here "infrastructure space becomes a medium of what might be called extrastatecraft – a portmanteau describing the often undisclosed activities outside of, in addition to, and sometimes even in partnership with statecraft" (Easterling, 2015). The symbolic elements of strategic coupling that justify the putting into practice of extrastatecraft are the discourses on corporate citizenship and shared value which also help to gain and maintain social legitimacy.

The second main finding is that for their strategic coupling projects the extractive industries make use of the global intelligence corps of sustainable urbanism. The international master planners provide a specific language and set of techniques that translate corporate rationales of material and symbolic strategic coupling into spatial designs and urban development projects. Through narratives of consensual decision making and ecological modernization the master planners aim at depoliticizing urban governance and demobilizing growing critiques of the inherent violence of the operationalization of city space. The master plans thus not only allow for strategic coupling in material and symbolic terms, but also in political terms. It is probably for this capacity of flexibly assembling different strategic needs that the master plan methodology has been packaged into a model and best practice and has started to be diffused around the world. A major role here is not only played by the travelling technocrats and persuasive practitioners of global sustainable urbanism, but also by international organizations as the OCDE. With its first Meeting of Mining Regions and Cities in Antofagasta in 2017, the OCDE provided an important platform for the interchange of ideas, policies, and spatial technologies such as the master plan model. The meeting was also a definite step in "worlding" (Roy & Ong, 2011) Antofagasta, that means, putting the city, its companies and urban planners on the global map of sustainable urbanism and advanced corporate citizenship. What this leads to is a widening gap between a "world-class" urbanism of spectacle, urban infrastructure designed for cultural consumption on the one hand, and a city with dire needs and environmental suffering, on the other. What I conclude in this context is that Antofagasta and its master plan represent a special type of peripheral urbanization and peripheral urbanism – one where urban processes, infrastructures, and governance dynamics are subordinated to the particular logics of capital accumulation by the global networks of resource extraction, sidelining the needs of a great number of urban dwellers, especially the poor.

ACKNOWLEDGMENT

This chapter is supported by the Academic Productivity Support Program, PROA VID 2019, University of Chile, and the Fondecyt de Iniciación proyect No. 11150789.

NOTES

1 Empirical research for this article has been conducted before the social mobilizations in Chile started in October 2019. In how far the social mobilizations and the resulting and ongoing elaboration of a new constitution – as well as the global COVID-19 pandemic – affect transnational resource extraction urbanism shall be discussed elsewehere.

2 At Centinela, Zaldivar, and Antucoya around 9,000 workers employed or subcontracted. In 2016 around 110,000 tons of copper cathodes, 180,000 tons of copper concentrate, and 213,000 ounces of gold were produced.
3 With shares of English Rio Tinto, the Japanese Jeco Corp and the International Financial Corporation, the latter belonging to the World Bank.
4 https://www.bhp.com/-/media/documents/media/reports-and-presentations/2017/170807_bhpinformedesustentabilidatabhpchile.pdf. BHP controls the open-pit copper mines Minera Escondida and Escondida Pampa Norte (which includes the Spence and Cerro Colorado mines).
5 http://www.nuevamineria.com/revista/bhp-billiton-exitos-economicos-y-conflictos/.
6 In 2008 the region received USD 126 million in public investment, only around 2% of national investment in regional development.

REFERENCES

Arboleda, M. (2015). In the nature of the non-city: Expanded infrastructural networks and the political ecology of planetary urbanisation. *Antipode, 48*(2), 233–251. https://doi.org/10.1111/anti.12175

Arboleda, M. (2016). Spaces of extraction, metropolitan explosions: Planetary urbanization and the commodity boom in Latin America. *International Journal of Urban and Regional Research, 40*(1), 96–112. https://doi.org/10.1111/1468-2427.12290

Arboleda, M. (2020). *Planetary mine: Territories of extraction in late capitalism*. London, New York: Verso.

Arias, M., Atienza, M., & Cadematori, M. (2013). Large mining enterprises and regional development in Chile: Between the enclave and cluster. *Journal of Economic Geography, 14*(1), 73–95. https://doi.org/10.1093/jeg/lbt007

Atienza, M., & Aroca, P. (2013). Potencialidad económica regional. In CORE Antofagasta (Ed.), *Region de Antofagasta: Pasado, presente y futuro* (pp. 413–419). Antofagasta.

Baker, T., & McGuirk, P. (2017). Assemblage thinking as methodology: Commitments and practices for critical policy research. *Territory, Politics, Governance, 5*(4), 425–442. https://doi.org/10.1080/21622671.2016.1231631

Bañados, R. (2018, Juni 22). *Vivir en medio del Chernóbil chileno.* The Clinic. http://www.theclinic.cl/2018/06/22/vivir-medio-del-chernobil-chileno/ (Accessed: December 2018).

BHP. (2016). *Informe de Sustentabilidad BHP Chile 2016.* Minera Escondida: Pampa Norte.

Brenner, N. (2014). *Implosions/explosions: Towards a study of planetary urbanization*. Berlin: Jovis Verlag.

Brenner, N., & Schmid, C. (2015). Towards a new epistemology of the urban. *City, 19*(2–3), 151–182. https://doi.org/10.1080/13604813.2015.1014712

Breul, M., & Revilla Diez, J. (2018). An intermediate step to resource peripheries: The strategic coupling of gateway cities in the upstream oil and gas GPN. *Geoforum, 92*, 9–17. https://doi.org/10.1016/j.geoforum.2018.03.022

Bridge, G. (2008). Global production networks and the extractive sector: Governing resource-based development. *Journal of Economic Geography, 8*, 389–419. https://doi.org/10.1093/jeg/lbn009

Bridge, G. (2009). The hole world: Scales and spaces of extraction. *New Geographies, 2*, 43–48.

Bridge, G., & McManus, P. (2000). Sticks and stones: Environmental narratives and discursive regulation in the forestry and mining sectors. *Antipode, 32*(1), 10–47. https://doi.org/10.1111/1467-8330.00118

Caldeira. T. (2017). Peripheral urbanization: Autoconstruction, transversal logics, and politics in cities of the global south. *Environment and Planning D: Society and Space, 35*(1), 3–20. https://doi.org/10.1177/0263775816658479

Coe, N., Hess, M., Yeung, H.W-C., Dicken, P., & Henderson, J. (2004). "Globalizing" regional development: A global production networks perspective. *Transactions of the Institute of British Geographers, 29*(4), 468–484. https://www.jstor.org/stable/3804369

Coe, N., & Yeung, H. (2015). *Global roduction networks: Theorizing economic development in an interconnected world*. Oxford/New York: Oxford University Press.

Cooperativa. (2018). *Antofagasta: Campamentos aumentaron en un 487% en los últimos seis años*. https://www.cooperativa.cl/noticias/pais/region-de-antofagasta/antofagasta-campamentos-aumentaron-en-un-487-en-los-ultimos-seis-anos/2018–05–20/155039.html (Accessed 10 June 2018).

Correa, F. (2016). *Beyond the city: Resource extraction urbanism in South America*. Austin, TX: University of Texas Press.

CREO Antofagasta. (n.d.). *Qué es CREO Antofagasta?* http://creoantofagasta.cl/que-es-creo-antofagasta/ (Accessed 5 October 2018).

Devenin, V. (2017). *Collaborative community development: A guide for managers*. Santiago: Network for Business Sustainability, Chile.

Easterling, K. (2015). *Extrastatecraft: The power of infrastructure space*. London, New York: Verso.

Ehrnström-Fuentes, M. (2016). Legitimacy in the pluriverse: Towards an expanded view on corporate-community relations in the global forestry industry. In *Ekonomi och samhälle – Economics and society – 304*. Helsinki: Hanken School of Economics.

Gonzalez, S. (2010). Bilbao and Barcelona "in motion": How urban regeneration "models" travel and mutate in the global flows of policy tourism. *Urban Studies, 48*(7), 1397–1418. https://doi.org/10.1177%2F0042098010374510
Harvey, D. (1982). *The limits to capital.* London, New York: Verso.
Herod, A. (2009). *Geographies of globalization: A critical introduction. Critical introductions to geography series.* Wiley-Blackwell: Oxford.
Jessop, B. (2002). Liberalism, neoliberalism, and urban governance: A state-theoretical perspective. *Antipode, 34*(3), 452–472. https://doi.org/10.1111/1467-8330.00250
Kaika, M. (2017). Don't call me resilient again! The new urban agenda as immunology ... or ... what happens when communities refuse to be vaccinated with "smart cities" and indicators. *Environment and Urbanization, 29*(1), 89–102. https://doi.org/10.1177%2F0956247816684763
Kanai, M. (2014). On the peripheries of planetary urbanization: Globalizing Manaus and its expanding impact. *Environment and Planning D: Society and Space, 32*(6), 1071–1087. https://doi.org/10.1068%2Fd13128p
Kanai, M., Grant, R., & Jianu, R. (2018). Cities on and off the map: A bibliometric assessment of urban globalisation research. *Urban Studies, 55*(12), 2569–2585. https://doi.org/10.1177/0042098017720385
Kitchin, R., Lauriault, T., & McArdle, G. (2015). Knowing and governing cities through urban indicators, city benchmarking and real-time dashboards. *Regional Studies, Regional Science,* 2(1), 6–28. https://doi.org/10.1080/21681376.2014.983149
Künkel, J. (2015). Urban policy mobilities versus policy transfer. Potenziale für die Analyse der Neuordnung. *sub/urban,* 3(1), 7–24.
Larner, W., & Laurie, N. (2010). Travelling technocrats, embodied knowledges: Globalising privatisation in telecoms and water. *Geoforum, 41*(2), 218–226. https://doi.org/10.1016/j.geoforum.2009.11.005
Leiva, R. (2017). Chile's Grupo Luksic and the "New Spirit of Capital." https://rca.ucsc.edu/images/Leiva-Chiles-Luksic-Economic-Conglomerate
Lukas, M., & Brueck, A. (2018). Urban policy mobilities und globale. Produktionsnetzwerke. Städtische Planung in Chile als Legitimationsinstanz extraktiver Industrien. *Suburban-Zeitschrift fuer kritische Stadtforschung,* 6(2–3), 69–90.
MacKinnon, D. (2013). Strategic coupling and regional development in resource economies: The case of the Pilbara. *Australian Geographer, 44*(3), 305–321. https://doi.org/10.1080/00049182.2013.817039
McCann, E., & Ward, K. (2011). *Mobile urbanism: Cities and policymaking in the global age.* Minneapolis, MN: Minnesota University Press.
McGuirk, J. (2014). *Radical cities: Across Latin America in search of a new architecture.* London, New York: Verso.

Montero, S. (2017). Persuasive practitioners and the art of simplification: Mobilizing the "Bogotá Model" through storytelling. *Novos Estudos CEBRAP, 36*(1), 59–75. https://doi.org/10.25091/s0101-3300201700010003

OCDE. (2013). *Estudios Territoriales de la OCDE: Antofagasta, Chile.* Recuperado el 29 November 2015. http://www.keepeek.com/Digital-Asset-Management/oecd/urban-rural-and-regional-development/estudios-territoriales-de-la-ocde-antofagasta-chile-2013_9789264209275-es#page5 (Accessed December 2018).

Porter, M., & Kramer, M. (2011). Creating shared value: How to reinvent capitalism – and unleash a wave of innovation and growth. *Harvard Business Review, 89*(1-2), 62–77

Rapoport, E. (2015). Globalising sustainable urbanism: The role of international masterplanners. *Area, 47*(2), 110–115. https://doi.org/10.1111/area.12079

Robinson, J. (2002). Global and world cities: A view from off the map. *International Journal of Urban and Regional Research, 26*(3), 531–554. https://doi.org/10.1111/1468-2427.00397

Roy, A., & Ong, A. (2011). *Worlding cities: Asian experiments and the art of being global.* Malden, Oxford: Blackwell Publishing.

Sanhueza, O. (2016). *Tribunal ratifica multa a terminal internacional de Antofagasta por contaminar el puerto.* https://www.biobiochile.cl/noticias/nacional/region-de-antofagasta/2016/10/06/tribunal-ratifica-multa-a-terminal-internacional-de-antofagasta-por-contaminar-el-puerto.shtml (Accessed: 10 June 2018).

Scherer, A., Rasche, A., Palazzo, G., & Spicer, A. (2016). Managing for political corporate social responsibility: New challenges and directions for PCSR 2.0. *Journal of Management Studies, 53*(3), 273–298. https://doi.org/10.1111/joms.12203

Swyngedouw, E., & Kaika, M. (2014). Urban political ecology: Great promises, deadlock … and new beginnings? *Documents d'Anàlisi Geogràfica, 60*(3): 459–481. https://doi.org/10.5565/rev/dag.155

Tironi, M., & Zenteno, J. (2013). "Licencia social para operar": Sostenibilidad y las justificaciones del RSE en la gran minería chilena. In J. Ossandón & E. Tironi (Eds.), *Adaptación: La empresa chilena después de Friedman.*

Vásquez, A., Lukas, M., Salgado, M., & Mayorga, J. (2017). Urban environmental (in)justice in Latin America: The case of Chile. In R. Holifield, J. Chakraborty, & G. Walker (Eds.), *The Routledge handbook of environmental justice.* London: Taylor & Francis.

Wilson, J (2013). The violence of abstract space: Contested regional developments in Southern Mexico. *International Journal of Urban and Regional Research, 38*(2), 516–538. https://doi.org/10.1111/1468-2427.12023

Wilson, J. (2016). The village that turned to gold: A parable of philanthrocapitalism. *Development and Change, 47*(1), 3–28. https://doi.org/10.1111/dech.12163

Yeung, H. (2009). Transnational corporations, global production networks, and urban and regional development: A geographer's perspective on multinational enterprises and the global economy. *Growth and Change: A Journal of Urban and Regional Policy, 40*(2), 197–226. https://doi.org/10.1111/j.1468-2257.2009.00473.x

13 Planetary Urbanization and Maquiladoras: Unveiling Abstract Space in Yucatán, Mexico

CLAUDIA FONSECA ALFARO

You would never know the village was there unless you already knew it was there. Driving fast on federal highway #176 you would assume there was an endless flatness of luxuriant green, wild, and uncultivated land all around, an in-between countryside between the towns worth naming along the way. However, if you knew where to look and where to turn, you would see the one-lane, paved road that led to San Pedro Chacabal, *comisaría*,[1] 765 inhabitants (INEGI, 2010). Without access to public transport, most residents had no other choice but to walk the eight kilometres that separated them from the next village, Ucí, in case they needed something as basic as a grocery store. However, if you followed the highway by car, the city of Motul – 23,240 inhabitants (ibid.) – was only ten minutes away. A regional centre, Motul is the provider of services for the people that do not or cannot travel to Mérida, Yucatán's capital thirty-one kilometres further on. For the residents of *comisarías* such as San Pedro Chacabal, Motul is an island of Yucatecan urbanity – a city which serves as a transportation node; hosts secondary and higher education centres; provides access to regional and nationwide retail chains, pharmacies and banks; and, offers "modern" food alternatives such as pizza and hamburgers. Most importantly, Motul is a centre of activity because it supplies jobs. The city is proud to be the home of Montgomery Industries[2] – the biggest employer in the municipality (1,800 workers on the payroll)[3] and the leading maquiladora factory in Yucatán.

This contribution is inspired by Roy's (2016) call to study settlements that do not display the "universal grammar of cityness" found in the Euro-American heartland and by Choplin and Pliez's (2015) "inconspicuous geography of globalization." The chapter zooms in to Motul, Yucatán – a state depicted as marginal and poor (cf. OECD, 2007;

COESPY, 2013) – to explore the relationship between maquiladoras and urbanization. The maquiladora industry in Yucatán boomed between 1990 and 2001, followed by a decline. Maquiladoras, a type of Export Processing Zone (EPZ) within the Mexican context, are manufacturing plants that operate under exceptional regulations and an advantageous tax regime within the bounds of a demarcated territory (Sklair, 1989; Farole & Akinci, 2011). Despite its exceptionality, the EPZ topography – in its maquiladora form or one of its many configurations[4] – is actually now common across the world. Heirs to medieval citadels, colonial trade ports, and Hanseatic cities (Bach, 2011) – and an outcome of capital's continuous search across the globe for the most advantageous and fluid business conditions – EPZs cater to global capitalism's need for "zones of exception" (Ong in Roy, 2011) and "diversity" in its labour subjects (Tsing, 2009). The modern EPZ, a product of the mid-twentieth century, exploded in numbers beginning in the 1970s as trade entered a new stage in globalization (Farole & Akinci, 2011). The International Labour Organization (ILO) reports that in 2006, there were 3,500 EPZs across 130 countries, employing 66 million workers (Singa Boyenge, 2007).[5] In the case of maquiladoras, the factories contributed in 2012 to 65% of all Mexican exports and employed 80% of all manufacture workers (Palencia Escalante, 2012).

Studies that explore the relationship between EPZs and urbanization or the built environment can be found in the literature. Grineski and Collins (2008) report environmental burdens caused by maquiladoras in Ciudad Juárez where the "peri-urban poor" are more exposed to the hazards of pollution. Landau (2005) describes the effects that maquiladoras have on urban sprawl. Hagemann (2015) studies the relationship between the built environment and export oriented factories in Turkey in what she calls the "effects of globalization in cities." Bach (2011) and Easterling (2012) reflect on the different configurations of EPZs and simply call the phenomenon, "the Zone." Both authors reflect on the types of "urban forms" emerging within the demarcated territories of the Zone. While Easterling (2014) argues the Zone "transplants" its logic to the urban creating an object which is "the city as zone and the zone as city" (Easterling, 2012), Bach (2011) calls this object the "Ex-City."

The purpose here is to explore the relationship between maquiladoras and urbanization through the lens of *planetary urbanization* (Brenner & Schmid, 2014, 2015b) and *abstract space* (Lefebvre, 1991) taking the case of Montgomery Industries and Motul. Following Bach (2011) and read as the Zone, maquiladoras are an example of the global flows of capital embedded in a place. I argue that, while it is fruitful to reflect on the exceptional characteristics of the Zone and what this means for the urban, this type of reflection obscures how the process of urbanization that materializes in the form of the Zone extends beyond the Zone itself.[6] Analysed through planetary urbanization and abstract space, I explore how the effects of the maquiladora topography – as an expression of contemporary capitalist urbanization – expand beyond its territory. In other words, while the maquiladora territory is the legally *enhanced* space of capitalism, the rest (meaning, the non-exceptional territory) is also the space of capitalism, an abstract space that attempts to alienate and fashion everyday life in the image of capital. This in turn uncovers a hidden urbanization: the infrastructural veins that protect and support the circulation of capital. In Bach's (2011) understanding, the Ex-City is a "vestige of the 'real' city and hence ex in the double sense of originating out of the city and exemplifying that which was once known as city." However, I argue the Ex-City is not a simple "urban" outcome of the Zone, or the logic of globalization; it is an outcome of capitalism itself and the subordination to its logic. In the case of Motul, the Ex-City is there, even if it is not as obvious. Under planetary urbanization, the whole planet has become the Ex-City – reminiscent of Lefebvre's worldwide urban fabric (Lefebvre, 2003), where space is fashioned in the image of capital. I support my argument with empirical material gathered particularly in Mérida and Motul during two fieldwork periods conducted in 2014 and 2015–2016, comprising a total of thirteen weeks.[7] Thirty-six semi-structured interviews were carried out with academics, government officials (local and at the state level), a maquiladora manager, maquiladora workers and ex-workers, and residents (the mayor, a journalist, business owners, and members of the public). Participant observation was also carried out. A literature review, newspaper records, and governmental reports serve to complement the material.

The structure of the chapter is as follows: In the first section, I unwrap the theoretical concepts of the Zone, planetary urbanization, and abstract space; I then go on to introduce the case of Montgomery Industries and Motul; and, in the third section, discuss how abstract space helps unveil Motul as the Ex-City. I conclude with the argument that, under planetary urbanization, the Ex-City does not need exceptional territory to exist.

The Zone, Planetary Urbanization, and Abstract Space

The Zone Becoming the Ex-City

One of the ways to reflect on the effects that global capitalism has on the local built environment is through the concept of the Zone. In its most basic form, the Zone exhibits an exceptional topography of business-friendly space where capital can breathe freer and flow faster. Bach (2011) explains that the logic of the Zone comes from its genealogy and "lies in capitalism's impetus to maintain strategically ambiguous spaces." The logic of the Zone establishes a clear division between its jurisdiction and the "outside"; in other words, between the Zone's exceptional territory and the rest of a country's "normal" territory. Both exceptional and normal spaces are established by the power of the state. Easterling (2012, 2014) sees the Zone as a "legal and economic instrument" that shelters activities that would otherwise not fit in the economic logic of the rest of a country, and that not only externalizes, but also diminishes the obstacles to profit. Obstacles to profit can exist in the form of taxes, wages, labour standards, trade unions, and availability of land (cf. Milberg & Amengual, 2008).

A global phenomenon, the Zone exhibits different configurations depending on where it is established: from fenced-in factories or demarcated industrial parks comprising a few hectares in places like Latin America to "hybrid Zone-cities" of fourteen million people such as Shenzhen, China (Bach, 2011; Farole & Akinci, 2011). In the cases of the hybrid Zone-cities, the logic of the Zone becomes a template that "rewires" the understanding of what a city is, affecting the way urbanization unfolds producing, for example, what Easterling (2014) calls a "Zone-metropolis." This is Bach's (2011) Ex-City, a place shaped by the logic of capital, emerging from the city but not a "real" city anymore. The Ex-City echoes Lefebvre's (2003) understanding of the "implosion-explosion" of the city. However, Bach and Easterling understand that the Ex-City or the Zone-metropolis are outcomes of globalization or global commodity chains while Lefebvre argues that all of urbanization is an outcome of capitalism itself. Bach and Easterling fail to understand the difference between a process and an outcome. This is where planetary urbanization and abstract space can deepen our understanding of the connections between the Zone and urbanization.

Planetary Urbanization

Mainly developed by Brenner and Schmid, the concept of planetary urbanization is Lefebvrean in origin, "Society has been completely urbanized" (Lefebvre, 2003, p. 1). It sounds like a contentious statement and gives the idea of one giant city spreading all over the globe. However, Lefebvre (ibid., p. 86) clarifies, the suggestion should not be conceptualized in terms of size or dimensions, but instead, it should be understood in terms of the "properties of the phenomenon." Large agglomerations centres remain relevant; rural areas, dispersed populations, somehow pristine and wild landscapes have not completely disappeared; and the concept does not suggest there is urban sprawl across the planet or global homogenization of a certain "city-ness" (Schmid, 2014). Brenner and Schmid's (2015b) main argument is a call to reconceptualize the notions of the urban and the city; to reflect on the significant difference between the thing – the outcome (i.e., a settlement) – and the process (i.e., urbanization) (Harvey, 2014). The "global process of capitalist urbanization" (ibid.) has reached a critical zone where the city has metaphorically imploded on itself and exploded tumultuously into the planet, making "countryside" things or even "pristine" places such as natural conservation areas part of a worldwide urban fabric (Lefebvre, 2003; Brenner & Schmid, 2011). In the words of Brenner and Schmid (2015b), planetary urbanization has mobilized "the operationalization of places, territories and landscapes, often located far beyond the dense population centers, to support the everyday activities and socioeconomic dynamics of urban life." This operationalization also includes the immobile infrastructures for transportation and communication – roads, canals, pipelines, waterways, railways, seaports, and airports – that allow capital to be produced and flow faster (ibid.). The authors, of course, do not disregard that the operationalization of hinterlands have a historic connection to capitalism (Brenner & Schmid, 2015a). What is worth noting is that, since the 1980s, there seems to be a new cycle of urbanization that relies on speed – the compression of space and time in the form of, for example, instant communication (Harvey, 2014) and the unification and integration of the market (Merrifield, 2013). Additional distinctive elements in this new cycle of urbanization is the active role of "extrastatecraft" actors (Easterling, 2014) and the alarming rate of environmental degradation (cf. Bellamy Foster & Clark, 2012). In other words, a new form of hinterland emerges. The expansive magnitude of the current cycle of capitalist urbanization has allowed cities to exist by stretching their tentacles – and impacts – farther and wider than ever before. Against this background, as Brenner and Schmid (2015b) argue, trying to understand the urban in terms of spatial dichotomies (e.g., urban/rural or city/countryside), a nominal essence, or degree of agglomeration is misleading. Instead, planetary urbanization might help unveil forms of urbanization that would otherwise be overlooked. Examples of empirical contributions that have already used planetary urbanization within the Latin American context include Arboleda's (2015) study of mining in the "non-city" of Huasco, Chile, that supplies raw materials to world markets; and Wilson and Bayón's (2016) case of the Manta–Manaus transport corridor across Ecuador and the Amazon in Brazil, part of a wider South American infrastructure project called IIRSA that is meant to bring economic development to the region.[8]

Abstract Space

The unfolding of planetary urbanization can be understood through abstract space. Reflecting on the concept's origin, Lefebvre's (1991) premise is that an economic system replicates itself through the production of space. Because space is a social product, capitalist urbanization can be understood as an outcome and precursor of capitalism. Following this line of thought, abstract space is the specific space of capitalism (Stanek, 2008), the "medium" where exchange can happen and where the world of commodities is spread out (Lefebvre, 1991). Just like the commodity, abstract space is fetishized and hides the human relations and labour embedded in it (Stanek, 2008). According to Wilson (2014), abstract space seems apolitical, a "neutral backdrop" when, in reality, it seeks to dominate, mould in the name of capital – while reducing the obstacles and the resistance it encounters. Through the expansion of abstract space, the everyday is shaped by conceived understandings of what space should be according to the rationality of capital[9] (Wilson, 2013a).

The Maquiladora in Motul

Montgomery Industries[10]

Montgomery Industries, headquartered in Hong Kong, is the outcome of several factors that unfolded in Mexico and Yucatán between the 1980s and 1990s: (a) the launch of the Henequen Restructuring Program and Comprehensive Development of Yucatán[11] in 1984 to replace the collapsing mono-agricultural exports model that had sustained the state since the nineteenth century; (b) the expansion of maquiladoras beyond the traditional enclaves along the border with the United States as national restrictions were lifted; and (c) Mexico's accession to the North American Free Trade Agreement (NAFTA) in 1994 (Canto Sáenz, 2001; Mendoza Fernández, 2008; Plankey Videla, 2008). The maquiladora industry was one of several proposals to create an "industrialized state" that would bring jobs and development to Yucatán after the collapse of the state's monocrop economy (Gobierno del Estado, 1983; Cervera Pacheco, 1987). Nevertheless, the Yucatecan government invested millions of pesos advertising abroad the "unique" business qualities that the state could offer, such as geographical proximity by sea to the east coast of the United States (Canto Sáenz, 2001) and not only a cheap but also non-unionized labour force (Ramírez, 1994). Advertising was supplemented with improvements and expansion of the infrastructure that already existed in the state (Canto Sáenz, 2001). At first, there were only a handful of maquiladoras in the 1980s, but an unexpected boom swept the state between 1990 and 2001. During this period, the number of factories exploded from 13 to 131, which translated into a meteoric growth of 1007%.[12] After the peak year of 2001 came a bust and a steady decline. As of 2019, there were 55 maquiladoras operating in the state (INEGI, 2020).

Montgomery Industries was inaugurated in 1995 within Motul Industrial Park, a piece of land of one hundred hectares at the edge of Motul, located within easy access to the old two-lane road that led to Mérida. In the early 2000s, a new four-lane federal highway – #176 – was constructed (see figure 13.1). Designed to circumvent obstacles such as villages, the highway halved the travel time between Motul and the capital. Subsequently, the transport time to Progreso – the seaport from which Montgomery ships its jeans and other denim garments – also decreased. The land where the industrial park is located used to be an *ejido*[13] which was expropriated by presidential decree in 1993 (Diario Oficial de la Federación, 1993). Despite its size, only three buildings exist within the industrial park, taking up around six hectares: Montgomery Industries, the former Mayan Palace maquiladora (no longer in operation since hurricane Isidore hit the peninsula in 2002), and Mayan Knits, a factory built in 2014 with a public–private investment of 300 million pesos[14] (SE, 2012) that in 2019 was still not operational (Diario de Yucatán, 2019). Motul Industrial Park does not exhibit any symbols to communicate the existence of the industrial park itself. No gates or fences establish the perimeter of the park, no symbols showcase the "industrial activity" happening there (see figure 13.1). The indistinguishable piece of land simply looks merged into the surrounding neighbourhoods and uncultivated land (see figure 13.2).

The existence of Motul Industrial Park might be invisible, but the presence of Montgomery Industries is not. In its heyday, Montgomery Industries had 6,000 employees working across three different shifts and could provide jobs not only to local residents but to people from the surrounding villages as well (Morales, García, & Pérez, 2000). During its highest production peak – before the declining years initiated by 9/11 and hurricane Isidore in 2002 and accelerated by the 2008 financial meltdown – the factory produced ten to twelve million units per year. Because of the sheer volume of production and a lack of measures to ensure the proper management of industrial waste, there have been

Figure 13.1 The indistinguishable Motul Industrial Park. On the left, Motul Industrial Park (the building is Mayan Knits). On the right side, highway #176. Photo by the author, 2016.

environmental impacts in Motul (Navarrete, 2008). The municipality lacks a sewage network, a wastewater plant,[15] and a waste management system (all solid waste is deposited in an open dump). It is unclear how Montgomery Industries currently takes care of its wastewater and solid industrial waste, but according to Navarrete (2008) at least until 2003, the factory deposited its wastewater in "receptive wells" thirty meters deep and unloaded its solid waste in the municipal dump. Navarrete (ibid.) reports a case from 2001 where the federal environmental agency (PROFEPA) discovered what appeared to be a lunar landscape of blue dust covering the surface of the municipal dump. On closer inspection, it was discovered that Montgomery Industries had discarded plastic bags filled with a blue dye and the bags had disintegrated under the heat of the sun. After assessing that the blue dust was not considered toxic under Mexican regulations – and following confirmation from municipal authorities that all types of industrial solid waste were allowed in the municipal dump – the case was closed after reaching a solution: Montgomery would use sturdier, "suitable

Figure 13.2 On the left, an empty lot in Motul Industrial Park. Across the street, the municipality's cultural centre and a private house. Photo by the author, 2016.

bags," and pledged to contribute financially to the construction of a landfill in the future.[16]

Nowadays, twenty-five years after its inauguration and despite its decreased workforce, Montgomery Industries is still impressively large. Outside the factory gates – in sharp contrast to the invisibility of the industrial park – there are clues in the form of bicycles, motorcycles, cars, and buses that give an idea into the amount of people working inside. Inside, the workers that cut, sew, wash, and pack the two million units that the factory now ships per year – mostly to the United States (87%), but also to Italy (5%), Japan (3%), and Hong Kong (2%) – continue to perform Marxist magic tricks while obtaining the meagre benefits that the factory provides. For example, brand new from the production line, garments can only be distinguished from each other by style, colour, and the type of fabric, or to the trained eye, an assessment of the number of operations that were carried out to sew the item. However, once workers place a brand's tag, a pair of jeans, for example, is suddenly worth 6800 yen; USD 20, USD 79 or USD 89 – despite the fact that all of them were manufactured in the same place by the same hands. The workers also continue to "travel the world" with their embedded labour, just like a government official – unaware of its raw irony – exclaimed when the factory opened, "[Montgomery Industries] will allow the people from Motul to travel the world with the 'Made in Mexico' brand" (DY, 1995b).

Despite the exhausting working conditions and low wages that the research participants report, Montgomery Industries continues to provide labour "benefits" that not so many other employers in Motul – not even the municipality – offer, even though they are legally obliged to. *Prestaciones conforme a la ley* are social security benefits such as (a) public healthcare and childcare, (b) a pension, (c) a potential mortgage from INFONAVIT,[17] (d) being entitled to *aguinaldo*,[18] (e) loans to buy goods and services from FONACOT,[19] (f) maternity leave, and (g) severance payment when dismissed. Montgomery Industries also offers some degree of stability to men in comparison to working as an *albañil*[20] – a job that is managed by contractors and therefore dependent on transitory projects. In the case of women, it offers a job alternative closer to home, as opposed to, for example, working as a domestic worker in Mérida, which would require commuting (not a nuisance but an economic burden).[21]

Motul: "The Industrial City" that Never Became One

Back in the 1990s, Motul was on the political map (DY, 1997), optimism was high throughout the state, and the maquiladora industry seemed to be a certain future. Headlines were filled with future plans for more and more maquiladoras, more and more jobs (cf. DY, 1995a, 1999, 2000). When Montgomery Industries arrived, there were expectations that Motul would become an "industrial city" (DY, 1995b). However, most plans never materialized. The maquiladora landscape throughout the state began to decline in the early 2000s, leaving industrial ruins behind. Within this context, Montgomery Industries' permanence in Motul is a success story. Research participants perceive a before-and-after-Montgomery effect that they can see reflected in the city. Motul has been an important settlement since the nineteenth century when the economy of the state relied on agriculture (Barceló Quintal, 2011), but the landscape has changed tremendously nevertheless. *Motuleños*[22] have substituted their bicycles for motorcycles – a lucky few from motorcycles to cars – and the horse-drawn carriages used as taxis around the

Figure 13.3 A *mototaxi* stand near Mercado Soriana Express, a national supermarket chain. Photo by the author, 2015.

city have been replaced by *tricitaxis* and *mototaxis* (see figure 13.3). New INFONAVIT neighbourhoods have been constructed and regional and nation-wide retail chains have arrived.

Discussion

The Zone: A Legal Magic Trick in Space and Time

Maquiladoras, production centres of commodities connected to global circuits of capital, are a clear example of global capitalism embedded in a place. In its rationality and architecture – exceptional territory where business-friendly regulations apply – the topography of the Zone seeks to represent frictionless abstract space. The logic of the Zone is built around the concept of temporality and location – allowing for a legal, magic trick in space and time. The magic trick happens in two ways. The regime of the Zone is structured so that raw materials and parts can come in to be processed by a local labour force and then shipped as a finished product. These objects temporarily "leave" Mexican territory where other regulations apply and instead exist within the Zone jurisdiction. This is the first trick in space and time. A labour force that would probably be unwanted in the commodity's final destination, the global North, does not need to move. Embedded in the commodity, labour moves without moving (reflected in the phrase "travel the world with the Made in Mexico brand"). This is the second trick. In the words of Wise and Cypher (2007), discussing maquiladoras in general, "what is actually taking place is the disembodied exportation of labor or, alternatively, that the workforce is being exported without requiring Mexican workers to leave the country." This (i.e., the immobile labour force) is emphasized as a competitive advantage by Yucatecan government officials and scholars in books, articles, and brochures where the myth of a stable, non-unionized employee is created (cf. García de Fuentes & Pérez Medina, 1996; Castilla R. & Torres G., 2000; Mendoza Fernández, 2008; SEFOE, 2010). In contrast, the commodity must be mobile. The raw material – fabric, official price tags, special branded buttons, and even detergent for some orders – and later on the finished product, can come and go as they please through the gates of the Zone. The capability of these items to cross borders and flow across the globe is impressive, and, in my view, their freedom to travel echo Cook's (2004) ethnography of Jamaican papayas. As a tropical fruit desired by consumers in the global North, papayas are global travellers, flown around the world. In contrast, the Jamaican papaya pickers and packers are not entitled to a visa. Just like the papaya labour force must remain immobile, the maquiladora labour force must do so as well within the Zone. In addition to the commodity, maquiladoras can also be mobile and hop around when the conditions start looking advantageous in other places. Judging by the number of EPZs across the world, there are plenty of Zone territories to choose from.

The Maquila: Abstract Space Extends beyond the Zone

The Zone represents a legal border between the host country and global production chains, an imaginary border between the local and the global, and a material border between what is maquiladora territory and what it is not. However, the Zone's border is not only porous, but an illusion between an "inside" and an "outside." The abstract space of the Zone extends beyond the Zone itself. The most obvious effects of the maquiladora are reflected in its name which, according to Sklair (1989), comes from *maquila*, the portion of flour that a miller would keep as payment for grinding somebody's grain. "Maquila" is a metaphor for what remains in the host country after the commodity has left the Zone – for example, wages (cf. Wise & Cypher, 2007), impacts on the built environment (cf. Landau, 2005), and environmental degradation (cf. Grineski & Collins, 2008). The logic behind abstract space produces the maquila leftovers. The pollution case in Motul reported by Navarrete (2008) – involving Montgomery Industries and the (improper) disposal of industrial solid waste and wastewater – illustrates the rationality of capital reflected in the ludicrous solution of simply using "sturdier bags" and summarized in a single phrase, "Water in Yucatán is quite cheap, treatment plants are not" (ibid.). Economic factors always take precedence over other concerns in a context where abstract space rules.

Abstract space seeks to be a medium for capital; therefore, it expands in a way useful to it. Following the logic of abstract space, immobile infrastructures for transportation and communication that would allow maquiladora capital to be produced and flow faster were prioritized by the government in the state: (a) the old harbour in the seaport of Progreso was improved so bigger cargo ships could moor, and its surface area was extended so that more containers could be processed (PuertosYucatán, n.d.; Gobierno del Estado, 2000); (b) the network of roads was expanded and modernized – such as the case of federal highway #176 – (Canto Sáenz, 2001); (c) the airport in Mérida was upgraded and a second one (Chichén Itzá International Airport) was constructed in the eastern part of the state (Gobierno del Estado, 2000); (d) natural gas pipelines were extended (ibid.); (e) electricity generation capacity was increased (ibid.); and (f) land was put aside in the form of industrial parks (Canto Sáenz, 2001). These projects were seen as "development detonators" that were "essential," even if significant monetary investment would be required (Gobierno del Estado, 2000). In contrast, other types of infrastructure continued to lack in the state, such as piped water, sewage systems, and healthcare facilities (cf. Gobierno del Estado, 1983, 2007; Cervera Pacheco, 1987).

Abstract space is easier to detect in the materiality of infrastructure – in the seaport where the commodity can be shipped, the road so the commodity can be transported, and the increased energy generation requiring pipelines and cables so the commodity can be manufactured. However, abstract space can also expand to places where nothing has been built; such is the example of Motul Industrial Park. Regardless of the fact that only three buildings exist within it, and that there is no enclosure, the industrial park is officially demarcated. Potential plots of land are waiting to be transformed into the Zone when, if ever, another maquiladora arrives. Uncultivated land, *monte,* is already potential abstract space, already put aside for a purpose. Something similar can be seen in the other industrial parks that were created and that stand almost empty in other parts of the state. At least four out of the nine industrial parks in Yucatán have an "urbanization rate" below 24%, and seven out of the nine have an occupancy rate below 44% (SEFOE, 2010). This echoes Arboleda's (2016) report of "spatial fragmentation" in Latin America where only a small percentage of the land under concession for mining projects is actually used. Nevertheless, concession holders are the ones that can decide what to do with the land (ibid.). Wise and Cypher (2007) argue that the "precarization and disaccumulation" that goes within the Zone is not only in terms of labour – subsistence wages, job-related injuries, and economic insecurity for the workers – but also in the form of land dispossession. In the case of Motul Industrial Park, the dispossession of land was operationalized by the state when the ejido was expropriated. Looking at a case of land expropriation carried by the state for the Zone, but in another context, India, Levien (2011) interprets this as an accumulation by dispossession.

The Ex-City of Motul: When the Exception Is the Norm

While the Zone is the exceptional, legally *enhanced* abstract space, the rest – the territory that is not demarcated as the Zone in Yucatán – is also shaped by the logic of abstract space; however, in a depoliticized way. Through his work in Chiapas, the south of Mexico, Wilson (2013b) describes how abstract space gets produced through a process that makes the act look like "a pragmatic response to economic necessity in the interest of the common good." This process of depoliticization that Wilson refers to is reminiscent of the state's "development detonators" and its efforts to promote maquiladoras. The before-and-after-Montgomery effect reported by the research participants has brought jobs closer to home, easier mobility, access to credit and loans, shopping alternatives, and improved living conditions in general. However, all this has occurred within a context where economic concerns have precedence over other factors (e.g., environmental and social ones). For example, the absurdity when the logic of abstract space takes prevalence over reality – the everyday – can be reflected in the pollution case occurred in Motul. A second example is the infrastructure built in the state to facilitate the maquiladora project. The positive impacts that might have trickled down to people in Motul (e.g., a shorter trip to Mérida) were secondary effects in a context were the rationale was to facilitate the workings of the maquiladora industry. The official raison d'être of the maquiladora industry was to create jobs and improve life conditions; however, the imperative has been the maquiladora industry itself.

Motul is not a Zone. However, Motul functions in a similar fashion to the Zone, not because the city is under special jurisdiction, but because it functions under the logic of abstract space. Less effective than the Zone, Motul also externalizes and diminishes the obstacles to profit (e.g., by offering lax labour and environmental standards). Motul is part of the "exoskeleton" (Bach, 2011) that protects and supports the circulation of capital throughout the world and in this sense becomes an Ex-City. The Ex-City of Motul forms part of the urban fabric – of the planetary urbanization that supports global capitalism – even if it is not as visible at first sight. Within a Mexican context, maquiladora production in the south of the country represents a small percentage in comparison to the production in the north (INEGI, 2017). Motul is, without a doubt, an important city within the Yucatecan context. However, at the global level, Motul is an ordinary city (Robinson, 2006), not a metropolis, megacity periphery or one of Saskia Sassen's famous global cities.

There are other Ex-Cities or inconspicuous places at the global level, even if locally there is nothing inconspicuous about them. Arboleda (2015) looks at the urbanization of a non-city, Huasco, a mining settlement in Chile that supplies natural resources to the circuits of capitalism. Mining in Huasco is voracious and obvious, "a bleeding wound in the surface of the Earth" (ibid.). Through his work, Arboleda attempts to unveil a place and a process that would otherwise only come to light if there were a disruption in the system. The hidden infrastructure that sustains the urban (fabric) only becomes visible when it does not function properly (e.g., a gas pipe that breaks, a water sewage system that stops functioning). Huasco and its urbanization, even with its bleeding wounds, only materialize when there is disruption. Motul is another one of these inconspicuous places, perhaps even more hidden. There is no monstrous machinery or voracious depletion. There is no feel of heavy industry. There are no obvious wounds in the Earth. Nevertheless, the modest Ex-City of Motul is part of the urban fabric.

The Planetary Ex-city

This contribution has used the readings of planetary urbanization (Brenner & Schmid, 2014, 2015b) and abstract space (Lefebvre, 1991) to focus on an example of the "inconspicuous spaces" (Choplin & Pliez, 2015) of global production chains and its connection to urbanization. The chapter has zoomed to the maquiladora industry in Yucatán and has presented the case of Motul, a city that hosts Montgomery Industries, the biggest maquiladora in the state. When they are understood through the concept of the Zone (Bach, 2011), maquiladoras are a legally enhanced abstract space. Taking into consideration the number and types of Zones that exist across the globe (Singa Boyenge, 2007; Farole & Akinci, 2011), the exceptional territory of the Zone is actually common. In the case of Motul, planetary urbanization helps reveal that abstract space extends beyond the maquiladora territory, uncovering an Ex-City where the everyday has been fashioned in the image of capital. However, under planetary urbanization, the Ex-City does not need exceptional territory to exist. The Zone might be a legally enhanced space of capitalism, but the "outside" – meaning the non-exceptional space – also follows the logic of abstract space and is part of the urban fabric that engulfs the planet. The planetary Ex-City is not a simple outcome of the Zone, or the logic of globalization; it is an outcome of capitalism itself and the subordination to its logic. Motul represents the example of abstract space conquering new territories that were not as accessible some decades ago (e.g., a structure as NAFTA was not present to facilitate it). Planetary urbanization has been able to reach a region that was not as integrated to it thirty years ago since the global process of capitalist urbanization has compressed space and time even more. However, the planetary Ex-City exhibits tensions: even though the exception of the Zone is actually the norm and abstract space engulfs the globe, global does not mean universal (cf. Roy, 2016). Despite its tendency to homogenize, global capitalism needs variety – in the form of homogenous difference (Fonseca Alfaro, 2018). Now, through the expansion of the urban fabric, even Motul – with all its particularities – can be operationalized to support the everyday life of an elsewhere, along with all the Huascos and Amazonian settlements of the world.

NOTES

1 In Yucatán, a settlement that belongs to a municipality's jurisdiction but that it is not the seat of the local government.
2 All factory names have been changed.
3 Information drawn from interviews with J. Becerril (Mérida, 24 September 2014) and B. Castilla Ramos (Mérida, 23 September 2014), academics at the Universidad Autónoma de Yucatán; and a manager at Montgomery Industries (Motul, 30 November 2015).
4 The EPZ regime has different names according to variations in size, concessions, subsidies, and regulations offered by the host country. "Maquiladora" is a term used to describe an EPZ in Mexico, and across Latin America and the Caribbean (Farole & Akinci, 2011). Other types of EPZs are Free (Trade) Zone, Exclusive Economic Zone, Special Economic Zone, among others (Singa Boyenge, 2007).
5 After a literature review, this reference continues to be the most comprehensive and recent report. See an ILO comment from 2017, which supports this assessment: https://www.ilo.org/global/about-the-ilo

/newsroom/news/WCMS_599888/lang-en/index.htm (Accessed 23 November 2020).

6 This chapter builds on arguments developed in my doctoral dissertation but further explores the concept of the Zone vis-à-vis the Ex-City against a background of planetary urbanization. For an extensive discussion of the Zone, see Chapter 7 in Fonseca Alfaro (2018).

7 Supported by a grant from the Swedish Society for Anthropology and Geography (SSAG).

8 For additional examples that operationalize planetary or extended urbanization within a Latin American context, see the work of Juan Miguel Kanai or Roberto Luís Monte-Mór.

9 Abstract space is the dominant form of space, not the only one. Because of its innate contradictions, cracks can appear within it, allowing *differential space* (a postcapitalist space) to emerge (Lefebvre, 1991).

10 Unless referenced, all facts about Montgomery Industries were drawn from information provided by a factory manager during an interview (Motul, 30 November 2015), a tour of the factory (Motul, January 15, 2016), and a presentation at *CANACO* (Mérida, 2 December 2015).

11 *Programa de Reordenación Henequenera y Desarrollo Integral de Yucatán*

12 Own analysis from the following sources: Cervera Pacheco, 1987; Canché Escamilla, 1998; Canto Sáenz, 2001; Canto Sáenz & Cruz Pacheco, 2004; INEGI, 2015, 2017.

13 An outcome of the Mexican Revolution, *ejido* is "a piece of land farmed communally under a system supported by the state" (Oxford English Dictionary). Reforms in 1992 allowed for its commodification (cf. Vázquez Castillo, 2004).

14 Around USD 16 million (27 July 2016).

15 This problem is widespread throughout Yucatán (CentroEure, 2014). where the bedrock, made of limestone, has been cited as a hindrance to construct the infrastructure that is necessary to address the issue, since it is considered either "technically unfeasible" or not "cost effective" (cf. Navarrete, 2008; COESPY, 2013).

16 Unclear if Montgomery Industries kept its promise. As of 2016, the municipality of Motul continues to have an open dump – despite an attempt to run a landfill for some years – according to information drawn from an interview with the editor of a local magazine (Motul, 18 January 2016), and conversations with the Director of the Department of Public Services and Urban Image (Motul).

17 Institute of the National Housing Fund for Workers (my translation).

18 A type of Christmas bonus that all workers are entitled to.

19 Institute for the National Fund for Employee Consumption (official translation).

20 Bricklayer.
21 This description is given according to the gendered division of labour in place in Motul.
22 A native or inhabitant of Motul.

REFERENCES

Arboleda, M. (2015). In the nature of the non-city: Expanded infrastructural networks and the political ecology of planetary urbanisation. *Antipode, 48*(2), 1–19. http://doi.org/10.1111/anti.12175.

Arboleda, M. (2016). Spaces of extraction, metropolitan explosions: Planetary urbanization and the commodity boom in Latin America. *International Journal of Urban and Regional Research, 40*(1), 96–112. http://doi.org/10.1111/1468-2427.12290.

Bach, J. (2011). Modernity and the urban imagination in economic zones. *Theory, Culture & Society, 28*(5), 98–122. http://doi.org/10.1177/0263276411411495.

Barceló Quintal, R.O. (2011). Los ferrocarriles en Yucatán y el henequén en el siglo XIX: El camino hacia el progreso. *Mirada Ferroviaria, 15,* 5–16.

Bellamy Foster, J., & Clark, B. (2012). The planetary emergency. *Monthly Review, 64*(7), 1–25.

Brenner, N., & Schmid, C. (2011). Planetary urbanisation. In M. Gandy (Ed.), *Urban Constellations* (pp. 10–13). Berlin: Jovis.

Brenner, N., & Schmid, C. (2014). The "urban age" in question. *International Journal of Urban and Regional Research, 38*(3), 731–755. http://doi.org/10.1111/1468-2427.12115.

Brenner, N., & Schmid, C. (2015a). Combat, caricature and critique in the study of planetary urbanization. *City, 19*(2/3). http://urbantheorylab.net/uploads/Brenner_Schmid_Richard%20Walker_2015.pdf.

Brenner, N., & Schmid, C. (2015b). Towards a new epistemology of the urban? *City, 19*(2–3), 151–182. http://doi.org/10.1080/13604813.2015.1014712.

Canché Escamilla, J.L. (1998). La industria maquiladora de exportación en Yucatán, México. *Comercio Exterior, 48*(4), 324–327. http://revistas.bancomext.gob.mx/rce/magazines/345/9/RCE9.pdf.

Canto Sáenz, R. (2001). *Del henequén a las maquiladoras: La política industrial en Yucatán 1984–2001.* México D.F.: Instituto Nacional de Administración Pública, A.C./Universidad Autónoma de Yucatán.

Canto Sáenz, R., & Cruz Pacheco, E. (2004). Las maquiladoras en Yucatán y el Plan Puebla-Panamá. *Comercio Exterior, 54*(4), 328–334. http://revistas.bancomext.gob.mx/rce/magazines/65/6/RCE6.pdf.

Castilla, R.B., & Torres, G.B. (2000). Mujeres en Yucatán: Nuevas figuras obreras a partir del modelo maquilador extranjero. *Dialectología y tradiciones populares, 55*(2), 197–235. https://doi.org/10.3989/rdtp.2000.v55.i2.446.

CentroEure. (2014). *Programa de desarrollo urbano regional de la zona metropolitana de Mérida (Municipios de Mérida, Progreso, Conkal, Kanasín, Ucú y Umán).* Mérida, Yucatán. http://www.seduma.yucatan.gob.mx/archivos/consulta_publica_du/proyecto_programa_regional_du.pdf.

Cervera Pacheco, V.M. (1987). *Sexto informe de gobierno.* Mérida.

Choplin, A., & Pliez, O. (2015). The inconspicuous spaces of globalization. *Articulo – Journal of Urban Research, 12.* http://articulo.revues.org/2905.

COESPY. (2013). *Plan estatal de desarrollo 2012–2018: Yucatán.* Mérida, Yucatán.

Cook, I. (2004). Follow the thing: Papaya. *Antipode, 36*(4), 642–664. http://doi.org/10.1111/j.1467-8330.2004.00441.x.

Diario de Yucatán. (2019). *Fábrica lleva cinco años sin operar, El Diario de Yucatán.* https://www.yucatan.com.mx/yucatan/fabrica-lleva-cinco-anos-sin-operar (Accessed: 23 March 2020).

Diario Oficial de la Federación. (1993). *DOF: 29/12/1993, Decreto por el que se expropia por causa de utilidad pública una superficie de 100–24–64 hectáreas de agostadero de uso común, de terrenos ejidales del poblado Muxupip, municipio del mismo nombre, Yuc. (Reg.-3010).* http://dof.gob.mx/nota_detalle.php?codigo=4818403&fecha=29/12/1993&print=true (Accessed: 27 January 2016).

DY. (1995a). Confirman que habrá nueva maquiladora de ropa en el parque industrial de Motul. *El Diario de Yucatán.* Año *LXXI,* Edición 200, 16 diciembre, 3 (Sección local).

DY. (1995b). En tiempo récord se hizo la planta Monty. *El Diario de Yucatán.* Año *LXXI,* Edición 51, 20 julio, Portada (Sección local).

DY. (1997). Plantearán hoy varios proyectos al Presidente: El recorrido del Dr. Zedillo abarcará cinco municipios. *El Diario de Yucatán.* Año *LXXII,* Edición 257, 13 febrero, Portada (Sección local).

DY. (1999). Tres maquiladoras generarán 22,700 empleos en dos años. *El Diario de Yucatán.* Año *LXXV,* Edición 325, 21 abril, 8 (Local).

DY. (2000). La industria textil de Valladolid tiene que "importar" trabajadores. *El Diario de Yucatán.* Año *LXXVI,* Edición 157, 3 novienbre, Portada (Interior del estado).

Easterling, K. (2012). Zone: The spatial softwares of extrastatecraft. *Places Journal,* June. https://doi.org/10.22269/120610.

Easterling, K. (2014). *Extrastatecraft: The power of infrastructure space.* London and Brooklyn: Verso.

Farole, T., & Akinci, G. (2011). *Special economic zones: Progress, emerging challenges, and future directions.* The World Bank.

Fonseca Alfaro, C. (2018). *The land of the magical Maya: Colonial legacies, urbanization, and the unfolding of global capitalism.* Malmö University.

García de Fuentes, A., & Pérez Medina, S. (1996). Factores de localización de la industria maquiladora: El caso de Yucatán, México. Yearbook. *Conference*

of Latin Americanist Geographers, 22, 17–30. https://www.jstor.org/stable/25765825
Gobierno del Estado, Y. (1983). *Plan de desarrollo estatal 83–88.*
Gobierno del Estado, Y. (2000). *Quinto informe de gobierno. Yucatán. 1995–2001.* Mérida, Yucatán.
Gobierno del Estado, Y. (2007). *Plan etatal de desarrollo* 2007–*2012.*
Grineski, S.E., & Collins, T.W. (2008). Exploring patterns of environmental injustice in the Global South: "Maquiladoras" in Ciudad Juárez, Mexico. *Population and Environment, 29*(6), 247–270. https://doi.org/10.1007/s11469-017-9806-3.
Hagemann, A. (2015). From flagship store to factory: Tracing the spaces of transnational clothing production in Istanbul. *Articulo: Journal of Urban Research, 12*. http://articulo.revues.org/2889.
Harvey, D. (2014). Cities or urbanization? In N. Brenner (Ed.), *Implosions/explosions: Towards a study of planetary urbanization*. Berlin: Jovis.
INEGI. (2010). *Censo de población y vivienda 2010, Instituto Nacional de Estadística y Geografía*. http://www.inegi.org.mx/est/contenidos/proyectos/ccpv/cpv2010/Default.aspx (Accessed: 31 March 2016).
INEGI. (2015). *Industria maquiladora de exportación, Banco de Información Económica*. http://www.inegi.org.mx/sistemas/bie/ (Accessed: 13 March 2015).
INEGI. (2017). *Estadística integral del programa de la industria manufacturera, maquiladora y de servicios de exportación*. Banco de Información Económica. http://www.inegi.org.mx/sistemas/bie/ (Accessed: 16 May 2017).
INEGI. (2020). *Industria manufacturera, maquiladora y de servicios de exportación (IMMEX), Banco de Información Económica (BIE)*. Instituto Nacional de Estadística y Geografía. https://www.inegi.org.mx/app/indicadores/?tm=0 (Accessed: 6 July 2020).
Landau, S. (2005). Globalization, maquilas, NAFTA, and the state: Mexican labor and "the new world order." *Journal of Developing Societies, 21*(3–4), 357–368. http://doi.org/10.1177/0169796X05058293.
Lefebvre, H. (1991). *The production of space*. Oxford: Blackwell Publishing.
Lefebvre, H. (2003). *The urban revolution*. Minneapolis: University of Minnesota Press.
Levien, M. (2011). Special economic zones and accumulation by dispossession in India. *Journal of Agrarian Change, 11*(4), 454–483. http://doi.org/10.1111/j.1471-0366.2011.00329.x.
Mendoza Fernández, M.T. (2008). *La industria maquiladora de exportación en el estado de Yucatán y el desarrollo regional*. Mérida: Ediciones de la Universidad Autónoma de Yucatán.
Merrifield, A. (2013). The urban question under planetary urbanization. *International Journal of Urban and Regional Research, 37*(3), 909–922. http://doi.org/10.1111/j.1468-2427.2012.01189.x.

Milberg, W., & Amengual, M. (2008). *Economic development and working conditions in export processing zones: A survey of trends.* Geneva: ILO

Morales, J., García, A., & Pérez, S. (2000). Impacto regional de la maquila en la península de Yucatán. In M.E. de la O Martínez (Ed.), *Globalización, trabajo y maquilas: Las nuevas y viejas fronteras en México* (pp. 311–344). México D.F.: Plaza y Valdés.

Navarrete, M. (2008). Garment maquiladoras in rural Yucatán: An environmental tale. *Journal of Latin American Geography, 7*(2), 105–132. http://doi.org/10.1353/lag.0.0014.

OECD. (2007). *OECD territorial reviews, Yucatan, Mexico.* Available at: http://www.oecd-ilibrary.org/urban-rural-and-regional-development/oecd-territorial-reviews-yucatan-mexico-2007_9789264037106-en.

Palencia Escalante, C. (2012). La industria de exportación, impulsora de la competitividad en México. *Comercio Exterior,* Septiembre, 10–12.

Plankey Videla, N. (2008). Maquiladoras. *International encyclopedia of the social sciences* (2nd ed.). Macmillan Reference.

PuertosYucatán. (n.d.). *History, Administración Portuaria Integral de Progreso S.A. de C.V.* http://www.puertosyucatan.com/ap/historia_en.htm (Accessed: 11 May 2017).

Ramírez, L.A. (1994). La política del desarrollo regional en Yucatán: Descentralización administrativa y empresariado. *Estudios demográficos y urbanos.*

Robinson, J. (2006). *Ordinary cities: Between modernity and development.* London: Routledge.

Roy, A. (2011). Slumdog cities: Rethinking subaltern urbanism. *International Journal of Urban and Regional Research, 35*(2), 223–238. http://doi.org/10.1111/j.1468-2427.2011.01051.x.

Roy, A. (2016). Who's afraid of postcolonial theory? *International Journal of Urban and Regional Research, 40*(1), 200–209. http://doi.org/10.1111/1468-2427.12274.

Schmid, C. (2014). The trouble with Henri: Urban research and the theory of the production of space. In L. Stanek, C. Schmid, & Á. Moravánszky (Eds.), *Urban revolution now: Henri Lefebvre in social research and architecture* (pp. 27–48). Farnham: Ashgate Publishing Limited.

SE. (2012). *Invierten 335 millones de pesos en Yucatán para detonar la creación de empleos.* Secretaría de Economía. http://www.2006-2012.economia.gob.mx/eventos-noticias/informacion-relevante/7625-boletin079-12 (Accessed: 21 March 2016).

SEFOE. (2010). *Guía del inversionista: Yucatán 2010.* http://www.sefoe.yucatan.gob.mx/esp/pdf/guia-inversionista-2010.pdf.

Singa Boyenge, J.-P. (2007). *ILO database on export processing zones* (Revised). ILO Working Papers, ILO.

Sklair, L. (1989). *Assembling for development: The maquila industry in Mexico and the United States* (2011 ed.). Oxon: Routledge.

Stanek, L. (2008). Space as concrete abstraction: Hegel, Marx, and modern urbanism in Henri Lefebvre. In K. Goonewardena, et al. (Eds), *Space, difference, everyday life: Reading Henri Lefebvre* (62–79). New York: Routledge.

Tsing, A.L. (2009). Supply chains and the human condition. *Rethinking Marxism, 21*(2), 148–176. http://doi.org/10.1080/08935690902743088.

Vázquez Castillo, M.T. (2004). *Land privatization in Mexico: Urbanization, formation of regions, and globalization in ejidos* [ebook]. New York: Routledge.

Wilson, J. (2013a). "The devastating conquest of the lived by the conceived": The concept of abstract space in the work of Henri Lefebvre. *Space and Culture, 16*(3), 364–380. http://doi.org/10.1177/1206331213487064.

Wilson, J. (2013b). The urbanization of the countryside: Depoliticization and the production of space in Chiapas. *Latin American Perspectives, 40*(2), 218–236. http://doi.org/10.1177/0094582X12470378.

Wilson, J. (2014). The violence of abstract space: Contested regional developments in Southern Mexico. *International Journal of Urban and Regional Research, 38*(2), 516–538. http://doi.org/10.1111/1468-2427.12023.

Wilson, J., & Bayón, M. (2016). Black hole capitalism. *City, 20*(3), 350–367. http://doi.org/10.1080/13604813.2016.1166701.

Wise, R.D., & Cypher, J.M. (2007). The strategic role of Mexican labor under NAFTA: Critical perspectives on current economic integration. *The ANNALS of the American Academy of Political and Social Science, 610*(1), 119–142. http://doi.org/10.1177/0002716206297527.

14 Rural Livelihoods, Urbanization, and Incomplete Population Transitions in Brazil

ALISSON F. BARBIERI AND RICARDO OJIMA

Introduction

In traditional rural spaces, such as the Amazon and the semi-arid Northeast, the extension of urban features (services, infrastructure, cultural values) and the diversification of rural livelihoods, including urban-oriented activities, have produced a myriad of urban forms beyond cities and towns that have required new definitions beyond the traditional categories of city/country and urban/rural (Barbieri et al., 2009a). These new rural-urban configurations have redefined the ways rural households have made their livings. In this vein, the allocation of farm household rural labour in urban activities has been a widespread strategy to diversify income and decrease dependence of dwindling farm production and resources (Barbieri, 2006).

At the same time, important population transitions have unfolded in Brazil. We define "population transitions" as the changes in the demographic, epidemiological, and urban profiles of the population over time. It is a response to the modernization process associated with the transition from a traditional, mostly agrarian society to an increasingly urban and industrial and/or services-oriented society (Rostow, 1960).

The demographic transition is characterized by a sharp decrease in fertility and population growth rates and an increasingly aging population; the epidemiologic transition, by changes in population morbidity and mortality profiles as a consequence of changing age structure; and the urban transition, by increasing concentration of population in urban areas. While in developed countries these transitions were accompanied (and driven) by structural socioeconomic developmental changes, this has not been the case in developing countries such as Brazil where developmental changes happened without the corresponding socioeconomic and policy adaptation to the changing demographic, epidemiologic, and urban population. Rather, they have perpetuated a context of profound spatial (re)configurations in Latin America since the second half of last century.

Based on the "Livelihoods" and the "New Rurality" approaches, this chapter discusses how urbanization in the Brazilian Amazon and the semi-arid Northeast has been related to the adoption of livelihood strategies that increasingly include non-farm activities (particularly towards urban areas and infrastructure development sites). The two approaches provide a relatively optimistic view in terms of improving economic and environmental sustainability for rural households. However, we discuss how incomplete population transitions – demographic, epidemiological, and urban – have implied important challenges for rural-urban livelihoods. These challenges may continue in the next decades not only due to the unfolding transitions. In fact, and repeating the rapid urban transition of the most developed areas in the country (in the South and Southeast), this region's urbanization has consisted in a shift from a society where the majority depends on traditional family agriculture to one with a predominantly urban population that largely depends on state welfare policy, particularly governmental cash-transfer programs. Furthermore, recent neoliberal policies in Brazil may exacerbate inequalities and conflicts that challenge the potentially positive impacts of urbanization and rural-urban mobility to foster development and socioeconomic improvements for the most vulnerable population, as well as to reduce the positive impacts of cash transfer programs on the livelihoods of the poorest.

Besides this introduction, the chapter has three sections. In the first section we discuss how challenges to rural-urban livelihoods in a context of incomplete population transitions can be addressed by two important theories: the New Rurality and the Livelihood approaches. In the second section we illustrate this theoretical discussion with two case studies in Brazil, adding the consequences of neoliberal policies as a challenge to achieve sustainable livelihoods. The third section provides a summary and conclusions.

Changing Rural-Urban Livelihoods and Incomplete Population Transitions

The rapid urbanization that occurred in Brazil and Latin America is about to take place in Africa and Asia. Hence, the Brazilian experience can be useful to avoid the same challenges that take place here (Martine & Ojima, 2013). This process has produced new socio-spatial configurations that cannot be easily defined as urban or rural and include, for example, urban sprawl, leapfrog urban development and peri-urban settlements.

These configurations cannot be understood without an appropriate assessment of how populations have adopted urban-based strategies. Kay (2008) reviews a strand of rural development studies in Latin America in the last decades known as "New Rurality," which highlights the importance of part-time farming and non-rural activities – particularly those towards urban areas – as essential component of rural livelihoods. Still according to Kay (2008), rural population may engage in multiple activities and have different degrees of insertion in the modern, urban markets as a way to adapt their livelihoods. Among these activities, two are particularly relevant to understand how growing urbanization and rural-urban connectivity impact rural livelihoods: (i) rural-urban linkages and shift to non-rural activities, which may require greater skills and capital and thus potentially higher returns to labour or, alternatively, may involve marginal, low productivity labour; these strategies can generate "peasant differentiation" and higher inequality; and (ii) increasing population mobility between rural and urban areas as well as the spread of services and urban-style values and features towards rural areas (Kay, 2008, pp. 923–926).

Kay (2008) also discusses the commonalities between the New Rurality and the Livelihoods approaches (regarding the last, see, e.g., Ellis, 1998; Bebbington, 1999): "The rural livelihoods approach, which emerged slightly earlier in the UK ... did not have a direct and immediate influence on the new rurality thinkers – although these two approaches share some commonalities such as their emphasis on the importance of non-agricultural activities in the countryside" (p. 920). According to Livelihoods Approaches, farm households develop over time the ability to derive their livelihoods from distinct sources, including mobility and urban employment strategies that usually require higher access to human and social capital (Sherbinin et al., 2008; Barbieri et al., 2009b). These strategies involve the perception and response to the social, economic, and political environments (see, e.g., Bebbington 1999; Sherbinin et al., 2008; VanWey et al., 2012) as well as the ability to access global commodity markets taking advantage of economic liberalization and macroeconomic conditions (Richards et al., 2012; Weinhold et al., 2013).

The two approaches have limitations to explain changing livelihoods in contemporary settings for two main interrelated reasons. First, they were developed in a context of relatively abundant labour supply and of relatively high rural fertility rates; thus, household strategies are taken in an environment of potentially higher diversification strategies.

Second, many rural areas in Brazil are facing structural changes regarding the advancement of industrial, service, and agribusiness sectors *in tandem* with profound changes – or transitions – in the demographic and epidemiological profiles of the population (Barbieri, 2013). Barbieri (2013) defines the demographic transition as a long-run, intergenerational process which

> aims to represent the implications of changes in vital events – mortality and fertility – on population growth, size, and age structure. An agrarian society evolves from a low population growth characterized by homeostatic equilibrium (with high mortality and fertility) to a modern, post-industrial society, again characterized by low growth and homeostatic equilibrium (however, with low fertility and mortality levels). Rapid population growth occurs between these two extremes when mortality decreases at a pace faster than fertility. (Lee, 2003)

Besides population growth, two dramatic features of the demographic transition are the concentration of the population in urban areas and population aging (Barbieri, 2013; Barbieri et al., 2016). In fact, the final stage of demographic transition implies below-replacement fertility linked to a predominantly older and urban population (Lee, 2003).

In parallel with urbanization linked to the advancement of demographic transition and changes in the population structure, population migration and circulation – mostly from rural to urban areas until the late 1980s, and between urban areas of different sizes (particularly from small to middle and large size urban areas) from the 1990s – has had a new momentum in Brazil. In less urbanized areas such as the Amazon and the semi-arid Northeast, rural-urban mobility has implied labour reallocation facilitated by the spread of more efficient communication and transportation systems, as well as infrastructure development such as roads, dams, etc. (Barbieri et al., 2016).

Population changes predicted by the theory of demographic transition are also related to changes in societal morbidity and mortality levels that may affect overall welfare levels and inequality. As suggested by Barbieri et al. (2013), the epidemiological transition in Latin America engenders the coexistence of increasing prevalence of chronic diseases typical of developed countries (associated with an aging population) and external causes linked to environmental quality (reflected in higher infant mortality and high maternal mortality and morbidity). These dual health burdens – of chronic and infectious diseases – have not been coupled with the necessary investments to reduce historical poverty, inequality, and a lack of access by a large share of the population to water and sanitation (Prata, 1992).

Barbieri et al. (2016) hypothesize that the combination of population aging and the lack of urban infrastructure may represent the reproduction or amplification of population vulnerability and inequality in the future, implying severe threats to rural and urban livelihoods. Nonetheless, as urban population grows, the trend is that the urban system becomes more fragmented, not only in spatial terms in a sprawl-like urban form (Hogan & Ojima, 2008), but also in a political and institutional way which fosters the creation of inequality access conditions to health and welfare in urban areas.

It is not only metropolitan areas and the biggest cities that have challenges to face. Half the Brazilian urban population is concentrated in big cities, but another half is dispersed over a huge territory, with low densities and a lack of an institutional framework to cope with vulnerability and to structure adaptation policies. Commuting also became more significant outside metropolitan areas and around big cities, so smaller cities face several regional development issues (Ojima & Marandola Jr, 2012). However, new population profiles brought by unfolding population transitions seem to be more present in the urban life of smaller cities where institutions may be more fragmented and, in many ways, absent.

Recent neoliberal welfare and socioeconomic policies in Brazil may also pose a new challenge for rural livelihoods by draining public transfers from the poorest. These policies will reduce pensions and other social benefits, limit public investments in education and health as well as in environmental conservation investments, and open forest reserves in the Amazon for mining, cattle ranching, and other infrastructure building projects, thus affecting the access to natural capital by rural households. Kay (2008) suggests that labour and wage degradation, the feminization of rural labour, and the need to engage in multiple activities (including non-rural) are merely demonstrations of the precarious rural livelihoods and the need to survive in a context of increasing neoliberalization and the advance of the global capitalism in rural areas.

Study Cases in Brazil: Amazon and the Semi-arid Northeast

Considering the entire country and its population distribution, Brazil is already urban. Since the 1970s there is a clear recognition of the complexity of the metropolization process and of the dynamics of the Brazilian urban network. Some regions have faced peculiar conditions in their urbanization processes, with consequent impacts on rural livelihood strategies that require an analysis that takes into account the timing and historical context of each region. Furthermore, in many regions, endemic problems (such as high level of child mortality) remain as new urban problems.

Figure 14.1 shows the percentage of population living in urban areas (Degree of Urbanization) in Brazil in 2010, highlighting the Legal Amazon and the semi-arid Northeast (IBGE, 2010). These two are regions created by the federal government as targets to some specific development policies and programs and were chosen due to their being the least developed areas in the country. The Brazilian Amazon is officially named "legal Amazon" and involves 5,217,423 km^2 and 61% of the Brazilian territory, while the Brazilian semi-arid involves 982,563 km^2 and 11% of the Brazilian territory.

Despite the problems in the legal and administrative definitions of "urban" in Brazil – which may encompass places with precarious urban-like services, infrastructure and population density (Barbieri et al., 2009a), figure 4.1 shows higher urbanization in the most consolidated and socioeconomically developed regions in the South, Southeast and Center West, and higher "rurality" in the study areas in the Legal Amazon and semi-arid Northeast.

Amazon

Several authors suggest that in order to understand urbanization in the Amazon and its consequences on population wellbeing and inequality, an assessment needs to be made of its articulations with rural changes, particularly in terms of flows of people and of economic activities (Barbieri et al., 2009a; Barbieri et al., 2009b; Barbieri & Monte-Mór, 2008; Monte-Mór, 2004).

Following the discussion by Barbieri (2006) on the articulations between urban and rural places in the Amazon fostered by the complex livelihoods strategies involving production and labour arrangements in both places, we suggest that urbanization in the Amazon reproduces the population profiles typical of demographic, epidemiological, and urban transitions verified in the older, consolidated urban areas in Brazil. These transitions may increase population vulnerability, inequality, and poverty as verified in other parts of the country.

In the 1960s, the military regime designed the concept of Rural Urbanism to guide settlement and colonization policies in the Amazon. The concept became symbolic of a settlement process that, from its inception, made explicit the rural-urban articulations and fostered population circulation as a key element of rural livelihoods and territorial occupation

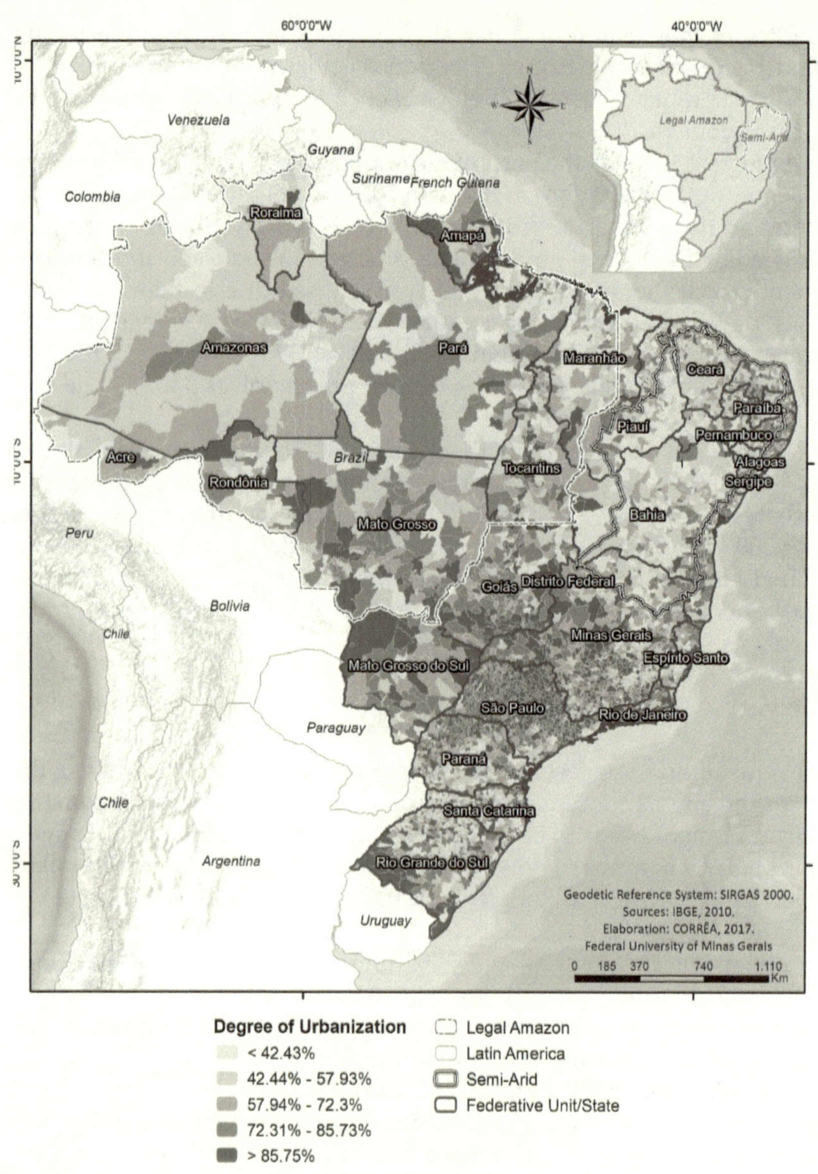

Figure 14.1 Percentage of population living in urban areas (degree of urbanization) in Brazil, highlighting the Legal Amazon and the Semi-arid Northeast – 2010. Map created by the authors.

(Barbieri, 2000). Rural Urbanism defines a hierarchy of places with distinct urban features, according to their scale in terms of population size and number of families. Communities were organized in rural villages, or "agrovilas" with a number of families ranging from 48 to 60 (Moran, 1981) or even 100 to 300 (INCRA, 1973) along the highway or side roads.These communities were served by health clinics, schools, warehouses, and one administrative centre (Moran, 1981; Oliveira, 1983). Each group of 20 to 28 rural villages (INCRA, 1973; Moran, 1981) would be connected to an "Agropolis" that contained more "urban" features like a telephone centre, post office, and telegraph, small agricultural cooperatives, high school, hotel, medical and dental clinic, sawmills, warehouses, and homes and offices for the colonization agents. Finally, a "Ruropolis," with about a thousand families, comprised two or three Agropolises and several rural villages within a radius of 140 km, still relying on small rural industries, larger facilities and hospitals (INCRA, 1973; Moran, 1981; Oliveira, 1983)."

Overall, Rural Urbanism as a strategy of territorial occupation linking rural and urban places in the Amazon was unsuccessful. Several factors contributed to the failure, such as the inability of colonists from Southern Brazilian to adapt to the Amazonian environment, the lack of institutional support (credits, technical assistance) and services as predicted in the hierarchy of places of Rural Urbanism, land tenure conflicts, the lack of economic sustainability of the projects (e.g., the lack of economies of scale for some services and agroindustries), as well as the inadequate planning in the construction of the *agrovilas* – for example, the barriers related to distance and difficulties of building a road network between rural plots and urban areas in the Amazon (Barbieri, 2000). One of the dramatic consequences of these failures was the reproduction, among the settled populations, of poverty and inequality and a lack of services and infrastructure in the origin areas. In this same vein, Browder and Godfrey (1997) mention that the lack of a more appropriate planning for rural villages, including attention to local realities and characteristics such as the soil type, topography, water supply, and distance between rural and urban lots, continue to lead to a reality of very poor and decaying infrastructure and the failure of many, if not most, official colonization projects implemented in the region (Browder & Godfrey, 1997). Browder and Godfrey (1997), for instance, refer to the adverse consequences of this settlement strategy in that "most of the agrovilas fell into a state of abandonment and disrepair, and one journalist referred to them as "rural slums" (p. 76).

Rural livelihoods involving urban articulations have deeply impacted livelihoods, exacerbating poverty and inequality in the following decades in many areas in the Amazon. Nowadays, the region has been a space for the expansion of capitalism through cattle ranching, agribusiness, and infrastructure building such as roads and hydroelectric plants, with corresponding drastic land use changes linked to conversion of primary forest (Barbieri et al., 2016; 2020). These transformations in the contemporary frontier have created a new momentum in Amazon urbanization and fostered urban labour markets. Barbieri et al. (2016; 2020) suggest that off-farm employment opportunities, together with cash transfer programs in Brazil (such as rural retirement and the *Bolsa Familia* program) create off-farm cash opportunities and decrease small colonists' dependency on farm production and natural capital. As an example, Barbieri et al. (2014), in an analysis of a survey in the rural area of the municipality of Machadinho in the Brazilian Amazon, show that 33% of all off-farm incomes are related to cash transfer programs from federal programs such as *Bolsa Familia* and rural retirement pensions. Nonetheless, the unfolding demographic transition in the Amazon, with decreasing fertility and an aging population which have decreased family labour pools in the most active working ages, will possibly be amplified and continue to challenge the sustainability of rural family agriculture. The persistence of infectious diseases combined with growing prevalence of chronic diseases and urban growth without adequate infrastructure (sanitation, precarious housing especially in peri-urban areas) have been and (in the absence of adequate policies) will likely continue to threat rural livelihoods, and create a new momentum in the reproduction of poverty and inequality in the region.

Semi-arid Northeast

The Brazilian Northeast Region (BNR) has traditionally been characterized as the main push-area of the Brazilian population – that is, an area with high rates of out-migration to other regions in the country, particularly the Southeast. This is explained in the literature on population mobility in Brazil (Moura, 1980; Camarano & Abramovay, 1999; Furtado, 1959) for conditions which range from environmental factors (drought, desertification, etc.) to low indicators of economic development such as infant mortality, life expectancy, economic dynamism, among others. In comparison to another Brazilian regions, the BNR is still where most of population lives in rural areas, but this feature has changed in recent years. About 50 years ago, the BNR population was mostly rural and living in small rural municipalities; nowadays they live in urban (but still small) municipalities, and this situation can be clearly noticed in the semiarid region. Economic data shows the trend of reducing participation of agriculture in regional GDP: in 2015, only 8.4% of the economic income in the semi-arid region was from agriculture.

Ab'Saber (1999) suggests that among all the regions with semiarid characteristics in the world, the Brazilian Northeastern Semi-arid is the most populated, with more than 23 million inhabitants in 982,563 km^2. Despite the unfavourable environmental conditions for agriculture and a lack of investments in industry or services, it still has an important share of population – more than 12% of Brazil's population. It seems to be contradictory, but the density is the result of unequal regional development policies which hinder a complete urbanization. Thus, rural population in the BNR seems to be a result of the inequity of land distribution, and not an agriculture vocation in the region.

Only in recent decades, as the population turned to living in urban areas, infrastructure reached some spaces inside the semiarid region. Fusco and Ojima (2017) show the huge commuting among cities of the semiarid region of Pernambuco State because of federal government investments in college education during the last 10 years. In the past, people had to migrate to state capitals to access superior education, but now they can stay and commute regionally. This type of mobility – which is typical for metropolitan and densely urbanized areas – became relevant to explain the regional development inside the semiarid region too. But the regional socioeconomic development delay remains the same. Most cities in the region rely heavily on federal government transfers, and more than 40% of the workforce does not have formal labour contracts. The *Bolsa Familia* Program and rural retirement share in the household budget increased and became another mechanism for coping with adverse conditions, helping people to avoid migration to other regions and, in some extent, attracting some to return to those places (Ojima et al., 2014).

Demographic transition in the region began later, but ran faster than the rest of the country. Associated with important flows of out-migration, the urban population here is becoming older than in other Brazilian regions that started their demographic and urban transitions earlier. The semiarid region is the last frontier of Brazilian urbanization process. Repeating the rapid urban transition of the most developed areas in the country (in the South and Southeast), this region's urbanization has consisted in a shift from a society where the majority depends on traditional family agriculture to one with a predominantly urban population that largely depends on state welfare policy. In contrast to the Amazon region, urbanization cannot result in a good path to improve social conditions in rural areas because there is limited potential for agriculture. Some regions suffer from an increasing desertification process. Thus, rural-urban connections need policies to indemnify the lack of other economic activities rather.

Although urban growth is not in itself the problem to be faced, we need to take a careful look so as not to allow the replication of forms of urban expansion in contexts of greater environmental and social vulnerability, including the persistence of a dual health burden, as is the case of the northeastern semiarid region. Late urbanization with a rapidly aging population, high levels of child mortality, malnutrition, and a low level of carrying capacity is a common scenario in the second most populated major region of the country. The recent shift into an austere economic policy and the economic effects of the COVID-19 pandemic can worsen the region's already serious social and economic conditions. As a matter of fact, the effects of the pandemic in the semi-arid region (as well as in the Amazon) became critical due to the lack of medical infrastructure and doctors in most of the municipalities.

Conclusions

In both study areas, theories such as the Livelihoods and the New Rurality are insufficient to inform policies aiming to reduce inequalities and increasing welfare in the least develop regions of Brazil if they do not consider three important aspects. The first refers to the nature of rural livelihoods in a context of increasing articulation to urban areas. The transformation of rural livelihoods in the Amazon – with rural-urban articulations both from off-farm livelihoods diversification including labour allocation in incipient urban areas – and the semi-arid Northeastern region of Brazil – with weak rural-urban connections and a huge dependence on governmental transfers (income transfers and public employment) – requires the design of policies that overcome the increasingly artificial boundaries between urban and rural. In this same vein, Barbieri et al. (2016) suggest that the assumption for successful rural livelihoods involves taking advantage of global commodity markets and economic liberalization and macroeconomic conditions, which is usually unmet, given that there is not a progressive and linear, but rather asymmetrical, and imperfect, integration of rural households into urban-based markets.

Second, rural-urban articulations have occurred in a context of population transitions (demographic, epidemiological, and urban) characterized by a growing older population without substantial improvements in their wellbeing. It is likely that there will be a change in morbidity and mortality patterns with the persistence of external causes in the poor places (e.g., vector-borne diseases and diseases related to poor infant nutrition) combined with the growing prevalence of chronic and degenerative diseases, and the concentration in urban areas without the appropriate infrastructure. These processes are even more dramatic in the poorer, non-structured (with precarious infrastructure, services, and integration to regional, national, and global markets) study regions in the Legal Amazon and the Semi-arid Northeast than in the rest of Brazil.

Third, and as discussed in Barbieri (2017), current neoliberal policies in Brazil may create additional challenges given its impact on (i) reduction in cash transfer programs towards the most vulnerable (the poorest elder and younger), (ii) withdrawal of labour rights and benefits which degrades urban labour conditions, (iii) the anticipated reduction in education and health investments (with dramatic impacts in the formation of human capital essential for better labour insertion and productivity in urban areas in the long run), and (iv) reduction in investments in conservation and environmental protection (decreasing the quantity and quality of ecosystem services and the stock of natural capital). This scenario may exacerbate inequalities and conflicts which counteract the potential benefits of urbanization and mobility (for example, in terms of wage labour, access to markets and urban services) to foster socioeconomic development for the most vulnerable populations (Barbieri, 2017).

Given these three aspects, we suggest the importance of taking a broader view of development policies aiming to overcome the challenges of population transitions in a context of growing rural-urban articulation as well as of neoliberal policies which downplay the role of the welfare state in in the Brazilian Amazon and Semi-arid Northeast. In the case of the Amazon, policies aiming to protect what is left of the rich tropical forests *and simultaneously* improving human welfare and rural livelihoods in areas of growing urbanization have virtually been ineffective. Various policies can alleviate poverty and achieve more sustainable development; but while remedial policies, such as better extension of welfare programs to the Amazon, could alleviate impoverishment in the short run, long-run policies are needed in a context of dramatic population transitions. These policies must be articulated with better planning of investments that may improve the welfare of local populations in frontier areas, particularly opening roads that improve accessibility to urban services and employment, and the maintenance of cash transfer programs towards the older and younger populations (Barbieri, 2006).

In the BNR, the lack of regional development has been always associated with water shortages. Since 2010, the region has suffered from one of the most severe droughts in history, and much of the poverty effects have been alleviated by welfare policies. The same policies that induced population mobility to urban areas in the past are now increasing regional mobility. In this sense, social policies avoided people to engage in long-distance migrations, but the region needs long term policies to include this new urban population. They have more access to some public services, like education, but with a lack of economic development and other public infrastructure services, they cannot assume the positive effects that an urban condition may offer. If these cities in the late urbanization transition do not achieve economic sustainability, they are likely to be exposed to environmental, health, and poverty vulnerabilities. It is clear that there is a fallacy in considering that most of the remaining urban growth in Brazil will occur in megacities (Martine et al., 2008); so, it is important to understand more fully this dispersed urbanization along the countryside can lead us to better answers to improve welfare to the urban poor. With a more urban population, even in smaller municipalities, new possibilities of population retention can potentially arise (Ojima, 2013).

ACKNOWLEDGMENTS

Funding for this chapter was provided by the Inter-America Institute (IAI), Project "LUCIA – Land Use, Climate and Infections in Western Amazonia" (CRNIII3036), and by the National Research Council – CNPq, Brazil, grants 447688/2014–6 and 306567/2016–4.

REFERENCES

Ab'Saber, A.N. (1999). Sertões e sertanejos: Uma geografia humana sofrida. *Estudos Avançados*, IEA/USP, São Paulo, *13*(36), 7–59. https://doi.org/10.1590/S0103-40141999000200002

Barbieri, A.F. (2000). *Uso Antropico da Terra e Malaria no Norte de Mato Grosso, 1992 a 1995.* [Master's Thesis]. Department of Demography. Belo Horizonte: Universidade Federal de Minas Gerais, 232 p.

Barbieri, A.F. (2006). *People, land, and context: Multi-scale dimensions of population mobility in the Ecuadorian Amazon.* Michigan: ProQuest/UMI.

Barbieri, A.F. (2013). Transições populacionais e vulnerabilidade às mudanças climáticas no Brasil. *Redes, 18,* 1–15. http://dx.doi.org/10.17058/redes.v18i2.3120

Barbieri, A.F. (2017). População e Mudanças Climáticas: (In)sustentabilidades e Desafios no Caso Brasileiro. [Paper presentation]. Seminar "Cedeplar: 50 years of History." Belo Horizonte, Brazil.

Barbieri, A.F., Brondizio, E., Costa, S., & Guedes, G.R. (2016). *Rural livelihoods and urbanization in the Brazilian Amazon* [Paper presentation]. Meeting of the Latin American Studies Association (LASA), New York.

Barbieri, A.F., Carr, D.L., & Bilsborrow, R.E. (2009b). Migration within the frontier: The second generation colonization in the Ecuadorian Amazon. *Population Research and Policy Review, 28*, 291–320. https://doi.org/10.1007/s11113-008-9100-y

Barbieri, A.F., Guedes, G.R., & Santos, R.O. (2020). Land use systems and livelihoods in demographically heterogeneous frontier stages in the amazon. *Environmental Development, 28.* [Online first]. https://doi.org/10.1016/j.envdev.2020.100587.

Barbieri, A.F., & Monte-Mór, R.L.M. (2008). Mobilidade populacional e urbanização na Amazônia: Elementos teóricos para uma discussão. In Rivero e Jayme Jr. (Ed.), *As Amazônias do Século XXI* (pp. 89–103). Belém: EDUFPA.

Barbieri, A.F., Monte-Mór, R.L.M., & Bilsborrow, R.E. (2009a). Towns in the jungle: Exploring linkages between rural-urban mobility, urbanization and development in the Amazon. In De Sherbinin et al. (Eds.), *Urban population and environment dynamics in the developing world: Case studies and lessons learned* (pp. 247–279). Paris: CICRED.

Barbieri, A.F, Santos, R.O., & Guedes, G.R. (2014). The migration, environment and development nexus in the frontier: A review of the literature based on empirical evidences from the Brazilian Amazon. Determinants of International Migration Conference, Oxford University.

Bebbington, A. (1999). Capitals and capabilities: A framework for analyzing peasant viability, rural livelihoods and poverty. *World Development, 27*(12), 2021–2044. https://doi.org/10.1016/S0305-750X(99)00104-7

Browder, J.O., & Godfrey, B.J. (1997). *Rainforest cities: Urbanization, development, and globalization of Brazilian Amazon*. New York: Columbia University.

Camarano, A.A., & Abramovay, R. (1999). *Êxodo rural, envelhecimento e masculinização no Brasil: Panorama dos últimos 50 anos*. IPEA, Rio de Janeiro.

Ellis, F. (1998). Household strategies and rural livelihood diversification. *Journal of Development Studies, 35*(1), 1. https://doi.org/10.1080/00220389808422553

Furtado, C. (1959). *A operação Nordeste*. Rio de Janeiro: Ministério da Educação e Cultura, Instituto Superior de Estudos Brasileiros.

Fusco, W., & Ojima, R. (2017). Educação e desenvolvimento regional: Os efeitos indiretos da política de descentralização do ensino superior e a mobilidade pendular no estado de Pernambuco. *Revista Brasileira de Gestão e Desenvolvimento Regional, 13*(1), 247–263.

Hogan, D.J., & Ojima, R. (2008). Urban sprawl: A challenge for sustainability. In G. Martine, G. McGranahan, M. Montgomery, & R. Castilla-Fernandez

(Eds.), *The new global frontier: Urbanization, poverty and environment in the 21st century* (1st ed., pp. 205–219). London: IIED/UNFPA and Earthscan Publications.

IBGE. (2010). Censo Demográfico de 2010. https://censo2010.ibge.gov.br/resultados.html.

Instituto Nacional De Colonização e Reforma Agrária – INCRA. (1973). *Urbanismo rural.* Brasília: Empresa Gráfica Gutenberg.

Kay, C. (2008). Reflections on Latin American rural studies in the neoliberal globalization period: A new rurality? *Development and Change, 39*(6), 915–943. https://doi.org/10.1111/j.1467-7660.2008.00518.x

Lee, R. (2003). The demographic transition: Three centuries of fundamental change. *Journal of Economic Perspectives, 17*(4), 167–190. https://doi.org/10.1257/089533003772034943

Martine, G., McGranahan, G., Montgomery, M., & Fernández-Castilla, R. (Eds.). (2008). *The new global frontier: Urbanization, poverty and environment in the 21st century.* Londres: Earthscan.

Martine, G., & Ojima, R. (2013). The challenges of adaptation in an early but unassisted urban transition. In G. Martine & D. Schensul (Eds.), *The demography of adaptation to climate change* (1st ed., pp. 138–157). New York: UNFPA.

Monte-Mór, R.L. (2004). *Modernities in the jungle: Extended urbanization in the Brazilian Amazon* [PhD Dissertation]. University of California at Los Angeles (UCLA).

Moran, E.F. (1981). *Developing the Amazon.* Bloomington: Indiana University Press.

Moura, H. (1980). *Migração Interna, Textos Selecionados.* Tomo I. Fortaleza, Banco do Nordeste.

Ojima, R. (2013). Urbanização, dinâmica migratória e sustentabilidade no semiárido nordestino: O papel das cidades no processo de adaptação ambiental. *Cadernos Metrópole (PUCSP), 15,* 35–54.

Ojima, R., Costa, J.V., & Calixta, R.K. (2014). Minha vida é andar por esse país ... : A emigração recente no semiárido setentrional, políticas sociais e meio ambiente. *REMHU (Brasília), 22*(43), 149–167. https://doi.org/10.1590/1980-85852503880004310

Ojima, R., & Marandola Jr, E. (2012). Mobilidade populacional e um novo significado para as cidades: Dispersão urbana e reflexiva na dinâmica regional não metropolitana. *Revista Brasileira de Estudos Urbanos e Regionais* (ANPUR), *14*(2), 103–116. https://doi.org/10.22296/2317-1529.2012v14n2p103

Oliveira, A.E. (1983). Ocupação humana. In Salati et al. (Eds.), *Amazônia: Desenvolvimento, integração e ecologia* (pp. 144–327). São Paulo: Brasiliense, CNPq.

Prata, P.R. (1992). The epidemiologic transition in Brazil. *Cadernos de Saúde Pública, 8*(2), 168–175. https://doi.org/10.1590/S0102-311X1992000200008

Richards, P.D., Myers, R.J., Swinton, S.M., & Walker, R.T. (2012). Exchange rates, soybean supply response, and deforestation in South America. *Global Environmental Change, 22*(2), 454–462. https://doi.org/10.1016/j.gloenvcha.2012.01.004

Rostow, W.W. (1960). *The stages of economic growth: A non-communist manifesto.* Cambridge University Press.

Sherbinin, A., VanWey, L.K., Mcsweeney, K., Aggarwal, R., Barbieri, A.F., Henry, S., Hunter, L.M., Twine, W., & Walker, R. (2008). Rural household demographics, livelihoods and the environment. *Global Environmental Change, 18*(1), 38–53. https://doi.org/10.1016/j.gloenvcha.2007.05.005

VanWey, L.K., et al. (2012). The ecology of capital: Shifting capital portfolios, context-specific returns to capital, and the link to general household wellbeing in frontier regions. In B. Kind, & K. Crews-Meyer (Eds.), *The politics and ecologies of health.* Routledge.

Weinhold, D., Killick, E, & Reis, E.J. (2013). Soybeans, poverty and inequality in the Brazilian Amazon. *World Development, 52*, 132–143. https://doi.org/10.1016/j.worlddev.2012.11.016

15 The Urbanization of Mexico's Countryside: A Socio-political Approach to Spatial Transformation

GABRIELA TORRES-MAZUERA

According to information from the United Nations, the year 2007 marks – around the world – the change from a predominantly rural society to an urban one (United Nations, 2005, *World Urbanization Prospects: The 2005 Revision Population).* Currently, most of the world's population lives in localities with over 2,500 inhabitants, and their income originates mainly from trade, services, and industry; hence its urban character (Davis, 2006). In Mexico, the outlook is similar: in 2010, only 23.2% of the population lived in localities with under 2,500 inhabitants, and 13.7% of the economically active population worked in activities related to agriculture, fishing or stockbreeding (INEGI, 2010).

The confirmation of the radical worldwide transformation from rural to urban is based on quantitative criteria such as the number of inhabitants per locality and the percentage of the economically active population in the primary sector. Yet, it tells us very little, or almost nothing, about the qualitative nature of such revolution.

Many villages that in quantitative terms could today be defined as rural exhibit social dynamics associated with an urban world; in parallel, those localities that would be categorized as urban, exhibit characteristics close to what is conceived as rural.[1] Hence, for some scholars, the array of rural settlements displaying a combination of social dynamics associated with the countryside and agriculture, with others attributed to what is urban and industrial, may be better defined in terms of a "new rurality."[2] This notion suggests a gap between an "old" and a "new rurality" related to the new social phenomena seen in the rural world: the insertion of rural inhabitants in non-agricultural economic activities; the increasing flexibility and feminization of rural work, whether agricultural or extra-agricultural; the increase in rural-urban interactions; and international migration. The notion is also useful to underscore a shift in the analytical approach. "New rurality" means a turning point in rural studies since it questions the dichotomous model that situates the countryside in opposition to the city, implicit in modernization theories that were the dominant paradigm during the twentieth century to explain social change within the rural world.[3]

From one vantage point, the notion of a new rurality is correct in bringing up that, what is referred as "rural" transcends agricultural activities; it is a good starting point for the analysis of social change in the contemporary rural world. However, it has the drawback of explaining little about social change from a socio-cultural and institutional standpoint.[4] Indeed, many studies using the new rurality concept have based their analysis on economic and socio-demographic indicators, focusing on the external factors driving social, economic, and political transformations in rural localities. They have, nevertheless, overlooked endogenous factors like redefining processes of rural inhabitants' social and political identities, as well as changes and tensions in their value systems. Even more important, they have disregarded formal structures and informal rules and procedures that give sense to social practices and the reconfigurations of power relationships associated with the new resources at stake in rural areas. The following questions have yet to be answered: What exactly does the new rurality, or to be more precise, the urbanization of the countryside, imply in terms of institutional change? Or, put in more concrete terms, what rearregments in family configuration, the community, the land-tenure system, religious groups, schooling, and the local government imply the urbanization of the countryside? Also, what kind of institutional change at the state level – and particularly related to those government institutions that materialize a national public policy address toward rural development (i.e., government agencies for rural development, and municipal government) – is involved in the new rurality?

My aim in this chapter is to briefly address some of these questions, in the context of San Felipe del Progreso (SFP), a municipality located in central Mexico (State of Mexico) that has endured a radical spatial transformation in the last four decades. To this end, I will summarize the main arguments developed in a previous work, where I proposed the notion of "urbanized rurality" as an alternative to "new rurality."[5]

A word must also be said about my socio-political approach to the urbanization of the countryside in Mexico. When I first arrived in San Felipe in 2002, as part of a research project on rural households' economic strategies, my initial interest was in the transformation of subsistence agriculture triggered by the economic opening to the global marketplace and the end of subsidies toward small farmers during the mid-1990s (changes brought about by provisions in the North American Free

Trade Agreement). However, the socio-political aspect of rural development caught my attention since my first fieldwork visit. More often than not, the people I was interviewing about the price drop of maize or the decay of agriculture production in the region ended up associating agrarian changes in the villages with the conflicts of power between rival families, or religious and partisan groups within and between the ejido settlements and the municipality seat. Local emphasis on conflict made me redirect my attention to the socio-political side of urbanization in a rural municipality.

Between 2004 and 2007, I went back and forth to San Felipe ejidos in short visits of two to four weeks, asking local authorities, officials from the municipal, state, and federal governments, Mazahua *ejidatarios,* rural inhabitants, men and women, about changes in agriculture, land tenure, and political life within ejidos and more broadly within the municipality. I relied on different qualitative techniques: participant observation, non-structured interviews, life histories, and focus groups.

What this experience left me with was the awareness that a deep understanding of social phenomena such as "the urbanization of countryside" needs to consider not only economic, demographic, and spatial reconfigurations, but also institutional change and the informal distribution of power. In Mexico, it implies focusing on two institutions of local governance in their intermingled and ever-changing relationship: the ejido and ayuntamiento.

In what follows I intend to describe the urbanization of rural Mexico as a process of change of those institutions. Section one is a brief account of how ejido and ayuntamiento have historically structured what we call "rural space" in Mexico, and their substantial recent transformation. The analysis of institutional settings requires considering not only rules, norms, and social practices, but also the way institutions construct actors and define their available modes of action (Scott, 1995, p. 27). In section two, I describe the transformation of political identities of people living in urbanized and de-agrarized ejidos. I contend that identitarian change in urbanize rurality is related to the new valuable ressorces at play, namely, the municipal budget for urban infrastructure and public services (roads, sewage, schooling, health clinics) that, in turn, transfigure the political landscape of the rural world. Lastly, in section three, I consider the changes in the public policy agenda in reference to rural development. My account encompasses the federal government neoliberal turn that inagurated a new conception of the peasantry. For more than seven decades (1910–80) the countryside and the peasants were the protagonists of national development in Mexico. Their social function, not only in the economic

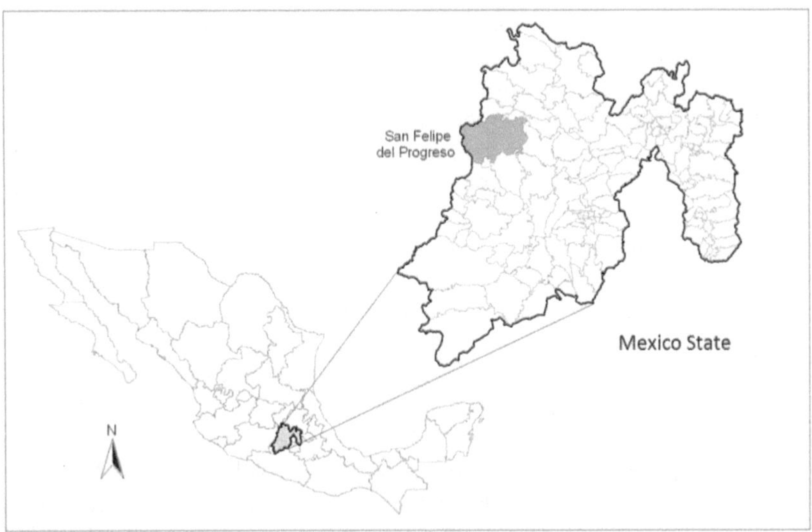

Figure 15.1 Location of the State of Mexico and San Felipe del Progreso in Mexico. Map created by the author.

realm, but more important, in their political participation, was evident and paramount. However, since the 1990s, its role in the national development agenda fades. Henceforth, peasants are conceived as residual factors of a global, competitive, and highly technological economy in which they no longer have a place, being described, in the official discourse, as inefficient and uncompetitive.

Institutions That Have Given Meaning to Modern Rural Mexico

The Ejido: The Peak and Decline of a Micro-Local Governance Institution

In modern Mexico, we can identify two local institutions with a colonial background and a modern "upgrading" that have modelled most of the rural order: the ejido and the municipal government, or *ayuntamiento*.[6] Both, with different functions, though fundamental in the local and micro-local governance of the countryside, have experienced considerable legal, socio-economic, and political changes since the 1970s. Let us begin with the ejido. The land restitution and redistribution (agrarian reform) initiated in 1915, in the aftermath of the Mexican revolution, generated a new type of legally recognized form of property, the ejido and the comunidad.[7] This agrarian reform program lasted more than 60 years and, when it ended in 1992, it had allocated half of the Mexican territory (103 million hectares) to almost four million peasants.

Over the course of their existence, ejidos and comunidades have significantly reconfigured the property relations, landscapes, and settlements of rural Mexico. In contrast to private property, both forms of communal land-holding were meant to provide a subsistence base for peasant families; they were not to become an economic commodity. Ejido and comunidad lands were therefore inalienable and indivisible. They entailed a set of obligations for ejido beneficiaries, like working the land with their own hands, living within ejido or comunidad villages, and participating in ejido or comunidad assemblies. Over time, ejidos and comunidades began to play an important political role as intermediaries between rural dwellers and the federal government (Warman 2001).

Ejidos were, ironically, the foundation of the urbanization of modern Mexico, as new settlements were created with reallocated lands (Leonard & Velázquez, 2009, p. 396; Warman, 2001). They also contributed to the expansion and strengthening of small villages granted with land that acquired a new political-administrative status as ejidos.

San Felipe del Progreso exemplifies the spatial reconfiguration of rural Mexico triggered by the land redistribution launched in 1921, with the first ejido granted in the municipality. Before land distribution took place, Mazahua peasants dwelled in four types of settlements according to their political jurisdiction: the municipal county seat (*cabecera municipal*), the Indian villages (*pueblos de indios*), the haciendas, and the ranches. The Mexican Revolution radically transformed the San Felipe landscape by introducing the ejido. In the first five decades of land reform (1920–70), people migrated from haciendas, ranches, Indian villages, and even from the county seat to the newly created ejidos. Ejidos granting was a way to take away power from local authorities living in the county seat of San Felipe del Progreso, in the hands of ranch owners, landowners, and, in general, local and regional elites that in many cases did not support the national government composed of revolutionary adherents. Therefore, the San Felipe county seat, the largest village within the municipality which, in 1900, had 2,433 inhabitants, by 1940, was left with only 717 inhabitants. In that same decade, 45 new urban zones were founded with the granting of ejidos becoming a pole of attraction for hacienda labourers (INEGI, Archivo histórico de localidades, 1900–2020). This growth was similar to that experienced at the national level, which, between 1920 and 1940, involved the creation of 40,000 new ejido settlements (Warman, 2001, p. 42).

By 1992, when the land reform was cancelled,[8] the total of 84 new settlements associated with ejidos in San Felipe del Progerso had been established, with a total of 38,586.32 hectares and 4,428 beneficiaries

Table 15.1 Historical origin of the fifty largest villages in San Felipe del Progreso, 2000

Ejidos granted to peasants from haciendas	63.41%
Colonial origin villages (*Pueblos de indios*)	17.07%
Hacienda settlements that became an ejidovillage after land granting	9.76%
New villages (neighbourhoods) derived from old ones	7.32%
Recently created villages (founded in the 1990s)	2.44%

Source: INEGI. Archivo histórico de localidades 1900–2020. https://www.inegi.org.mx/app/geo2/ahl/; RAN, Padrón e Historial de Núcleos Agrarios. Available at: https://phina.ran.gob.mx/index.php.

(RAN, Padrón e Historial de Núcleos Agrarios). In 2000, 63.41% of the 50 largest villages of the municipality were created altogether with new ejidos. Only 17% were originally villages prior to ejido creation, but they also benefitted from the land granting or restitution (see table 15.1).

Since its creation as a form of land tenure, the post-revolutionary ejido functioned as a multifunctional institution that regulated virtually all areas of community life in the granted villages including agricultural production and credit, access to land and natural resources, provision of services, and political participation. Ejidos were organized around an assembly in which all ejido holders could participate and was chaired by an ejido board. The ejido assembly organized community work or tasks for the activities that benefitted the ejido and urban settlement, like building a road or path or dredging irrigation canals. This assembly also decided where the community school or church were to be built.

In ejido villages, the ejido board acted as the local authority and representative before federal agencies, particularly the Department of Agrarian Reform and the Department of Agriculture, which had greater presence in rural areas. Ejidatarios were usually members of the largest Mexican union, the National Peasant Confederation, a peasant organisation closely bounded to the Institutional Revolutionary Party (PRI).[9]

The rural geopolitical order centered around the ejido began to crumble after the 1970s. An array of demographic, economic, sociocultural, and political transformations had an eroding impact on the ejido as a local micro-government body. Let us begin by describing the demographic driver in the case study.

Between 1970 and 2000 San Felipe del Progreso population increased from 87,173 to 177,287, which led to the concomitant growth of ejidos and villages. In that same period, the percentage of people living in settlements with a population under 2,499 decreased from 92.30% to 47.48% (INEGI, 1970; 2001) (see table 15.2).

Table 15.2 San Felipe del Progreso demographic transformation between 1970 and 2000

San Felipe del Progreso	1970	1990	2000
Total population	87,173	155,978	177,287
Population percentage living in villages with 2,499 inhabitants or fewer	76.25%	70%	47.48%

Source: INEGI, 1970, 1990, 2001.

Table 15.3 San Felipe del Progreso population growth in five ejidos, 1940–2000

	Settlement origin	1940	1960	2000	2000 neighbourhoods*	Total population 2000
Emilio Portes Gil	ejido	892	1,611	3,076	1,330	4,506
Dolores Hidalgo	ejido	637	1,053	3,033	s/b	3,033
Santa Ana Nichi	pueblo	847	1,681	1,526	1,738	3,264
San Lucas Ocotepec	pueblo	434	650	3,361	s/b	3,361
San Pedro el Alto	pueblo	1,258	2,396	3,195	4,230	7,425

Source: INEGI. Archivo histórico de localidades 1900–2020. https://www.inegi.org.mx/app/geo2/ahl/.

* The neighbourhoods that once depended on mother-towns (ejidos) are recorded as separate localities in the 1990s Population Census. Thus, to analyse the population increases we included the neighbourhoods' population today, coverted in independent localities.

The vocation of ejidos as human settlements was confirmed in the 1970s when the population of most ejidos grew rapidly. However, this growth rendered the coordination and implementation of public works concerning the village life – previously organized by the ejido assembly – inefficient.

The population growth of ejidos also led to the emergence of new neighbourhoods settled within the ejido lands, but far from the urban centre which sought autonomy from the ejido assembly once they had more than 700 inhabitants. (This because according to the local legal framework, an urban settlement with 700 or more inhabitants may have its own authority and may negotiate directly with the municipal government regarding access to resources for urban development [San Felipe del Progreso, 2008].)

Since the 1990s, ejido board members, who used to be the only local authority in the ejido village, started to be replaced by a new authority: the "municipal delegate." This novel mediation appeared following the empowerment of the municipal government (see next section), its main role being to request public works from the municipal president.[10]

Another factor that weakened the ejido assembly as a micro-local governance institution was the waning of agriculture activity and the increased number of people working outside the ejido domain. According to data from a survey conducted in 2000 in the region of Alto Lerma, which includes San Felipe del Progreso, the percentage of the population working in the service and industry sectors amounted to 71.24% of the economically active population (Orozco, 2005).

Today, ejido assembly participation is in decline. The end of production subsidies and credits to small-scale and peasant agricultural activity since the 1990s in Mexico, as well the decreasing incorporation of San Felipe del Progreso inhabitants in that economic sector, is the reason why ejidatarios and peasants have stopped participating in the ejido assemblies According to the ejido president board, interviewed in 2004, ejido assemblies are now held sporadically and with very little assistance.

The Counterpart: Municipal Government Empowerment

Since the 1990s, municipal governments' remarkable budget increase of 170%, directed mainly at the construction of urban infrastructure, services, housing improvement (metal sheets and cement), and public works (schools, sports courts and fields, health clinics, street paving, housing improvement), significantly transformed the existing power relations within the rural municipalities (Enciclopedia de los municipios y delegaciones en Mexico, 2010). It also had important consequences on local political organization to the extent that municipal governments must now negotiate upwards, with the state government (the governor), and downwards, with the villages under their jurisdiction.[11] This negotiation has increased the municipal government's power at the local arena, to the extent that the municipal president now decides how to distribute new federal resources directed mainly to urban development among the ejidos/rural villages.

Once more, the demographic transformation of San Felipe del Progreso village, where the municipal government is located, is an indicator of the municipal government's empowerment. After a sharp population decrease in the 1930s, due to rural immigration related to the Mexican Revolution, San Felipe del Progreso demographics began expanding in the 1990 (see table 15.4).

Table 15.4 Population growth of San Felipe del Progreso village (municipal seat), 1900–2000

Year	1900	1930	1960	1970	1990	2000	2010
Population	2,433	665	836	991	1,818	3,512	4,350

Source: INEGI. Archivo histórico de localidades 1900–2020.

The population growth of the municipal seat goes hand in hand with the expansion of the municipal government. In 1980, the number of San Felipe del Progreso municipal government officers was 57, which increased to 150 by 1994, and to 500 by 2000 (San Felipe del Progreso, 1980; San Felipe del Progreso, Municipal archives, 1970–2000).[12] The increase in municipal bureaucracy is not an extraordinary trait unique to San Felipe del Progreso. According to Grindle (2007), in 2000, the number of municipal staff members at the national level grew 45% compared to 1990. This rise is due to the enlarged budget and the surge of tasks that local governments must thereafter carry out. Between 1998 and 2005, the municipal government of San Felipe del Progreso increased its municipal budget by 172.8% (Enciclopedia de los municipios y delegaciones en Mexico, 2010; San Felipe del Progreso, 1980; 2009).[13]

The growth of municipal resources led to an administrative restructuring within the municipal government; below, we sum up the main changes.

In the 1980s, in San Felipe del Progreso, the "official receiver" (*síndico*) was – after the municipal president or mayor – the most important position since it had to deal with all matters regarding municipal crimes. However, since the 1990s, there is a municipal judge within a new office branch of the Mexico State Public Ministry in charge of those tasks. Likewise, the civil registry, which was previously the municipal government's main source of income and was managed by the mayor, today functions as a regular and independent service department, and is not the most important one. Finally, in 1997, the municipal secretary position was created to address matters pertaining rural villages within the municipal jurisdiction, previously run by the mayor. These changes, which apparently have eliminated functions and therefore power from the municipal president, hide the true transformation experienced by municipal governments in the urbanized rurality.

Currently, the economic vitality of the municipality comes from urban works carried out in rural villages thanks to the federal government (in 2006 urban works accounted for 69.1% of the total municipal budget) (San Felipe del Progreso, 2009).[14] The importance of municipal urban development has required the creation of a new department of Public Works and Development that concentrates a good part of the municipal government staff and budget. The most desired position of trust at present is the public works manager. A new management area has also been created to oversee resources from a federal fund (Ramo 33) destined for the construction of urban infrastructure. This restructuring demonstrates the new direction taken by the municipal government' administration, nowadays mainly aimed at municipal urban development (Information collected during 2005 field research).

Social and Political Identities of Urbanized Rural Space

The urbanization of rural localities in central Mexico not only implies a structural transformation in the relationship among the ejido and the municipal government, but also new patterns of social differentiation within the rural villages linked to – also new – social and political identities.[15]

To analyse social differentiation, it is useful to think in terms of opposing identity categories that are not limited to class or ethnic affiliation but are a mixture of production relations and other systems of social relations independent of the production process that go beyond the community sphere. In the municipality of San Felipe del Progreso one of the identity pairs that sheds light on the socio-cultural reconfiguration associated with the urbanization of the countryside is that of professionals (*profesionistas*) versus peasants-ejidatarios (ejido land holders). Professionals, as a socio-political group, make sense when compared to the peasant/ejidatarios category that once was the leading group with the most power and control over ejidos settlements. Unlike other identity categories in San Felipe del Progreso, such as Mazahua Indians and mestizos, the peasant/ejidatarios and professional identities are forms of identification and self-identification in different realms of daily life in ejidos and municipalities that since the 1990s have become established as political identities. Current opposition between these categories point to different generational cohorts with contrasting values regarding what is a valuable life in ejidos.

People who today identify as peasants in the ejidos of San Felipe usually belong to the generation born between 1930 and 1950, who inherited ejido parcels from the first ejido holders and who grew up in a time when working the land was the main *raison d'être* for the ejido. Their area of social relations is limited – in general – to the locality (many of them speak Spanish as a second language, Mazahua being their mother tongue), and their links are reactivated in village festivities and family celebrations. Many of these people migrated to Mexico City at a certain point in their lives in search of work but returned to their village where they invested their savings. Some of them grow corn as a social obligation related to the idea that "the land belongs to those who work it," rather than as a compulsory source of income or food supply for the household.

In contrast, children of this generational cohort, born after the 1960s, have taken on non-agricultural activities, both inside and outside ejidos, and make up the group of "professionals." The new generation, generally, speaks Spanish fluently, and many of its members had access to high school, thanks to scholarships from the National Indian Institute in the 1980s. Access to middle school enabled many of them to become professionals – namely, elementary, or middle education teachers, nurses, police officers, physicians, and lawyers. Largely, this group does not hold land in ejidos, except for those who have inherited it from their ejido-holder parents or grandparents. For this group, land is seen as a form of investment that contributes to family assets, particularly when they can build their homes; some of them invest in subsistence agriculture that they value because of the quality of the corn harvested for family consumption.[16]

Professionals are role models for many ejido inhabitants, since they are the new rural entrepreneurs who open businesses in the locality, migrate to the United States, or take on municipal government positions or the delegate position at ejidos.

The transformation of political institutions in the rural area is closely linked to a transformation of political identities. The opposition that currently exists between peasant/ejidatarios and professionals categories reveals the transformation in the structuring of power relations in the ejidos of San Felipe. Previously, ejido land and access to the ejido board were at stake and only accessible to a person who was a peasant-ejido holder; whereas today, the municipal government and control over the provision of urban services are only accessed by those who identify as professionals with enough education and cultural capital.

Space and Rural Development Seen by the State

A comprehensive description of the new rural spatiality must consider the production of space from public discourses and policies. In Mexico, the de-agrarization and urbanization of the rural world is associated with an ideological shift of the political elite. Since the late 1980s a neoliberal turn was promoted in legal reforms such as the Constitutional

Article 27 reform, in 1992, that ended the agrarian distribution and, with it, peasants' demands and political claims associated with it. This change had an impact on policies aimed at agricultural activity and rural producers, as well as on rural development. From the productive perspective, we see the creation of a new official discourse directed to the rural world that is no longer based on the demand of land but focuses on the problem of production and productivity. The "agrarian question" that until 1980 implied an unending debate about production relations between the urban and the rural world, the creation and transformations of social classes, particularly that of the peasant, and the role of rural development within national development policy, has been replaced by a purely technical and productive question: How to become more competitive in the global market of cash crops and livestock products? The new programs and subsidies to agriculture have been, ever since, directed not at peasants but at competitive and efficient farmers in the private and business sectors (on very few occasions coinciding with the ejido sector). This change happens simultaneously with the emergence of a new form of state intervention directed at the rural world, which no longer mainly targets peasants and ejido holders but, rather, other social groups, including women, migrants, the elderly, and ethnic groups. At present, state intervention policies aimed at what is conceived as "rural space," are mostly aimed at the provision of urban services, the increase in the quality of life of rural inhabitants (health, education, food), and the insertion of rural households into a system of consumption controlled by the state or by the market that – in general terms – tends to eliminate self-sufficiency. Similarly, the objectives and presence of those government ministries (departments) with intervention in rural areas have changed: the Department of Agriculture focuses most of its budget in supporting agro-industry producers, at the expense of the peasant and ejido sector; the Department of Agrarian Reform that up until the 1990s had a strong presence in ejidos and communities, withdrew from this field, changed its name and objectives (now the Department for Agrarian, Urban and Territorial Development), redirecting them toward urban development. Lastly, the department with the greatest presence in the rural areas today is the Department of Social Development with a program of wide national coverage dedicated to fighting against poverty and focused on women and youngsters. I present a small sample of this transformation, taking as an example the changes in the Mexico State Department of Rural Development and the new rural demands in San Felipe del Progreso.

Between 1970 and the early 1980s, the federal government conceived a rural development policy for the region and the rest of the country that provided subsidies for fertilizers and other agricultural supplies, such as herbicides or pesticides, the establishment of a guaranteed price for corn and other basic grains in the Mexican diet, and the purchase, distribution, and sale of basic grains via storage warehouses and rural stores established in regions deemed marginal. In Mexico State, a policy parallel to the federal policy was developed by the Institute of Agricultural and Livestock Development (DAGEM),[17] created with the objective of increasing the areas where corn was grown, expanding the intensive use of supplies and improved seeds, and providing basic grains for growing urban areas during those years. Until the 1990s, resources from the federal government were channelled through the State of Mexico Department of Rural Development (SEDAGRO).

In 2004, SEDAGRO offered a set of productive projects that were part of the *Alianza contigo* program.[18] Three programs were offered: PAPIR, which supported productive projects proposed by individual or group producers, particularly women, youngsters, the elderly or Indigenous people; a training program via workshops to prepare preserves, industrialize pork, and sewing; and lastly, a rural organization program (PROFEMOR) that offered administrative and accounting advice to groups of producers. The PAPIR program was the most welcomed one in San Felipe del Progreso in 2004, and although one of its objectives was agriculture productive reconversion and the introduction of cash crops, few projects had been proposed towards this result. Since 2000, people of the municipality had been interested in receiving financing, not for agriculture projects but, rather, mostly for services such as Internet cafes, car repair shops, and stationery stores; women and the elderly proposed sewing workshops, industrial sewing machines, production of processed food, flowers, and handicrafts. This preference reflects the transformation of the rural world and is proof of the way in which countless projects and programs aimed at rural development have consolidated the urban vocation of San Felipe del Progreso.[19]

Conclusions

In this chapter, I have identified two contrasting spatial arrangements that give an idea of the change experienced in Mexico's rural world. To define them, I used the notion of ejido hegemony that allows me to describe a spatial, economic, political, social, and cultural configuration organized around ejidos that involved the formation of social groups associated with the identity categories of peasant and ejido holders. Until the late 1970s, these were the main actors in power disputes in a rural municipality such as San Felipe del Progreso. The resources at stake were associated with the ejido as a productive unit, and political participation was controlled by the ejido assembly. Ejido holders were a privileged group in terms of access and control of resources at a local level, but ever since the 1970s, they began to lose political and social cohesion. New social and political groups, such as professionals, have emerged in the ejidos of San Felipe del Progreso, ending the ejido hegemony and giving rise to a new territorial order whose gravity centre is the municipal government.

Thanks to federal contributions, the municipal government has managed to consolidate its role as provider and administrator of local urban development. In San Felipe del Progreso, the main economic and political dynamics today have to do with the urbanization of rural villages, whether in the construction of urban and housing infrastructure or services. The centre of urbanized rurality is the municipal government that has become the main mediator between federal and state governments on the one hand, and the localities distributed in its territory on the other, with representatives of the municipal government, the delegates, having replaced ejido officers.

The ethnographic approach to the urbanization of the rural world in Mexico sheds light on some of its unexpected consequences. Today, rural villages are mostly the site of reproduction of the economically active population working in non-agricultural activity sectors in other regions of the country or even outside of Mexico. This reproduction of the labour force involves the expansion of mechanisms and forms of the provision of consumption goods for rural households, a task in which the state, and more specifically municipal governments, are increasingly involved.

NOTES

1 An example is Larissa Lomnitz's work "Como sobreviven los marginados" (1975) where we see the continuation of social dynamics such as kinship networks, reciprocities, and compadrazgos (the reciprocal relationship existing between a godparent and the godchild's parents) that thrive in Mexico City's marginal neighborhoods, very similar to those in rural towns.
2 See Arias (1992); Carton de Grammont (1995, 2004, 2009); Ruiz Rivera and Delgado (2008); Llambí (1996); Kay (1996); Pérez (2001); and Teubal (2001).
3 See for instance, Arthur Lewis's thesis (1954). It is important to note that since the 1960s, these theories have been very much criticized. Authors like Cardoso and Falleto (1969), Gunder Frank (1969), and G. Germani (1971) developed the concept of over-modernization to explain the dysfunctional

results of modernization in developing countries. However, these authors focused on studying the effects of dependent development in cities or economic sectors, and failed to analyse in detail and from a qualitative perspective the effects of over-modernization in the rural world.

4 New rurality has also been analysed from a political sociology standpoint by some Mexican scholars as the end of peasant corporatism associated with the National Peasant Confederation and the emergence of new organized groups with novel demands. However, such approaches are overshadowed by the vast literature on new rurality from an economic or geographic conceptualization viewpoint (Carton de Grammont, 1995; Mackinlay, 2004; Mackinlay & Otero, 2006; Harvey, 1990, Haubert & Torres, 2003).

5 See Torres-Mazuera Gabriela (2012), La ruralidad urbanizada en el centro de México. Reflexiones sobre la reconfiguración local del espacio rural en un contexto neoliberal. Mexico: UNAM.

6 The municipality is recognized by the 1917 federal Constitution as the smallest unit of administrative and political governance with the *ayuntamiento* or municipal government as its main governing institution. The size of municipalities around Mexico varied considerably, as well as the number of villages under their jurisdiction. In the Mexico State, where San Felipe del Progreso is located, municipalities are, usually, wide territories with more than 50 villages under the municipal government jurisdiction.

7 The ejido and community were created in 1917 after the Mexican Revolution as a response to peasants' demands for land restitution and allocation. The land was granted to a group of no less than 21 peasants who received individual plots of land for farming, in addition to a communal parcel for livestock or/and forestland, and a small plot for families to build their houses (Warman, 2001). Ejidos and communities spread all over Mexico, and in 2017, there were over 32,082 amounting to 99.8 million hectares, representing 51% of the Mexican territory.

8 The end of Agrarian reform in México occured in 1992 when Article 27 of the Mexican 1917 Constituion was reformed under a neoliberal backlash.

9 Founded in 1929 as the National Revolutionary Party, it adopted its present name, Institutional Revolutionary Party (PRI), in 1946. PRI leaders controlled the federal government without interruption, from the party's creation through to 2000.

10 According to state law, there are three types of sub-municipal authorities in the State of Mexico, namely the delegates, deputy delegates and citizen councils. The delegates and deputy delegates are authorities representing rural populations at the municipal government.

11 This transformation is the result of a descentralization policy launched in the late 1980s consisting in the increase of federal budget. Before that, the ejido assemblies used to organize all the infrastructure provision, which was scarce and precarious, by themselves with little aid from the state.
12 This increase in the number of officers, of over 700% between 1980 and 2000, contrasts with the growth if municipal population in a similar period (1970–2000), which was a little over 200% (87,173 people in 1970 to 177,287 in 2000).
13 Which contrasts with the insignificant participation of tax to the municipal budget that in 2005 was only 0.7% (INAFED, 2005).
14 Thus, in the last 10 years, six building material stores and two construction companies have emerged.
15 For analytical purposes, I differentiate a social identity from political one. A social identity is not always political in that it is not activated for strategic purposes of collective action. A social identity that becomes political implies the formation of a new social group that claims certain political rights linked to the identity characteristics that differentiate it from other groups.
16 For an analysis of this preference, see Appendini, Cortes, and Díaz Hinojosa, 2008.
17 In 1976, Dagem was replaced by the Coordinating Commission of Agricultural and Livestock Development of the State of Mexico (Codagem), which unlike the former that depended directly on the state government, has legal personality and state and federal fiscal resources (Vizcarra Bordi, 2002).
18 The Zedillo administration (1994–2000) created Alianza para el campo to promote rural development. This program was aimed at "viable" producers and sought to establish the means to increase productivity and profitability "based on free decisions by agricultural producers." Applying the principles of co-responsibility and participation, the program focused on specific projects in which the federal government, state government, and producers participated. During the Fox administration, the program name changed to Alianza contigo and covered more, opening its support to projects not necessarily directed to agricultural production.
19 For an exhaustive review of the different rural development projects in the municipality and their unexpected effects, see Torres-Mazuera (2012).

REFERENCES

Appendini, K., Cortes, L., & Díaz Hinojosa, V. (2008). Estrategias de seguridad alimentaria en los hogares campesinos: La importancia de la calidad del maíz y la tortilla. In Kirsten Appendini y Gabriela Torres-Mazuera (Eds.),

¿Ruralidad sin agricultura? Perspectivas multidisciplinarias de una realidad fragmentada (pp. 103–128). México: El Colegio de México.
Archivo municipal de San Felipe del Progreso. Ramo Presidencia. 1970–2000.
Arias, P. (1992). *Nueva rusticidad mexicana*. México: Conaculta.
Cardoso, F.H., & Falleto, E. (1969). *Dependency and development in Latin America*. Berkeley: University of California Press.
Carton de Grammont, H. (1995). Nuevos actores y formas de representación social en el campo. In J.F. Prud'homme (Coord.). *El impacto social de las políticas de ajuste en el campo mexicano* (pp. 105–168). Plaza y Valdés.
Carton de Grammont, H. (2004). La nueva ruralidad en América Latina. *Revista Mexicana de sociología, 66*, 279–300. https://doi.org/10.2307/3541454
Carton de Grammont, H. (2009). La desagrarización del campo mexicano. *Convergencia, 16*(50), 3–55.
Davis, M. (2006). *Planet of slums*. London, New York: Verso.
Enciclopedia de los municipios y delegaciones en Mexico. (2010). http://www.inafed.gob.mx/work/enciclopedia/EMM15mexico/index.html
Frank, A.G. (1969). *Latin America: Underdevelopment or revolution*. New York: Monthly Review Press.
Germani, G. (1973). *Latin modernization, urbanization, and the urban crisis*. Boston: Little Brown.
Grindle, M.S. (2007). *Going local: Decentralization, democratization, and the promise of good governace*. Princeton and Oxford: Princeton University Press.
Harvey, N. (1990). *The new agrarian movement in Mexico 1979–1990*. London: University of London.
Haubert, M., & Torres, G. (2003). *Hacia un nuevo pacto Estado-campesinos? Desenlace de la crisis y ajustes al modelo de desarrollo* [Presenté au Séminaire AMER]. Guadalajara.
INEGI. (1970). *IX Censo General de Población*. México: Instituto Nacional de Estadística y Geografía.
INEGI. (1990). *XI Censo General de Población y Vivienda*. México: Instituto Nacional de Estadística y Geografía.
INEGI. (2001). *XII Censo General de Población y Vivienda*. México: Instituto Nacional de Estadística y Geografía.
INEGI. (2010). *Censo de Población y Vivienda 2010*. México; Instituto Nacional de Estadística y Geografía.
INEGI. *Archivo histórico de localidades 1900–2020*. México: Instituto Nacional de Estadística y Geografía. https://www.inegi.org.mx/app/geo2/ahl/
Kay, C. (2008). Reflections on Latin American rural studies in the neoliberal globalization period: A new rurality? *Development and Change, 39*(6), 915–943. https://doi.org/10.1111/j.1467-7660.2008.00518.x

Leonard, E., & Velázquez, E. (2009). El Istmo posrevolucionarios (1): Agrarismo oficial y fronteras internas en la construcción de los espacios rurales. El reparto agrario y el fraccionamiento: construcción local del Estado e impugnación del proyecto communal. In E, Velázquez, et al. (Coords.). *El Itsmo mexicano: Una región inasequible. Estado, poderes y dinámicas espaciales* (siglo XVI–XXI, pp. 395–398). México: Casa chata.

Lewis, A. (1954). Economic development with unlimited supplies of labor. *Manchester School of Economic and Social Studies, 22*(2), 130–191. https://doi.org/10.1111/j.1467-9957.1954.tb00021.x

Llambí, L. (1996). Globalizacion y nueva ruralidad en América Latina: Una agenda teórica y de investigación. *La socieidad rural mexicana frente al nuevo milenio, 1*, pp. 55–58. Plaza y Valdés, INAH, UAM, UNAM.

Lomnitz, L. (1975). *Cómo sobreviven los marginados.* México: Siglo XXI.

Mackinlay, H. (2004). Los empresarios agrícolas y ganaderos y su relación con el Estado. *Polis: Investigación y análisis sociopolítico y psicosocial,* 2(04), 113–144.

Mackinlay, H., & Otero, G. (2006). Corporativismo estatal y organizaciones campesinas: Hacia nuevos arreglos institucionales. In Gerardo Otero (Coord.), *México en transición: Globalismo neoliberal, Estado y sociedad civil.* H. Cámara de Diputados LIX Legislatura, Simon Fraser University, la Universidad Autónoma de Zacatecas y Miguel Ángel Porrúa.

Orozco, María Estela. (2005). Articulación de economías domésticas al desarrollo regional del Alto Lerma, México. *Papeles de Población, Nueva Época, 11*(46).

Pérez, E. (2001). Hacia una nueva visión de lo rural. In Giarracca N. (Ed.), *¿Una nueva ruralidad en América Latina?* Buenos Aires: Clacso.

RAN, Padrón e Historial de Núcleos Agrarios. https://phina.ran.gob.mx/index.php/ (Accessed 01 August 2018)

Ruiz Rivera, N., & Delgado, J. (2008). Territorio y nuevas ruralidades: Un recorrido teórico sobre las transformaciones de la relación campo-ciudad en el marco de la globalización. *EURE: Revista latinoamericana de estudios urbano regionales, 102*, 77–96.

San Felipe del Progreso. (1980). *Radiografía municipal,* Toluca.

San Felipe del Progreso. (2008). *Bando municipal, 2008.*

San Felipe del Progreso. (2009). *Plan de desarrollo municipal 2006–2009.*

San Felipe del Progreso. Municipal archive, 1970–2000.

Scott, R. (1995). *Institutions and organizations.* London: Sage.

Teubal, M. (2001). Globalización y nueva ruralidad en América Latina. In Giarracca N. (comp.) *¿Una nueva ruralidad en América Latina?* Buenos Aires: Clacso.

Torres-Mazuera, G. (2012). *La ruralidad urbanizada en el centro de México: Reflexiones sobre la reconfiguración local del espacio rural en un contexto neoliberal.* México: UNAM. Colección: La pluralidad cultural en México.

United Nations. (2005). *World urbanization prospects: The 2005 revision population*. https://population.un.org/wup/

Vizcarra Bordi, I. (2002). *Entre el taco mazahua y el mundo: La comida de las relaciones de poder, resistencia e identidades*. Gobierno del Estado de México. Toluca: Université de Laval.

Warman, A. (2001). *El campo mexicano en el siglo XX*. México: Fondo de Cultura Económica.

Conclusion: Peripheral Urbanization: Current Trends, Methodological Advances, and the Decolonization of Urban Theory

NADINE REIS AND MICHAEL LUKAS

The aim of this book was to reconnect to the tradition of theorizing urbanization from the margins. As we clarified in the introduction, underlying our methodology is an understanding of (sub)urbanization as an abstracted concept that refers to a *process* rather than a specific form of settlement (Harvey, 1996; Brenner & Schmid, 2015, p. 165; Keil, 2018, p. 11). This implies that for any study of urbanization globally, and in Latin America today, it is not sufficient to focus analysis on *cities*. As the contributions in this book have shown, it is ever more evident in the global periphery in the age of neoliberal capitalism that urbanization is a territorial process that reaches far beyond the limits of the classic urban settlements. Overall, urbanization in twenty-first-century Latin America takes on manifestations that are qualitatively different from those of the largest part of the twentieth century. One of the major manifestations of this is the increasing importance of suburbanization and the development of urban hinterlands in all its contradictory forms (Keil, 2017). In Latin America, these socio-spatial transformations have been taking place in the context of renewed extractivism of primary resources, the expanding frontiers of the agro-industries ("agro-extractivism"), and the establishment of *maquila* and *maquila*-like export production plants. This neo-extractivism (Gudynas, 2009) is based on and gives rise to massive new infrastructural projects, such as road networks, waterways, ports, airports, energy plants, and hydrological infrastructure – many realized in the past two decades and others under construction or planned – which shape the specific pathways of urbanization in the region (Bebbington, 2009; Wilson & Bayon, 2017). In analytical terms, this implies that the study of urbanization must also take into account the structural transformation of rural livelihoods, since the specific pathways and characteristics of depeasantization have immediate relevance for the urbanization process in shaping not only the migration patterns of labour, but also the urbanization of the countryside itself. It is in this regard that the megacity has ceased to be the only and even the major site of study to understand urbanization in Latin America – a claim that inspires the title of this book.

This book thus started as an experiment to investigate, from an interdisciplinary perspective, the natures, causes, implications, and politics of current urbanization processes in Latin America through exploring the usefulness of the notion of peripheral urbanization as boundary concept (Mollinga, 2008, p. 24). In this sense, the aim of this book was to advance theory building on urbanization in Latin America based on a set of empirical case studies from various countries in the region that cover different theoretical and disciplinary approaches. Therefore, this edited collection includes perspectives and approaches from the fields of geography, anthropology, sociology, urban studies, agrarian studies, urban and regional planning, including academics, journalists, practitioners, and scholar-activists. The book sets out in part 1 with three chapters that epitomize important ways of seeing and studying urbanization from the margins in Latin America today and that are also key points of reference throughout this book: First, Teresa Caldeira's conception of peripheral urbanization as the principal way of urban space production in the Global South; second, Raúl Zibechi's understanding of the urban peripheries as crucial territories of resistance to neoliberal capitalism through communities engaging in collective work; and Martín Arboleda's work on planetary urbanization and the metabolic interdependence between built and unbuilt environments and city and non-city, thus surpassing the recently dominant epistemology of urban studies, where cities were considered the only morphological embodiment of urbanization.

This conclusion aims to generate a synthesis of the different conceptual approaches and case studies from a wide range of countries in Latin America through the perspective of peripheral urbanization. The first part of this chapter summarizes the key trends of urbanization in Latin America as identified by the authors of this book. In the second part, we present two propositions that are both the major conclusions of this investigation, and guidelines for a future research agenda. The first point is that we must understand urbanization processes in Latin America in the twenty-first century as peripheral in a world-systemic sense, that is, as still dependent on the specificity of the region's integration into the capitalist world system. This signals the continued relevance, but also the need for revision, of dependency theory, a hitherto highly productive theoretical agenda in and from Latin America. Our second main point is that to overcome structuralist and Eurocentric accounts and decolonize urban theory (Farrés & Matarán, 2014; Vainer, 2014), we must take seriously the agency of the ordinary, subaltern and insurgent people and the specific urban forms and life they produce, as well as the situatedness of knowledge. In the outlook of this conclusionary chapter, we argue that the concept of peripheral urbanization has the potential to do so, as it is a relational concept that shifts attention to places, processes, practices, and bodies of knowledge that have often been neglected in urban theory in the past – the "peripheral" in all its senses – but are fundamental to understanding urbanization and contesting hegemonic knowledge on the urban and urbanization.

Current Trends of Urbanization in Latin America

Inclusion of Suburban Spaces into the Financialized Capital Cycle

The contributions in part 2 of this book show that the ways in which urban peripheries are integrated into capitalist relations of production have been under transformation. They reveal the specific forms and manners in which peri-urban spaces and the population of the Latin American *hábitat popular* (Connolly, 2013) have been included into financialized accumulation processes and the new economies of extractivism in Latin America.

First, urban peripheries in Latin America are deeply heterogeneous and fragmented spaces that are increasingly shaped by global finance capital and real estate companies. Across Latin America there has been an expansion of new forms of settlements in urban peripheries. These have not come into existence in the "conventional" way established since the period after the 1950s – that is, the purchase or occupation of land and autoconstruction by the low-income population – but through the activities of private capital backed by neoliberal housing and land policies of the World Bank and national governments (Rolnik, 2013; Lukas, 2014; Sanfelici & Halbert, 2015; Soederberg, 2015). The contributions of Salazar, Reis and Varley on Mexico City and Cáceres on Santiago de Chile show that private actors have had a major impact on urban space production since the end of the 1990s. While fenced residential areas for the wealthy have been a ubiquitous phenomenon in Latin America since the 1980s (Caldeira, 2000), the novelty of the recent phenomenon is that (a) finished, often uniform homes in ready-made "urban" neighbourhoods are sold to the customer; (b) developments are located in the urban periphery, often in far distance from public traffic infrastructure; (c) sellers of these homes target the low- and middle-income population; and (d) the new housing policies do not only involve a privatization of planning and construction, but the financialization of the supply and demand side of housing, that is, the financing of real estate actors through international financial markets and the securitization of mortgages backed by the state. The withdrawal of the state from housing policies has come along with the promise of the private sector to provide the working classes with homes in safe neighbourhoods, with good infrastructure and a modern urban lifestyle. In Chile, this promise has only partly been fulfilled. Most importantly, Cáceres shows that the higher residential standard of the inhabitants of satellite towns in the periphery of Santiago is only of precarious nature: it comes at the price of increased financial vulnerability, and many cannot afford to hold on to their new homes in the face of high prices for privatized basic urban infrastructure. In Mexico, the quality of life in the new suburban *conjuntos urbanos* in terms of urban infrastructure is often worse than in autoconstructed areas. Moreover, millions of houses stand empty as inhabitants could not afford to pay back the credits they had taken on for overpriced homes. Salazar, Reis, and Varley show that the presence of private real estate in Mexico's urban periphery has severely increased land prices on the informal market, on which the low-income population still relies. This has contributed to the increase of autoconstruction in other areas of the megaregion of Mexico City.

Second, the contributions by Salazar, Reis, and Varley, and Mason-Deeze indicate that the urban poor have become included into extractive forms of capital accumulation by the financial sector. The financialization of popular life and economy marks one of the latest frontiers of the extractive operations of capital in Latin America and must be seen as part of the new logics of extractivism (Gago & Mezzadra, 2017). In the Mexican case, the financialization of housing policy has delivered immense and secure profits for international financial investors over more than a decade, which were based on the extraction of wealth from the working class through usurious credit contracts. The contribution by Mason-Deeze reveals that finance capital has also entered Argentina's consolidated autoconstructed areas. In the working-class neighbourhoods of Buenos Aires, the financial sector has successfully taken over community organized solidarity economies that had emerged in the urban periphery after the last financial crisis through the expansion of consumer credit to informal workers. This has increased the dependency on credit to meet their basic needs, made their livelihoods highly vulnerable, and created conflicts among communities and within families. Mason-Deeze argues that this process must be seen as part of the economies of extraction (Gago & Mezzadra, 2017) under neoliberal capitalism in Latin America.

Overall, these contributions show that urban peripheries and their residents have become inserted into new forms of consumption through credit. The ground for this inclusion of the popular classes into financialized capital accumulation was set by the peripheral states' failure to supply adequate housing, urban infrastructure, and consumer goods to the majority of the population. As we argue further below, this must be seen in the context of external dependency of the Latin American economies. Moreover, the contributions point to a new role of the state, which has been reconfigured to facilitate and back the activities of finance capital in order to extract value from the urban poor.

Urban Peripheries as Sites of Counter-Hegemonic Politics

Recent research has emphasized that the production and governance of urban space in Latin America is very much based on the practices and strategies of its residents, an aspect that has often been overlooked in overly political economy-focused approaches (Crossa, 2009; Gilbert & De Jong, 2015). The authors in part 3 of our book contest the idea that the homogenizing and fragmentizing processes induced by global political-economic structures are consolidated. They show that urban development is contested not only in terms of the fragile legitimacy of neoliberal urban policies, but also regarding actually existing counter-hegemonic practices from the grassroots. Several of our contributions focus on the agency of residents, laying emphasis on the opportunities for political deliberations that continue to exist within or despite the capitalist urbanization process. In particular, our authors emphasize the specific role of urban peripheries as sites of counterhegemonic practices.

Golda-Pongratz's perspective is that of an urban planner. Following the thinking tradition of John Turner, she argues that a key task of urban planning consists of supporting local initiatives in the building up of a community spirit through the reconstruction of urban memory, understood through the concept of the "palimpsest" – the multi-layered historical traces inscribed in self-built urban space. She explores this possibility in the periphery of Lima, where rural Andean traditions of shared economies and mutual aid continue despite neoliberal urban policies and global cultural influences. Here, teachers and social workers seek to re-connect local residents to the space they inhibit through old footage and interviews with original rural migrants on the self-construction of their neighbourhoods. This process raises awareness for the achievements of former generations who autoconstructed the area, reconnects the younger generations with their territory and carries the potential to re-define the identity of Lima Norte. Hence, the ultimate aim of such an approach to urban planning relates to the re-activation of the urban periphery's community-based heritage as a motor for urban renewal from the grassroots.

While much of the political dynamics and community spirit in Lima Norte has been lost over the years in an already established urban area, Brazil's recent history is characterized by a new scale and dynamics of organized land occupations for housing. Tonucci and Castriota show that in spite of their precarious condition, recent occupations are characterized by practices of community sharing and solidarity, collectively producing new commons and thus, pointing towards alternative urban modernities. These involve squares for community meetings, markets and festivities, community kitchens and daycare facilities, and common gardens, all of which often transcend the boundaries of public/private space. However, Tonucci and Castriota emphasize that in order to become real alternatives to capitalist urbanization – most manifest in the processes of regularization and land titling, followed by increasing land prices, land speculation, and the loss of the commons, as also exemplified by the history of Lima Norte – broader structural changes such as the legal recognition of collective property are needed. The challenge for urban social movements is thus to institutionalize *commoning* practices on a higher level.

The case studies by Moeller and by Schiavoni and Felicien indicate that there are already successful examples of enduring practices of commoning in Latin America. In particular, the two cases show the crucial role of agrarian and food distribution practices in the institutionalization of urban solidarity economies. Moeller emphasizes the relevance of persisting precolonial practices of peri-urban dwellers in Bolivia for what she terms "peripheral resistant adaptation." In line with Zibechi (this volume), she argues that urban peripheries carry specific potential for resistance due to the enabling conditions of informality. This is illustrated by the Indigenous *ayllu* practices of peri-urban residents in Cochabamba, which are based on communal land ownership and the maintenance of reciprocal relationships between urban and rural kin. Moeller describes how the urban periphery is characterized by a fragmented mix of urban and rural worlds, including agrarian production, industrial stockpiling, and residential and commercial exchange. Overall, she argues that peri-urbanism has the potential to overcome the urban/rural split that is so fundamental to capitalist urbanization, and that Andean heritage-based practices play a key role in the success of such "peripheral resistant adaptation" in Bolivia.

As in other Latin American cities, there are strong rural remnants in everyday practices around food in the urban peripheries of Caracas. Schiavoni and Felicien's contribution shows how these practices, while borne out of survival necessity, constitute a fertile ground for sustainably overcoming the urban/rural split and achieving food sovereignty. In Venezuela, grassroots-based practices of food production and distribution are recognized by the state through popular power laws. The creation of "communal corridors" that link up urban and rural movements organized in *comunas* facilitates the direct partnership of rural food producers and urban consumers, while *comunas* in urban peripheries of Caracas make use of their people power and infrastructure for food processing and distribution, as well as urban agriculture. In this way, the emergence of numerous cross-territorial *comunas* fundamentally challenges the longstanding urban-rural divide and may lay the ground for food sovereignty in Venezuela. Urban peripheries play a key role in this process since connections to the countryside are deeply embedded in people's collective consciousness.

In summary, the contributions of part 3 of our book indicate that urban peripheries in Latin America are sites of resistance to neoliberal capitalism: as in the past, newly occupied areas are major sites of communal organizing, but principles of solidarity economies also continue to exist in established urban areas due to heritage-based practices. These studies thus point to the case of decolonial theorists that Latin American cultures (as those of Africa and Asia) "have been partly colonized, but most of the structure of their values has been excluded – *disdained, negated and ignored* – rather than annihilated ... This disdain, however, has allowed them to survive in silence, in the shadows, simultaneously scorned by their own modernized and westernized elites (Dussel, 2012, p. 42). In particular, our authors showed that it is organizing around food supply, processing, and distribution, bridging the city and the countryside, where the logic of neoliberal capitalism may be sidestepped. Community organizing in urban peripheries has a specific potential to overcome the urban/rural split due to often still existing connections of their inhabitants with the countryside and the fragmented nature of peri-urban space, including residential areas, communal and agricultural land. As argued by Zibechi, these neighbourhoods or translocal entities identify as communities not primarily based on common property, but on their *collective work* (Zibechi, this volume). This raises the question as to which extent such forms of collective work require certain forms of institutionalization, in the state or at another societal level – in particular regarding communal land ownership.

Changing Socio-spatial Patterns of Urbanization and De-peasantization

Between 1950 and 1990, the dominant phenomenon of urbanization in Latin America that needed to be explained was metropolitanization: the fast growth of the region's metropolises through rural-urban migration (Slater, 1977). The contributions in our book – especially but not only in part 4 – show that the new pattern of urban growth in Latin America involves the growth of medium-sized cities on the one hand, and the emergence of mega-regions on the other. Traditional metropolises continue to grow, but small- and medium-sized cities exhibit much higher growth rates throughout Latin America (UN, 2014, 15ff.). As discussed in the contributions by Arboleda, Barbieri and Ojima, Durán et al., and Lukas, this new pattern has emerged in the context of the new natural resources extractivism that has had a severe impact on Latin American economies since the 1990s, particularly in South America (Gudynas, 2009; Acosta, 2013; Svampa, 2019). New towns have emerged and formerly small towns in areas of natural resources extraction have turned into sizeable cities within a very short period of time, as exemplified by the Ecuadorian case. Durán et al. establish a clear connection between the "explosive urban development" and agro-extractivism in the Ecuadorian coastal region during the neoliberal period. They point out that land concentration and the dispossession of smallholders are major causes driving the migration of formerly rural population to the new urban centres. These towns and cities have emerged within regions characterized by the renewed upswing of the "traditional" banana industry, but especially with the push for the expansion of other industrially cultivated crops such as oil palm, corn, and teak. The neoliberalization of agricultural policy and trade liberalization played a key role in the re-concentration of land in favour of the agro-industry. Many small- and medium-scale farmers and cooperatives were forced to give up cultivation in the course of accumulated debt with exporting companies. Moreover, as pointed out by Arboleda, the expansion of agro-industrial monocultures as well as open-cast mining in Latin America is often facilitated by the violent dispossession of smallholders, which is enacted through armed private security forces and a militarized state. Increasing violence in the countryside thus plays a significant role for urban growth in South America.

It is important to note that there are similarities between the South American cases and the Mexican case. Mexico, Central America, and the Carribean play a different role in the new international division of labour (Grinberg, 2010). Natural resources extraction has also significantly expanded in this region (Bury & Bebbington, 2013, p. 43f.), but the recent economic history has primarily been characterized by a massive growth of the *maquila* or *maquila*-like manufacturing industry, particularly in Mexico. This industry is based on the exploitation of a cheap and disciplined labour force for the production of textiles, electronics, and other consumables, mostly for export to the US. However, most of the needed inputs are imported[1] and the enterprises are often transnationalized. Hence, there are little spin-off effects for the host country, and a large share of profits does not remain in the country (Reis, 2019). As in the case of primary resources extraction, these economic structures are constituted spatially. In her case study on Motul, a medium-sized city in Southern Mexico, Fonseca shows that the urbanization process coming along with the development from an agriculture- to a manufacturing-based economy in the global periphery is not only characterized by urban growth per se, but by the comprehensive expansion of abstract space as conceptualized by Lefebvre (1991): Even if industrial facilities have not been built (yet) and industrial parks stand almost empty in many parts of Yucatán, the peninsula has transformed into a territory where dispossession, environmental degradation, and the lack of basic public infrastructure in the new urban peripheries is normalized for the sake of the yet-to-come maquila industry.

While the socio-spatial patterns of growth have shifted from a unilateral growth of metropolises to smaller cities, this has not solved the main problems associated with urban growth in Latin America since the 1950s. Several contributions in this book (e.g., Barbieri & Ojima; Durán et al.; Fonseca; Lukas; Salazar, Reis, & Varley) show that recent urban space production has mostly happened in the same way as in the metropolises, that is, through processes of land occupation and autoconstruction. As in the older urban peripheries, these areas often lack urban infrastructure, in particular adequate access to water and sanitation. Their inhabitants are those most affected by the environmental degradation coming along with natural resources extractivism and industrial export production zones, like the contamination of air, water, and soil, causing serious health damages. As Barbieri and Ojima point out, the population in newly emerging urban peripheries in Latin America is carrying a double health burden: higher infant mortality and high maternal mortality and morbidity caused by environmental degradation; and the increasing prevalence of chronic diseases such as diabetes, associated with an ageing population. The effects of recent neoliberal policies and reductions in public spending due to austerity measures – the cutting of labour rights and budgets for cash transfer programs, education, health, and environmental protection programs – is exacerbating this situation. Strikingly, this is not only the case in autoconstructed areas, but also in "formally" produced housing, as shown by Salazar, Reis, and Varley. Mexico's social housing policy, tied to international financial markets, has produced millions of new homes for the working class, but in many cases the infrastructural situation is comparable or worse to that in areas of autoconstruction. Yet, as shown by Lukas (this volume), the new emergence and growth of cities related to extractivism also follows another logic than that of autoconstruction. In the Chilenean city of Antofagasta, a major global hub in the Global Production Network of the copper industry, multinational mining corporations have entered the sphere of urban planning with the aim of transforming Antofagasta into a global "model city." The contribution argues that this is part of the "strategic coupling" (Coe & Yeung, 2015) of urban space and institutions to the needs of lead firms of the extractive industries. By turning urbanism into spectacle and technocratic procedures, it de-politicizes the violent nature of planetary urbanization and embodies the privatization of urban governance. Hence, while we are seeing the return of land takings and the growth of marginal settlements in the context of de-peasantization, violent dispossession and extractivism on the one hand, peripheral urbanization in Latin America is also characterized by the increasing participation of private corporations in urban planning (also see Salazar, Reis, and Varley; Caceres, this volume).

The growth of small- and medium sized cities in Latin America is linked to another dimension of extractivism: the massive expansion of large infrastructural systems throughout the continent to facilitate the circulation and export of raw materials and manufactured products, including roads, airports, seaports, energy, and water infrastructure. As pointed out by Arboleda, primary commodity production is a main driver of the "explosion of space," whereas cheap labour extractivism in Mexico can be added to the drivers of this process (see the contribution by Fonseca). The case of the Ecuadorian coastal zone, situated on the northwest of the "Initiative for the Integration of the Regional Infrastructure of South America" (IIRSA, see the contribution of Arboleda) project, illustrates how small- and medium-sized cities fulfil certain functions in these infrastructural networks for resources extraction. Durán et al. show how these emerging cities function as mediators between the flow of goods, information, innovation, administration, urban-rural relationships on the one hand, and integrate local, regional, national, and global relationships on the other. Observing the developments in other parts of the continent, such as the regions of Mexico City and Lima (see Salazar, Reis &, Varley; Golda-Pongratz), this gives rise to the hypothesis that the mega-region as "polycentric agglomeration of cities and their lower-density hinterlands" (Florida et al., 2008, p. 5) constitutes one of the spatial dimensions of neoliberal capitalism not only in the global economic centre regions of North America, Europe and China, but also in the global periphery of Latin America (cf. Harrison & Hoyler, 2015). Arboleda takes up this phenomenon by arguing that "the explosion of space" in Latin America is characterized by a contradictory tension between spatial homogenization (in the form of multiscalar state spatial strategies) and territorial fragmentation (in the form of fixed capital allocations and state-led spatial segregation). Hence, "urban peripheries" are increasingly located not only at the margins of metropolises, small- and medium-sized cities, but as nodes within the various urban centres of emerging mega-regions.

Moreover, our authors have shown that the urbanization of the Latin American countryside is progressing economically, culturally, and politically. First and foremost, this is manifest in the increasing integration of urban and rural labour. Since the 1970s, the majority of households in the Latin American countryside have undergone a transition from a primary dependency on traditional family agriculture to a livelihood structure based on multiple income sources that transcend the urban-rural divide (see Barbieri & Ojima, and Moeller, this volume; Kay, 2008). The pace of depeasantization has increased with the neoliberalization of agrarian policies and the renewed surge of extractivism since the 1990s. As Barbieri and Ojima show for the case of Brazil, this process has recently occured in both the Amazon, where agro-extractivism is rampant, and the peripheral northeast of Brazil, which is characterized by unfavourable environmental conditions for agriculture. The expansion of agribusiness and urban infrastructure has created new off-farm employment opportunities in the Amazon. However, as Kay (2008) warns, migration and the need to engage in multiple activities, including wage labour, are first and foremost demonstrations of the precarious rural livelihoods and the need to survive in the context of increasing neoliberalization. In Mexico, the urbanization of the countryside is closely tied to the abandonment of subsidies for small-scale agriculture since the 1990s (see the chapters by Torres-Mazuera and Salazar, Reis, & Varley). Even in the peri-urban areas of Mexico City, many who can afford to hold on to their land continue to cultivate maize as an insurance strategy due to the instability of wage labour income (see Salazar, Reis, & Varley). Others who have migrated to cities derive part of their income out of wage labour in the countryside, that is, in the agro-industry or in the mining sector, as it is the case in Ecuador's coastal zone (Durán, et al.) and in the Chilean desert (Lukas). However, the increasing integration of the rural workforce into wage labour throughout Latin America has not necessarily been going along with substantial increases in living standards. As Barbieri and Ojima point out, a large percentage (around one-third in the Amazon) of the population in the Amazon and the northeast of Brazil depends on the government's (conditional) cash transfer programs – a trend visible for the region as a whole. In Latin America and the Caribbean, 21.5% of the population lived in households receiving such transfers in 2013 (ECLAC, 2015, p. 70).

As the contribution by Torres-Mazuera shows, access to higher education plays a major role in what she terms "urbanized rurality." While the older generation in Mexico's ejidos still cultivates maize, household income sources in the Central Mexican village she studies are largely based on the younger generation's economic activities in non-agricultural sectors in other areas or outside of Mexico, and locally, on "professional" employment, that is, as teachers, nurses, police officers, physicians, lawyers, or – not last – local state officers. Torres-Mazuera argues that this transformation of livelihoods has given rise to a power shift in rural political institutions, from the ejido, which was based on a rural peasant identity, to the municipal authority, which is based on a population identifying as "professionals." She discusses how the main task of the local government consists of the provision of consumption goods for the urbanized rural population. Similar to the Brazilian case, this is especially related to the provision of social goods for women, youngsters, the elderly, and Indigenous groups on the one hand, and the provision of urban infrastructure such as health, education, and housing to the general public on the other.

Overall, the contributions in part 4 have shown that urbanization in Latin America constitutes a complex and contradictory process of simultaneous homogenization and fragmentation. On the one hand, urbanization results in homogenization: the urban has become an all-embracing condition throughout Latin America. This is especially manifest in the urbanization of livelihoods, emerging in the context of export-oriented development models linked to an intensified process of depeasantization and the subsequent expansion of small and medium-sized cities and towns. However, economic growth is often jobless, and thus the neoliberal period in Latin America has come along with a massive growth of the informal sector. Moreover, migrant remittances, (conditional) cash transfer programs, and consumption credit have come to play a key role in the reproduction of the labour force in many countries. Hence, one of the most important characteristics of the recent phase of urbanization has been the transformation of the state-society relationship. In the period of ISI-led developmentalism, between 1940 and 1980, the state attempted to slow down the dismantling of small-holder agriculture through a system of guaranteed prices, subsidies, land reforms etc., and the stability of rural livelihoods still played a key role for political legitimacy. While this was only partly successful already at the time, resulting in increasing rural-urban migration and metropolitanization, the disintegration of rural livelihoods took on a new quality since the 1990s. In a context of jobless growth in the export economy and a far-reaching destruction of the local economy, a major role of the neoliberal capitalist state for ensuring the reproduction of the labour force consists not only in the provision of urban infrastructure for the growing urban population, but in the facilitation of consumption. The latter has particularly been enabled by the state through massive-scale cash transfer programs on the one hand, and the vast inclusion of Latin America's popular sector into relations of credit and debt. Consequently, three decades of neoliberal policy in Latin America have forcefully contributed to shifting the battleground for political legitimacy from the rural to the urban.

On the other hand, the paradox of urbanization is that it also implies increasing heterogenization and fragmentation: The expansion of industrialized agriculture and the "emptying" of rural areas of peasant population comes along with an increasing rural/urban split. As Merrington noted on the history of capitalist development in Europe, it involves "a conquest over the countryside, which becomes "ruralized," since it by no means represented in the past an exclusively agricultural milieu. From being a centre of all kinds of production, an autonomous primary sector that incorporates the whole of social production, the country becomes "agriculture," that is, a separate industry for food and raw materials, separated in turn into various specialized types of farming, districts etc." (Merrington, 1975, p. 72). Recent development in Latin America has borne witness to Marx and Engels's insight that the increasing rural/urban split is a crucial expression of capitalist development. In Latin America, the countryside becomes not only "agriculture," but more broadly, the territory of extraction, including (potential) spaces of natural resources and cheap labour extraction.

Peripheral Urbanization and the De-colonization of Urban Theory

What can we learn with regard to urban theory based on our insights of urbanization processes in Latin America? And how useful is the concept of peripheral urbanization in order to explain (and contest?) urbanization in Latin America and push forward the agenda of decolonizing urban theory? We believe that two major conclusions can be drawn.

First, the contributions in our book point to the continued relevance of Ánibal Quijano's insight that both the territorial and the social dimension of urbanization in Latin America cannot be understood outside of the historical specificity of the region's integration into the capitalist world system, that is, its peripherality (Quijano, 1968). In the 1970s, authors such as Manuel Castells, David Slater, John Walton, and Jeffrey Kentor coined the concept of peripheral urbanization in reference to dependency theory, explaining the phenomenon of metropolitanization, that is, the enormous interregional, rural-urban and intra-city inequalities emerging in twentieth century Latin America, as a result and function of the expansion of the dependency relation (Castells, 1977, p. 39ff.; Slater, 1977; Walton, 1977; Kentor, 1981). Similarly, the trends and patterns of urbanization identified in this book must be analysed in connection with the specific, subordinated role of Latin American economies in global capital accumulation. Since the 1970s, economic development in Latin America has increasingly been shaped by international financialization and the pressures emerging in the context of the 1980s debt crises. In the course of the subsequent so-called debt restructuring programs, international creditors enforced far-reaching neoliberal reform packages including, for instance, the dismantling of state support measures for small-scale agriculture, the privatization of state housing agencies and policies, the opening of capital accounts and the free convertibility of currencies, and the support of export-oriented economic policies. Due to the dependency of national currencies on international capital flows, autonomy on policy-making in Latin America has been seriously constrained by the potential effects of policy reforms on the movement of such flows (López González, 2013; Reis, 2019; Alami, 2018). Massively increased and volatile capital flows, combined with a lack of government regulation of the activities of banks and financial institutions, caused devastating financial crises in the 1990s. In dealing with these crises, the political pathways of Latin American states can be broadly divided into two categories: on the one hand, those countries that have continuously conformed to the requirements of Western-dominated multilateral and financial institutions and adapted their economic and financial policies to their needs, such as Mexico, Brazil, Chile, and Peru; and on the other hand, those who, with the help of strong social movements, put left-wing governments to power in the beginning of the 2000s, with the aim of opposing external dependency, such as Ecuador, Bolivia, and Venezuela. However, so far none of the two ways has been successful in breaking external dependency. Quite the contrary, debt levels are at a record high and have yet again massively increased with the COVID-19 crisis, whether debt is primarily with international financial markets (as in the case of the first group) or with Chinese financial institutions (as in the case of the second). In this context, all over Latin America, the neoliberal period has seen the breakthrough of export-oriented extractivism to generate increasing capital inflows, geared to the "[extraction] of as much as possible of high-demand resources (be it land, water, minerals, forests, agricultural products, oil reserves, cheap and disciplined labour, or others) at lowest cost within shortest period of time" (BICAS, 2016), and facilitated by huge investments into large-scale infrastructure for transport, energy and water. Moreover, pushed by the World Bank/ Wall Street agenda of "financial inclusion," many countries have seen massive increases in household indebtedness. As financial investors see the poor as reliable payers of high interest, and thus an investment vehicle resilient to economic crisis (Soederberg, 2013, p. 606), it is to be feared that the COVID-19 crisis will lead to the further ratcheting up the power of finance capital over the poor (Reis, 2020).

The dispossession of rural population, the urbanization of the countryside, the immense growth of small- and medium-sized towns and cities – that is, the enormous increase of the industrial reserve army – the often devastating living conditions in the new urban peripheries, and the extraction of value from the urban *clase popular* through the spread of consumption culture financed by credit ("financial inclusion") must thus be seen in the context of the specific role of Latin America and its working classes in changing global modes of capital accumulation. This indicates that urbanization in Latin America is *peripheral,* in the sense that it is conditioned (not determined!) by processes and decisions taken in the centres of global capital accumulation, and by the continuing extraction of wealth from the peripheral countries of the South by the countries of the centre. This points to the continued relevance of dependency theory and the original formulation of the peripheral urbanization concept for the analysis of urbanization in Latin America. A key insight of the Marxist strand of dependency theory was that the hierarchical relation between centre and periphery not only manifests as a macroeconomic phenomenon, but first and foremost in the dependency of the internal class relations in the periphery, hence at the level of relations of production (Quijano, 1968, p. 80; Marini, 1973; Bambirra, 1978). One of the tasks of contemporary urban theory is thus to engage with the latest debates on dependency in Latin America (see for instance Féliz, 2016; Osorio, 2016; Sotelo Valencia, 2017; Antunes de

Oliveira, 2019; Reis & Antunes de Oliveira, forthcoming) and examine the co-production of dependency and urbanization processes. In particular, such efforts must draw attention to the ways in which Latin American ruling classes have responded to new global geopolitical conditions and modes of accumulation, and how this manifests territorially. A case in point is the new power of Latin American real estate firms and holding companies over urban space production (see Cáceres, this volume; Lukas, Fragkou, & Vasquez, 2020; Lukas, this volume).

At the same time, our second main insight is that if analyses of urbanization are to be meaningful for emancipatory social struggles in Latin America and elsewhere, there is a need to overcome the flaws of structuralist, capital centric and Eurocentric theories of peripheral urbanization (see introduction). As postcolonial urbanists have pointed out, we must pay attention to the everyday practices that people employ to manage their existence, and the kinds of urbanism that this "urban collective life" produces (Bhan et al., 2020), to overcome Eurocentric ontologies and epistemologies. In Latin America, the existing and diverse community-based practices investigated by our authors show that despite the imposition of financialization and extractivism, hegemonic control over space is not absolute. Notwithstanding the neoliberal logics of capital accumulation and the World Bank's agenda of "entrepreneurialism," which have attempted to absorb autonomous forms of urban development for decades (Davis, 2006, p. 72), the subaltern continue to produce new urban space and forms of urban life through "occupancy urbanism" (Benjamin, 2008) and/or everyday practices (Crossa, 2009). Whether these practices have counter-hegemonic intentions or consequences, or generate themselves exclusionary and oppressive social forms (cf. Harvey, 1996, p. 59), and/or actually contribute to the stability and legitimacy of the capitalist state (Bayat, 2000, p. 545), are key questions for critical research. However, while ordinary people resist hegemonic institutions, occupy land, create new *commons*, and revive old or build up new institutions and infrastructure for overcoming the urban/rural split to gain food sovereignty, in the Latin American context it is crucial to note that colonial and pre-colonial practices, modern and non-modern, solidary and exploitative relations, and formal and informal structures overlap, together forming "motley" geographies (Gago, 2017). Latin American societies are "motley" in being characterized by "the parallel coexistence of multiple cultural differences that do not extinguish but instead antagonize and complement each other. Each one reproduces itself from the depths of the past and relates to others in a contentious way" (Rivera Cusicanqui, 2012, p. 105). As several of our authors have pointed out, urban peripheries play a key role in the vindication, mobilization, and reinvention of heritage-based practices due to the continuing presence of collective memories and the dominance of informal social structures (see Zibechi, Moeller, Golda-Pongratz, and Schiavoni & Felicien). A second major task of critical urban studies is thus to engage with the question of if and how the insights of the different strands of scholarly traditions and agendas in critical urban studies can be integrated, to account for both the power of global structures and the local specificity of urbanization in Latin America and the Global South, considering the situatedness of knowledge. In the Latin American context, this especially means to engage with the contributions of decolonial theory and its critique (cf. Kogan Valderrama, 2020).

Outlook

In the introduction, we identified three broad, partly overlapping, theoretical approaches or scholarly agendas in Latin American urban research that have guided this book:

1) an approach inspired by Marxist political economy, analysing urbanization processes in Latin America in relation to the global capitalist system,
2) a postcolonial approach, focusing on the role of everyday practices and politics of the subaltern in the production of urban space and life, and questioning the conditions of knowledge, and
3) an approach of global sub/urbanisms, arguing for an empirical and analytical focus on urban peripheries in order to understand current processes of urbanization.

The challenge that arises for urban theory is if and how these various strands of critical urban theory can be brought together to learn from and enrich each other, to inform our understanding of urbanization, urban life, and urban social movements in but also beyond Latin America. We suggest that the concept of peripheral urbanization is well positioned to open up such a debate, as notions of "periphery" and "peripheral" are present in all three strands of critical urban theory. One of the strengths of the concept of peripheral urbanization derives from its ability to explore the interlinkages, contradictions, and complementarities between the different conceptualizations of the peripheral. For example, the global sub/urbanisms approach views the peripheral as the space distinguished from the urban centre, while arguing that peripheral places "are, in many ways, the hotbeds of new political ruptures and eruptions themselves" (Keil, 2018, p. 197). Similarly, recent approaches inspired by critical political economy such as that by Arboleda (this volume) emphasize the urgency to look at those territories outside the city to understand urbanization, while the role of transnational capital in the homogenization and fragmentation of these spaces has so far been methodologically and analytically prioritized over the role of everyday practices of urban people in the production of these spaces. This stands in contrast to the conception of the peripheral by some authors in the postcolonial tradition. For instance, Simone refers to the peripheral not as the physical space outside or on the margins of the city but rather as a "range of fractures, discontinuities of 'hinges' disseminated over urban territories" (Simone, 2010, p. 45). In general, there is a tendency in postcolonial urbanism to refrain from spatial understandings of the peripheral and conceptualize it as a "way of producing space that can be anywhere," and in which the practices and the knowledge of the subaltern play a key role (Caldeira, this volume).

Across the different groups of scholars studying peripheral urbanization at present, there are, thus, those that tend to take on a more actor-oriented approach, focusing on the concrete local practices of urban space production that have often been overlooked in the past, and those that have, generally, a more structure-oriented approach, looking at how capital unfolds unevenly across space. We believe that peripheral urbanization can be useful as a boundary concept for urban theory if we understand the various meanings of the peripheral as referred to by the different disciplinary abstractions and methodological approaches not as mutually exclusive, but as intimately related dimensions of reality that also define each other (cf. Mollinga, 2008, p. 24). For example, the socio-spatial nature of capital accumulation in the periphery (both in a topological and a relational sense) not only shapes forms of resistance to these, but also (and perhaps especially) the cultural and material aspirations of many people who strive for a lifestyle that matches Western ideas of urbanized modernity and is to a large part equated with the ability to consume. At the same time, postcolonial urbanists' emphasis on local specificity, agency, possibility, and different forms of knowledge should remind those focusing on understanding the socio-spatial mechanisms and manifestations of capitalist development in the periphery that much urban space and many ways of urban living are not exclusively produced by the logic of the capitalist market, but by different societal logics. Decolonial theory is of particular relevance here, as it is further advanced in resolving these tensions, but has been hardly explored in urban studies so far. Building on the most elaborated versions of dependency theory, which emphasized the significance of internal power structures in the periphery for understanding peripheral (and global) capitalism (Marini, 1973; Bambirra, 1978), decolonial theory puts emphasis on the multiple hierarchies of class, gender and race that were imposed on peripheral societies with their colonization, and the mutual conditionality of modernity and coloniality (Grosfoguel, 2011; Quijano, 2000; Dussel, 2012). In the words of Ramón Grosfoguel, "what arrived in the Americas in the late fifteenth century was not only an economic system of capital and labor ... but a broader and wider entangled power structure that an economic reductionist perspective of the world-system is unable to account for" (Grosfoguel, 2011, p. 8). From this perspective, we propose the future agenda of reformulating the concept of peripheral urbanization based on an understanding of peripherality through the concept of the "colonial power matrix" (Grosfoguel, 2011). The "European modern/colonial capitalist/patriarchical world-system" (Grosfoguel, 2007, 2011) is closely related to the production of urban space and the construction of knowledges about the periphery through the binaries of modernity (Farrés & Matarán, 2014, p. 37; Patel, 2016, p. 7).

The argument we are putting forward by employing the boundary concept of peripheral urbanization is that while there remains the risk that the different ontological and epistemological propositions of the various strands of critical urban theory cannot always and easily be united, critical urban theory as a whole should aim to look for the complementarities between the different scholarly perspectives and agendas. After all, the common ground shared by all authors contributing to this volume is the shift of attention to places, processes, practices, and bodies of knowledge that have often been neglected in urban theory in the past – the "peripheral" in all its senses – but are nevertheless fundamental to understanding urbanization and contesting hegemonic knowledge on the urban and urbanization. Therefore, in our view the decolonization of urban theory (Farrés & Matarán, 2014; Vainer, 2014) consists in the development of theory based on the empirical analysis of everyday experiences and practices of urban (rural/urban) dwellers *in relation* to the multi-scalar territorial processes that are shaped but not determined by the colonial power matrix in the twenty-first century. This requires a combination of the rich spectrum of methodologies available to social research, including quantitative and qualitative methods and macro as well as micro perspectives such as political economic analyses, class analyses and ethnography; and, not least, adopting a "decolonial attitude"

(Farrés & Matarán, 2014, p. 8). In this way, it may become possible to grasp the "stratified" reality (Sayer, 1992, 118ff.) of urbanization and integrate hitherto largely separated thinking traditions.

Moreover, we believe that the potential of the concept of peripheral urbanization lies in the possibility of a reconnection to and revitalization of the Latin American paradigm of the *hábitat popular*, referring to "any type of construction where 'the people' (*el pueblo*) or the poorer classes of society live" (Connolly, 2013, p. 514). This would pave the way not only for the revitalization of a historical stage of academic research "where ontological, epistemological and methodological positions predominated that were compromised to the transformation of the social reality and where theoretical debates mattered beyond the academic curriculum" (Connolly, 2013, p. 551, translation by the authors). It would also facilitate the reconnection between the Latin American/ Spanish-speaking and Anglophone (and also the German- and French-speaking, for instance) scientific debates on urban development, which has substantially weakened with the fall of the "Grand Theories" in the end of the 1980s despite the bloom of postcolonial thinking in the Anglo-Saxon academic world (Connolly, 2013, p. 543ff.), and which is urgently needed today.

Finally, by choosing the distinctively relational concept of "peripheral urbanization," emphasis is given to the inherently global nature of the capitalist process of urbanization and the need to "learn from the South" if we are to understand and contest capitalist development in South *and* North. As David Slater concisely pointed out three decades ago, "in so much of today's critical geography, the West is implicitly viewed as a self-contained entity, as if somehow it can be apprehended and comprehended of and by itself. There appears to be little realization that it is in the outskirts that the system often reveals its true face, or as Cocks (1989, p. 4) defines it, in her book on the "oppositional imagination," "the political advantage in looking at peripheries and extremities is that power is exposed in what it drives from the center of life to the edges, and in what it incites as its own antitheses" (Slater, 1992, p. 314).

In this book we began to summarize the Latin American tradition of theorizing urbanization from the margins. Our authors reflected on the transformations of urban hinterlands, the production of urban space through neo-extractivism, and the urbanization of the countryside. By using the term "peripheral urbanization," one implicitly seems to accept the premise that besides "peripheral" urbanization, there are urbanization processes of the "centre" that differ from those in the periphery or the peripheral ones in some way. "Peripheral urbanization" is necessarily a relational term. Caldeira acknowledges this by stating that the modes of urban space production in the Global South differ from those of the North Atlantic (Caldeira, this volume). This raises a series of questions regarding the geographical reference points of the concept. With respect to this volume, it is clear that further comparative studies including different regions from the Global South or the global periphery are needed in order to identify the commonalities and differences of current forms and mechanisms of urbanization as identified in this book.

Moreover, neoliberalism has brought forth urban phenomena, above all a simultaneousness of fragmentation and homogenization of urban space, that show similarities in the Global South and North. This is evident for instance with a view on the growing gap between rich and poor, increasing segregation, and the global assimilation of consumption cultures. Large urban areas in the Global North, such as the old industrial centres of the American Rust Belt or the German Ruhr Area, are characterized by decay, widespread poverty, and an informal survival economy – and perhaps, also by newly emerging solidarity economies. With migrant farm labour camps in southern Europe and tent cities in Berlin or Los Angeles, squatter settlements without urban infrastructure cannot be said to be solely a phenomenon of the Global South. With Grosfoguel, we could thus conclude that peripheral urbanization refers, thus, to a territorial process that operates within the structures of global coloniality, which are shaped by the colonial power matrix of the twenty-first century: "In this sense, there is a periphery outside and inside the core zones and there is a core inside and outside the peripheral regions" (Grosfoguel, 2007, p. 220). Peripheral urbanization then is a relational and cosmopolitan concept and connects a spatial dimension – as in its addressing of different scales, places, and locales such as urban peripheries, operational hinterlands, and the Global South – to a political, economic and cultural dimension – which refers to the processes of peripheralization and marginalization, the ubiquitousness of the "New South" (Scholz, 2000), as well as the ubiquitousness of centre and periphery.

The questions that arise are in which way the "periphery" exists on different scales, and what can be learned from peripheral societies with respect to recent developments in the centres of the global economy. This book has been a first attempt to bridge the different theoretical and methodological approaches to studying peripheral urbanization in Latin America existing today. We hope it can serve as a basis for further empirical, conceptual, and political progress.

NOTE

1 In Mexico, 97% of inputs used in maquilas are imported (Guíllen, 2012, 68f).

REFERENCES

Acosta, A. (2013). Extractivism and neoextractivism: Two sides of the same curse. In M. Lang & D. Mokrani (Eds.), *Beyond development: Alternative visions from Latin America* (pp. 61–86). Amsterdam: Transnational Institute (TNI).

Alami, I. (2018). On the terrorism of money and national policy-making in emerging capitalist economies. *Geoforum, 96,* 21–31. https://doi.org/10.1016/j.geoforum.2018.07.012

Antunes de Oliveira, F. (2019). The rise of the Latin American far-right explained: Dependency theory meets uneven and combined development. *Globalizations* (published online). https://doi.org/10.1080/14747731.2019.1567977

Bambirra, V. (1978). *Teoría de la dependencia: Una anticrítica.* Mexico. ERA.

Bayat, A. 2000. From "dangerous classes" to "quiet rebels": Politics of the urban subaltern in the Global South. *International Sociology, 15*(3), 533–557.

Bebbington, A. 2009. The new extraction: Rewriting the political ecology of the Andes? *NACLA Report on the Americas, 42*(5), 12–20. https://doi.org/10.1080/10714839.2009.11722221

Benjamin, S. (2008). Occupancy urbanism: Radicalizing politics and economy beyond policy and programs. *International Journal of Urban and Regional Research, 32*(3), 719–729. https://doi.org/10.1111/j.1468-2427.2008.00809.x

Bhan, G., Caldeira, T., Gillespie, K., & Simone, A.M. (2020). The pandemic, southern urbanisms and collective life. *Society & Space, 3.* https://www.societyandspace.org/articles/the-pandemic-southern-urbanisms-and-collective-life

BICAS. (2016). *Agro-extractivism inside and outside BRICS: Agrarian change and development trajectories* [Call for papers]. The 4th international conference of BICAS. https://www.tni.org/es/node/23006

Brenner, N., & Schmid, C. (2015). Towards a new epistemology of the urban? *City, 19*(2–3), 151–182. https://doi.org/10.1080/13604813.2015.1014712

Bury, J., & Bebbington, A. (2013). New geographies of extractive industries in Latin America. In A. Bebbington & J. Bury (Eds.), *Subterranean struggles: New dynamics of mining, oil, and gas in Latin America* (pp. 27–66). University of Texas Press.

Caldeira, T.P.R. (2000). *City of walls: Crime segregation, and citizenship in São Paulo.* University of California Press.

Castells, M. (1977). *The urban question: A Marxist approach.* MIT Press.

Cocks, J. (1989). *The oppositional imagination: Feminism, critique and political theory*. Routledge, Chapman and Hall.
Coe, N., & Yeung, H. (2015). *Global production networks: Theorizing economic development in an interconnected world*. Oxford University Press.
Connolly, P. (2013). La ciudad y el habitat popular: Paradigma latinoamericano. In B.R. Ramírez Velázquez & E. Pradilla Cobos (Eds.), *Teorías sobre la ciudad en América Latina* (pp. 505–562). Universidad Autónoma Metropolitana.
Crossa, V. (2009). Resisting the entrepreneurial city: Street vendors' struggle in Mexico City's historic center. *IJURR, 33*(1), 43–63. https://doi.org/10.1111/j.1468-2427.2008.00823.x
Davis, M. (2006). *Planet of slums*. London: Verso.
Dussel, E.D. (2012). Transmodernity and interculturality: An interpretation from the perspective of philosophy of liberation. *TRANSMODERNITY: Journal of Peripheral Cultural Production of the Luso-Hispanic World, 1*(3). http://dx.doi.org/10.5070/T413012881 Retrieved from https://escholarship.org/uc/item/6591j76r
ECLAC (Economic Commission for Latin America and the Caribbean). (2015). *Inclusive social development: The next generation of policies for overcoming poverty and reducing inequality in Latin America and the Caribbean*. https://repositorio.cepal.org/bitstream/handle/11362/39101/4/S1600098_en.pdf
Farrés. Y., & Matarán, A. (2014). Hacia una teoría urbana transmoderna y decolonial: Una introducción. *Polis* [Online], *13*(37), 339–361.
Féliz, M. (2016). Transformations in Argentina's capitalist development since the neoliberal age. *World Review of Political Economy, 7*(3), 350–362.
Florida, R., Gulden, T., & Mellander, C. (2008). The rise of the mega-region. *CESIS Electronic Working Paper Series, 129*. https://static.sys.kth.se/itm/wp/cesis/cesiswp129.pdf
Gago, V. (2017). *Neoliberalism from below: Popular pragmatics and baroque economies*. Duke University Press.
Gago, V., & Mezzadra, S. (2017). A critique of the extractive operations of capital: Toward an expanded concept of extractivism. *Rethinking Marxism, 29*(4), 574–591. https://doi.org/10.1080/08935696.2017.1417087
Gilbert, L., & De Jong, F. (2015). Entanglements of periphery and informality in Mexico City. *International Journal of Urban and Regional Research, 39*(3), 518–532. https://doi.org/10.1111/1468-2427.12249
Grinberg, N. (2010). Where is Latin America going? FTAA or "twenty-first-century socialism"? *Latin American Perspectives, 30*(170), 185–202. https://doi.org/10.1177%2F0094582X09351713
Grosfoguel, R. (2007). The epistemic decolonial turn: Beyond political-economy paradigms. *Cultural Studies, 21*(2–3), 211–223. https://doi.org/10.1080/09502380601162514

Grosfoguel, R. (2011). Decolonizing post-colonial studies and paradigms of political-economy: Transmodernity, decolonial thinking, and global coloniality. *TRANSMODERNITY: Journal of Peripheral Cultural Production of the Luso-Hispanic World, 1*(1). https://doi.org/10.5070/T411000004

Gudynas, E. (2009). Diez tesis urgentes sobre el nuevo extractivismo. In Centro Latinoamericano de Ecología Social (Ed.), *Extractivismo, Política y Sociedad*. Quito.

Guíllen, A. (2012). Mexico, an example of the anti-development policies of the Washington consensus. *Estudos Avancados, 26*(75), 57–75. https://doi.org/10.1590/S0103-40142012000200005

Harrison, J., & Hoyler, M. (2015). Megaregions: Foundations, frailties, futures. In J. Harrison & M. Hoyler (Eds.), *Megaregions: Globalization's new form?* (pp. 1–28). Edward Elgar.

Harvey, D. (1996). Cities or urbanization? *City, 1*(1–2), 38–61. https://doi.org/10.1080/13604819608900022

Kay, C. (2008). Reflections on Latin American rural studies in the neoliberal globalization period: A new rurality? *Development and Change, 39*(6), 915–943. https://doi.org/10.1111/j.1467-7660.2008.00518.x

Keil, R. (2017). Extended urbanization, "disjunct fragments" and global suburbanisms. *Environment and Planning D: Society and Space, 36*(3), 494–511. https://doi.org/10.1177%2F0263775817749594

Keil, R. (2018). *Suburban planet: Making the world urban from the outside*. Polity Press.

Kentor, J. (1981). Structural determinants of peripheral urbanization: The effects of international dependence. *American Sociological Review, 46*(2), 201–211. https://doi.org/10.2307/2094979

Kogan Valderrama, A. (2020). Una crítica desde los territorios al giro decolonial en América Latina. Mapuexpress. Colectivo de Comunicación Mapuche. https://www.mapuexpress.org/2020/10/21/una-critica-desde-los-territorios-al-giro-decolonial-en-america-latina/

Lefebvre, H. (1991). *The production of space*. Blackwell Publishing.

López González, T.S. (2013). Deuda pública interna y estabilidad monetaria en México: El dilemma de la política fiscal con liberalización financiera. In M. Campos (Ed.), *Globalización financiera y nuevo marco institucional: Referencias obligadas para la política fiscal y financiera* (pp. 34–63). Universidad Nacional Autónoma de México.

Lukas, M. (2014). *Neoliberale stadtentwicklung in Santiago de Chile: Akteurskonstellationen und machtverhältnisse in der planung städtebaulicher megaprojekte*. Band 125 von Kieler geographische Schriften.

Lukas, M., Fragkou, M.C., & Vásquez, A. (2020). Hacia una ecología política de de las nuevas periferias urbanas: Suelo, agua y poder en Santiago de Chile.

Revista de Geografia Norte Grande, 76, 95–119. http://dx.doi.org/10.4067/S0718-34022020000200095
Marini, R.M. (1973/1991). Dialéctica de la dependencia. http://www.marini-escritos.unam.mx/024_dialectica_dependencia.html
Merrington, J. (1975). Town and country in the transition to capitalism. *New Left Review I/93*, September/October 1975, 71–92. https://newleftreview.org/issues/i93/articles/john-merrington-town-and-country-in-the-transition-to-capitalism
Mollinga, P. (2008). The rational organisation of dissent: Boundary concepts, boundary objects and boundary settings in the interdisciplinary study of natural resources management. *ZEF Working Paper Series, 33*. https://www.zef.de/uploads/tx_zefportal/Publications/wp33.pdf.
Osorio, J. (2016). *Sistema mundial, intercambio desigual y renta de la tierra.* Ciudad de México: Universidad Autónoma Metropolitana.
Patel, S. (2016). A decolonial lens on cities and urbanisms: Reflections on the system of petty production in India. *Asia Research Institute*, ARI Working Paper No. 245.
Pradilla Cobos, E. (1982). Autoconstrucción, explotación de la fuerza de trabajo y políticas del Estado en América Latina. In E. Pradilla (Ed.), *Ensayos sobre el problema de la vivienda en América Latina.* UAM-Xochimilco.
Quijano, A. (1968). Dependencia, cambio social y urbanización en Latinoamérica. *Revista Mexicana de Sociología, 30*(3), 525–570.
Quijano, A. (2000). Colonialidad del poder, eurocentrismo y América Latina. In E. Lander (Ed.), *La colonialidad del saber: Eurocentrismo y ciencias sociales: Perspectivas latinoamericanas* (p. 246). CLACSO.
Reis, N. (2019). A farewell to urban/rural bias: Peripheral finance capitalism in Mexico. *The Journal of Peasant Studies, 46*(4), 702–728. https://doi.org/10.1080/03066150.2017.1395856
Reis, N. (2020). The "Bank of Welfare" and Mexico's moral economy. In D. McDonald, T. Marois, & D. Barrowclough (Eds.), *Public banks and COVID-19: Combatting the pandemic with public finance* (pp. 171–192). Municipal Services Project (Kingston), UNCTAD (Geneva), y Eurodad (Brussels).
Reis, N., & Antunes de Oliveira, F. (Forthcoming). Peripheral financialization and the transformation of dependency: A view from Latin America. *Review of International Political Economy.*
Rivera Cusicanqui, S. (2012). Ch'ixinakax utxiwa: A reflection on the practices and discourses of decolonization. *The South Atlantic Quarterly, 111*(1), 95–109. doi.org/10.1215/00382876-1472612
Rolnik, R (2013). Late neoliberalism: The financialization of homeownership and housing rights. *International Journal of Urban and Regional Research, 37*(3), 1058–66. https://doi.org/10.1111/1468-2427.12062

Sanfelici, D., & Halbert, L. (2015). Financial markets, developers and the geographies of housing in Brazil: A supply-side account. *Urban Studies, 53*(7): 1465–1485. https://doi.org/10.1177%2F0042098015590981

Sayer, A. (1992). *Method in social science: A realist approach.* (2nd Ed). Routledge.

Scholz, F. (2000). Perspektiven des Südens im Zeitalter der Globalisierung. *Geographische Zeitschrift, 88*(1), 1–20. https://www.jstor.org/stable/27818864

Simone, A. (2010). *City life from Jakarta to Dakar: Movements at the crossroads.* Routledge.

Slater, D. (1977, May 11–13). *Towards a political economy of urbanization in peripheral capitalist societies: Problems of theory and method with illustrations from Latin America* [Paper presentation]. Prepared for the Zone/Fagligt joint workshop on imperialism and the spatial analysis of peripheral capitalism, Amsterdam, The Netherlands.

Slater, D. (1992). On the borders of social theory: Learning from other regions. *Environment and Planning D: Society and Space, 10,* 307–327. https://doi.org/10.1068%2Fd100307

Soederberg, S. (2013). Universalising financial inclusion and the securitization of development. *Third World Quarterly, 34*(4): 593–612.

Soederberg, S. (2015). Subprime housing goes south: Constructing securitized mortgages for the poor in Mexico. *Antipode, 47*(2), 481–499.

Sotelo Valencia, A. (2017). *Sub-imperialism revisited: Dependency theory in the thought of Ruy Mauro Marini.* Leiden: Brill.

Svampa, M. (2019). *Neo-extractivism in Latin America: Socio-environmental conflicts, the territorial turn, and new political narratives.* Cambridge University Press.

UN. (2014). *World urbanization prospects: Highlights.* United Nations Department of Economic and Social Affairs. https://esa.un.org/unpd/wup/Publications/Files/WUP2014-Highlights.pdf (August 2018).

Vainer, C. (2014). Disseminating "best practice"? The coloniality of urban knowledge and city models. In S. Parnell & S. Oldfield (Eds.), *The Roudledge handbook on cities of the Global South* (pp. 48–58). Routledge.

Walton, J. (1977). The international economy and peripheral urbanization. In N. Fainstein & S. Fainstein (Eds.), *Urban policy under capitalism* (pp. 119–35). Sage.

Wilson, J., & Bayon, M. (2017). Fantastical materializations: Interoceanic infrastructures in the Ecuadorian Amazon. *Environment and Planning D: Society and Space, 35*(5), 836–854. https://doi.org/10.1177%2F0263775817695102

Contributors

Martín Arboleda is an assistant professor of sociology at the Universidad Diego Portales, Santiago de Chile. His research explores the role that primary commodity production performs in the political economy of global capitalism. His fields of interest include global political economy, critical social theory, and agrarian studies. He is the author of the books *Planetary Mine: Territories of Extraction under Late Capitalism* (Verso, 2020) and *Gobernar la utopía: sobre la planificación y el poder popular* (Caja Negra Editora, 2021). His research on the political economy of resource extraction in Chile and Latin America has been published in several scholarly journals. He is currently working on a long-term research project on the globalization of agro-food systems, as well as on the social and political history of agrarian planning in Latin America.

Alisson F. Barbieri is an economist and demographer with a PhD in city and regional planning from the University of North Carolina at Chapel Hill. He is currently associate professor at the Department of Demography and researcher at the Center for Regional Planning and Development (CEDEPLAR), both at the Universidade Federal de Minas Gerais (UFMG) in Brazil, and professor in the graduate program in Sustainable Development at the Universidade Federal Rural do Rio de Janeiro (UFRRJ), Brazil. He is also one of the coordinators and member of the Scientific Committee of the Brazilian Research Network on Climate Change (Rede Clima). His research has focused on migration, population-environment-development interactions, and urban and regional planning.

Manuel Bayón has an MA in urban studies at FLACSO-Ecuador. He is a critical geographer with an MSc in Human Rights at UNED. In 2014, his thesis dealt with the territorial growth of Quito, capital of Ecuador, and the relationship between urbanization and Indigenous communities. Between 2014 and 2016, he was a researcher at CENEDET, directed by David Harvey, where he studied the planetary urbanization of the Amazon in the emblematic projects of the Citizen Revolution. He is FLACSO-Ecuador researcher in Contested Cities Ecuador and Contested Territories Amazonía. He is part of the Critical Geography Collective.

César Cáceres Seguel is a geographer and PhD in urban planning. He is a professor of urban geography at the Viña del Mar University, and researcher at the Regional Center for Social Inclusion (CRIIS). He has worked in issues related to neighbourhood development plans, social programs, and urban planning. His research focuses on studies on urban informality, socio-spatial segregation, neighbourhood planning, touristification, and gentrification. He has published scientific articles in Spain, the United States, Colombia, Argentina, Mexico, and Chile. He is currently working on a research project (FONDECYT) on the gentrification and touristification process in Valparaíso, Chile.

Teresa Caldeira is chair and professor of City and Regional Planning at the University of California, Berkeley. Her research focuses on predicaments of urbanization and reconfigurations of spatial segregation and social discrimination, mostly in cities of the Global South. She has been studying the relationships between urban form and political transformation, particularly in the context of democratization. Her work is interdisciplinary, combining methodologies, theories, and approaches from the different social sciences, and especially concerned with reshaping ethnographic methods for the study of cities. Teresa Caldeira's book *City of Walls: Crime, Segregation, and Citizenship in São Paulo* (University of California Press, 2000), won the Senior Book Prize of the American Ethnological Society in 2001. In 2012, she received a Faculty Mentor Award, Graduate Assembly, University of California, Berkeley, and was named a Guggenheim Fellow.

Rodrigo Castriota is an economist with a PhD in urban and regional development at the Centre for Development and Regional Planning (CEDEPLAR) at the Federal University of Minas Gerais (UFMG), Brazil. He is a member of the research group URBANO – Laboratório de Estudos em Urbanização, Planejamento e Outras Economias and has taught and published on the themes of urban theory, extended and planetary urbanization. His main research interests include urbanization processes in contemporary Amazonia, decolonial thought, and the relations between neo-extractive dynamics and urbanization.

Gustavo Durán has a PhD in architecture and urban studies from the Pontifical Catholic University of Chile. From 2013 he has been a research professor at FLACSO Ecuador, in the Department of Public Affairs, with dedication to the master's program in urban studies. His area of specialization is in the study of the Latin American city, with emphasis on the processes of new marginality and urban exclusion. In Bogotá, Quito, and Santiago, Chile, he has carried out studies on public household services, popular housing, and social housing, and socio-territorial trajectories of urban employment.

Ana Felicien is a Venezuelan agroecology activist and researcher. She has a master's degree in Tropical Ecology at the University of the Andes. She works at the Ecosystems and Global Change Laboratory of the Venezuelan Institute of Scientific Research (IVIC). She is a volunteer professor at the National Indigenous Experimental University (UNEIT).

Claudia Fonseca Alfaro holds a PhD in urban studies and is a postdoctoral fellow at the Institute for Urban Research (IUR), Malmö University, Sweden. A postcolonial and critical scholar, her research explores how global capitalism and urbanization unfold at the local level in the inconspicuous places of the Global South.

Kathrin Golda-Pongratz is a professor at the UIC Barcelona School of Architecture and associate professor at the Department of Urbanism at the Polytechnic University of Catalonia (BarcelonaTech). She holds a PhD in architecture and urban planning from Karlsruhe Institute of Technology. Her research and teaching focuses on urban memory, urban culture(s) and public space, postcolonial urbanization, non-formal urbanism and place-making strategies. Her experience expands into curating and film-making. Current research and exhibition projects centre on urban memory and visual imaginaries of self-built neighbourhoods in Lima and Barcelona. She is a member of the German Association of Latin American Research (ADLAF), the German Academy of Urban and Regional Planning (DASL), and the Academia Europaea (AE).

Michael Lukas is an assistant professor at the Department of Geography at the Universidad de Chile in Santiago de Chile. He is a geographer from the Freie Universität Berlin and holds a doctorate in geography from the Christian-Albrechts Universität zu Kiel, Germany. His research focuses on the intersections of international urban political economy, urban political ecology and power relations in urban governance and planning in Latin America. His research on Latin American urbanization and planning processes has been published in several scholarly journals. Currently he is a researcher in the international Contested Territories network.

Liz Mason-Deese holds a PhD in geography from the University of North Carolina at Chapel Hill. She has taught geography at the University of Mary Washington and George Mason University. She has also translated numerous books, including *Neoliberalism from Below* (Duke University Press, 2017) and *Feminist International* (Verso Books, 2020) by Verónica Gago, and translated and edited the volume *Feminicide and Global Accumulation* (Common Notions, 2021). She is currently working on a project to create a feminist counter-map of debt, exploring the links between the domestic space, debt, and global financial flows, funded by the Antipode Foundation. She is also a member of the Counter Cartographies Collective and the *Viewpoint Magazine* editorial collective.

Jonathan Menoscal has an MA in urban studies at FLACSO-Ecuador. He is an environmental and management risks engineer. He has been an environmental consultant in several projects for the municipality of Quito and several private companies between 2012 and 2015. He was a researcher at Flacso Ecuador between 2015 and 2017, working on the environmental impact of urban sprawl in the coastal region of Ecuador and proposing various projects for disaster risk management and resilience in vulnerable communities. In 2018 he worked in the National Secretariat for Planning and Development, specifically on the drawing up of the Inter-sectoral Agenda for the Sea, the Prospective Binational Ecuador-Colombia Plan, the updating of the guides for the Development and Land Management Plan, among others. Currently he is a researcher on the UKRI-GCRF Urban Disaster Risk Hub – Tomorrow's Cities.

Hannah-Hunt Moeller holds a Master of City Planning from Massachusetts Institute of Technology (MIT) and a Master of Science from the University of Michigan in Architecture. Currently, she practices as a designer and urban planner. Her work spans the worlds of research and design of the built environment. She consulted for Mexico City in 2020, promoting environmental risk reduction management strategies and worked as a researcher from 2018 to 2020 to manage climate adaptation in coastal United States. Her architectural portfolio includes public utility buildings and infrastructure for the water utility Denver Water and Portland's public transit authority TriMet.

Ricardo Ojima is a sociologist and demographer with a PhD from the University of Campinas, Brazil. He is currently head of the Department of Demography and professor in the graduate program in demography at the Federal University of Rio Grande do Norte (UFRN). He is also the former president of the Brazilian Association of Population Studies (ABEP, 2017–2020). His research has focused on the areas of population and environment, migration, spatial mobility, and urbanization.

Nadine Reis is a professor at the Centre for Demographic, Urban and Environmental Studies (CEDUA) at El Colegio de México, Mexico, and member of the Mexican National Researchers' System (SNI I). She holds a PhD in development sociology from the University of Bonn, Germany, and a Diplom in geography from Freie Universität Berlin, Germany. Her current research focuses on financialization, urbanization, and development in Latin America, especially Mexico. Her research topics include financial dependency, rural-urban relations, water governance, microfinance, and public banking. She has published in *The Journal of Peasant Studies*, *Globalizations*, and *Water Alternatives*, among other journals.

Clara Salazar is a professor at El Colegio de México's Centre for Demographic, Urban, and Environmental Studies (CEDUA) and member of the Mexican National Researchers' System (SNI III). She analyses peripheral urbanization, focusing on the legal condition of land property, the poor's strategies to access to the land, and the role of the government in titling. Currently, she focuses on analysing the legal reforms in Mexico to explain the effects of land privatization in urban expansion, in which the real estate sector has been participating actively. She has published four books, as author, co-author, and coordinator, as well as contributing around 50 chapters in books and articles in specialized magazines and journals. Her last book, which she also edited, provides a framework for peripheral urbanization in Latin America and has been translated into English for publication by Wiley-Blackwell, in association the Society for Latin American Studies, in 2017.

Christina M. Schiavoni is a PhD researcher focused on food sovereignty issues at the International Institute of Social Studies (ISS) in The Hague, Netherlands. Earlier, she was engaged in food sovereignty organizing and advocacy at the global level, including involvement in the International Planning Committee on Food Sovereignty (IPC) and the Civil Society Mechanism of the United Nations Committee on World Food Security, and in the US, where she helped found the US Food Sovereignty Alliance and the Food Sovereignty Prize. She has published in the *Journal of Peasant Studies, Globalizations, Third World Quarterly, Monthly Review, Development,* and various popular outlets. She is the co-editor of *The Politics of Food Sovereignty: Concept, Practice and Social Movements.*

João Tonucci is a professor of urban and regional economics at the Center for Development and Regional Planning (CEDEPLAR), School of Economic Sciences (FACE), Federal University of Minas Gerais (UFMG), and a researcher at Observatório das Metrópoles, Brazil. He holds a bachelor's degree in economics, a master's in architecture and urbanism and a PhD in geography. He was a visiting scholar at the City Institute, York University, Toronto, and is currently a Visiting Fellow at the Centre for Urban Studies at the University of Amsterdam. In addition to having worked as a planner, João researches, teaches, and publishes on critical urban theory, property markets, land policy, metropolitan planning, housing informality, popular economies, urban commons, and the right to the city.

Gabriela Torres-Mazuera has been a research professor at the Centre for Research and Higher Studies in Social Anthropology (CIESAS, Mexico) since 2011. She holds a PhD in social sciences from the Institute of Economic and Social Development at the University of Paris 1 Sorbonne-Pantheon (France). Her research focuses on contemporary rural and Indigenous Mexico, with a particular interest in the legal and political resistances and collaborations of Indigenous and peasant groups to processes of privatization and commodification of resources (land, seeds, water, forests); the governance of agrarian and Indigenous societies in contexts of legal and institutional change; and the lags and discrepancies between state legislations, Indigenous customary law, social practices, and individual action.

Ann Varley is a professor of human geography at UCL (University College London). She works on urban land and housing; informality and formalization; and gender, families, and households in Latin America. Her books include *Illegal Cities: Law and Urban Change in Developing Countries* (edited with Edésio Fernandes, Zed, 1998, and translated into Spanish with support from PROMESHA, the Programme for Training in Socio-Environmental Improvement, 2004) and *Landlord and Tenant: Housing the Poor in Urban Mexico* (with Alan Gilbert, Routledge, 1991). She has been a visiting lecturer in Mexico, Chile, the US, Denmark, Norway, and Lebanon, and co-edits the *Bulletin of Latin American Research.* Her work has been recognized by a residency at the Rockefeller

Foundation's Bellagio Study Center, and the award of the Royal Geographical Society's Busk Medal.

Raúl Zibechi is journalist, writer, militant, and one of Latin America's leading political theorists. He has written extensively on urban social movements in Latin America in Spanish and English. Two of his books available in English are *Dispersing Powers: Social Movements as Anti-State Forces* (AK Press, 2010), and *Territories in Resistance. A Cartography of Latin American Social Movements* (AK Press, 2012).

Index

GLOBAL SUBURBANISMS

Series Editor: Roger Keil, York University

Published to date:

Suburban Governance: A Global View / Edited by Pierre Hamel and Roger Keil (2015)

What's in a Name? Talking about Urban Peripheries / Edited by Richard Harris and Charlotte Vorms (2017)

Old Europe, New Suburbanization? Governance, Land, and Infrastructure in European Suburbanization / Edited by Nicholas A. Phelps (2017)

The Suburban Land Question: A Global Survey / Edited by Richard Harris and Ute Lehrer (2018)

Critical Perspectives on Suburban Infrastructures: Contemporary International Cases / Edited by Pierre Filion and Nina M. Pulver (2019)

Massive Suburbanization: (Re)Building the Global Periphery / Edited by K. Murat Güney, Roger Keil, and Murat Üçoğlu (2019)

The Life of North American Suburbs: Imagined Utopias and Transitional Spaces / Edited by Jan Nijman (2020)

In the Suburbs of History: Modernist Visions of the Urban Periphery / by Steven Logan (2020)

Beyond the Megacity: New Dimensions of Peripheral Urbanization in Latin America / Edited by Nadine Reis and Michael Lukas (2022)

www.ingramcontent.com/pod-product-compliance
Lightning Source LLC
LaVergne TN
LVHW090146080826
844660LV00013B/688/J

* 9 7 8 1 4 8 7 5 0 9 1 0 1 *